D0422580

Estimating for the Building Trades

JOSEPH STEINBERG is the Chairman of the Construction Technology Department of the New York City Community College. His broad experience in the building trades includes architectural and structural draftsman, chief estimator, construction superintendent, expediter, and consultant. He is a lecturer and co-author of *Practices and Methods of Construction* and *Construction Estimating.*

MARTIN STEMPEL has been teaching construction technology since 1947. He has broad experience as an estimator and consultant in all phases of construction. Previous to this he was a general construction superintendent, project manager, and in U. S. Coast Guard construction. He is co-author of *Practices and Methods of Construction.*

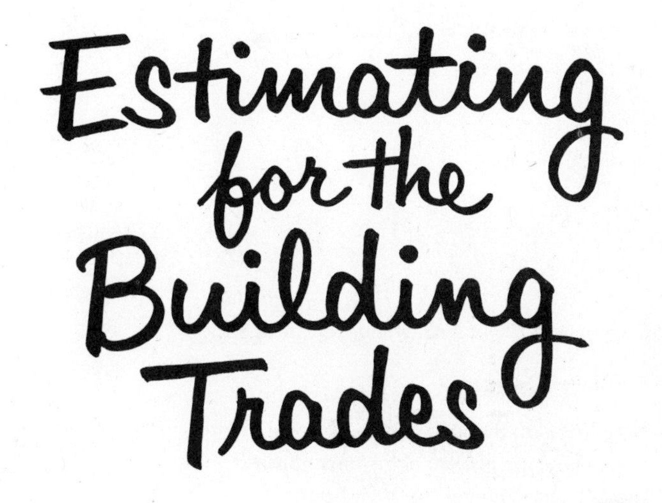

JOSEPH STEINBERG

Professor of Construction, Construction Technology Department, New York City Community College

MARTIN STEMPEL

Associate Professor of Construction, Construction Technology Department, New York City Community College

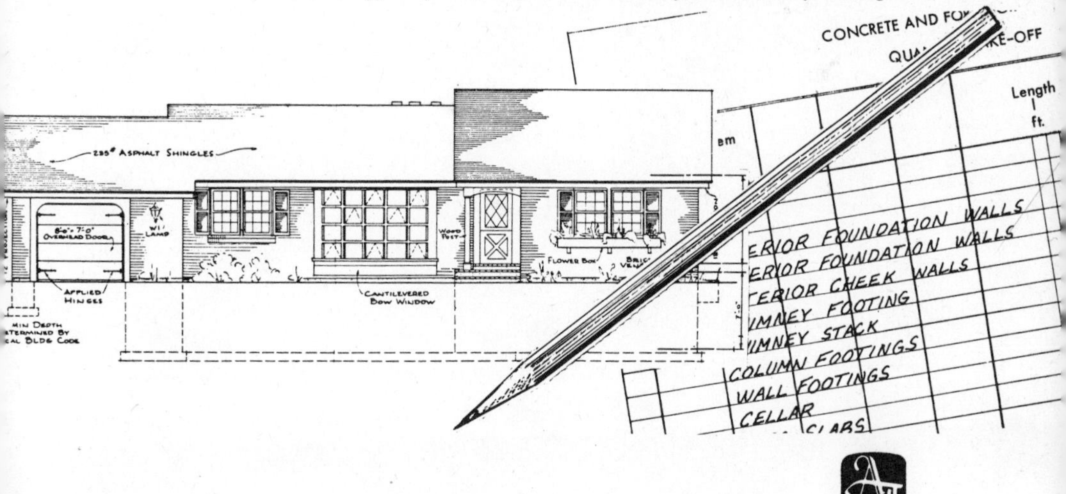

published by

AMERICAN TECHNICAL SOCIETY

Library of Congress Card Catalog No.: 65-26829

Printed in the United States of America

Illustrations on the cover are by

courtesy of Eugene Dietzgen Co.

PREFACE

The demand for competent building trades estimators is increasing. Thousands of new homes, apartments, and commercial and public buildings are being built each year. Graduates of estimating courses have little trouble finding well-paying jobs in the estimating field.

Estimating is defined in the dictionary as a means used to evaluate, to fix a cost, and to calculate a worth—all in an approximate manner. Contractors and estimators avoid "approximating" as much as possible, that is, they attempt to arrive at a total cost or a unit cost more by accurate methods than by guessing. A contractor needs accurate construction cost and overhead expense estimates in order to determine how low he can bid for a job and still be able to make a fair profit.

This book was written with the hope that the basic information on the estimating process would aid the student, contractor, and builder in making quick, accurate quantity take-offs and labor-time estimates. The authors' broad experience in practical estimating and teaching were used to develop the step-by-step methods used in this book. Illustrations were chosen to show details which do not always appear on plans and details which could cause trouble for the inexperienced estimator.

The emphasis is on accurate analysis of the working drawings and specifications. Contracts have been lost because the submitted bids were too high or too low. Why? In many cases, later examinations of work sheets have shown that the quantity take-offs were wrong, thus leading to errors in pricing and rejected bids. Careless and hasty study of specifications and working drawings had led to inaccurate quantity take-offs.

Although material and labor costs vary among different regions, and for this reason no actual costs are given, the detailed estimating *methods* contained herein *will* apply everywhere.

The chapters on "Specifications and Drawings," "Plan Reading," and "Square and Cubic Measure" will be refresher material for some readers but some of it will be new information for other readers. This material must be understood before estimates are attempted because it is fundamental to all estimating.

PREFACE

A set of house plans is included and used for many of the sample estimates throughout the book. Estimating larger structures involves the same basic methods used for these plans. The student who can estimate materials and labor for this house with little or no difficulty will have little difficulty estimating larger structures once he has had some practical estimating experience.

The early chapters are detailed and emphasize estimating *methods*. The remaining chapters emphasize the *materials* peculiar to each of the building trades.

This book replaces *How to Estimate for the Building Trades* which was successfully used by thousands of students for nearly 30 years. We wish to express our appreciation to Gilbert Townsend, J. Ralph Dalzell, and James McKinney for their pioneering efforts.

JOSEPH STEINBERG
MARTIN STEMPEL

ACKNOWLEDGMENTS

The authors and publisher wish to acknowledge the assistance rendered by the individuals, trade associations, publications, and companies listed below. The illustrations provided by them were of great value in the preparation of this book.

Allenform Corp.
American Concrete Institute
Anaconda Wire & Cable Co.
Associated Equipment Manufacturers
Bestwall Gypsum Co.
Caterpillar Tractor Co.
Celotex Corp.
Chicago Estimating Service
Crescent Insulated Wire & Cable Co.
Eugene Dietzgen Co.
E. L. Bruce Co.
Engineering News-Record
General Cable Corp.
Henry Furnace Co.
Herman York, Architect
Keuffel & Esser Co.
Linde Air Products Co.
Metal Lath Manufacturers Assoc.
National Board of Fire Underwriters
National Gypsum Co.
Portland Cement Assoc.
Practical Builder
Republic Steel Corp.
Structural Clay Products Institute
Taco, Inc.
United States Gypsum Co.
University of Illinois, Small Homes Council
Warren Webster & Co.

Our special thanks to Mr. Herman York, for providing architectural plans for the sample residence used in the text, and to the Caterpillar Tractor Company for their permission to adapt material from their manual on excavating equipment costs.

CONTENTS

1

ESTIMATING AND THE ESTIMATOR

The successful builder depends on accurate estimates of construction costs. Estimating these costs is an exacting process based on a thorough knowledge of the various trades involved in the construction industry.

What is estimating?

Estimating for the building trades is divided into two major phases: *quantity take-offs* and *labor pricing*. An estimate includes the costs of raw materials, mechanical equipment, and labor necessary for a construction job. When bidding for a contract, the contractor's overhead and profit are added to the estimate. It is almost impossible for one individual to do the quantity take-offs and labor estimates for all the various trades which are involved in the construction of a building. There are men in the construction industry who specialize in estimating one or two similar trades (e.g. electrical estimators and plumbing estimators). These specialists combine their estimates in order to determine the total construction cost of a building.

General contractors usually employ these various specialists to do all their estimating and pricing. Sometimes a general contractor works in close cooperation with a group of subcontractors who estimate their own costs and then submit the costs to the general contractor. Subcontractors usually include the cost of *material, labor, machinery, overhead,* and *profit.* A general contractor compares these estimates with his own rough estimates which are based on his experience and judgment — and determines a total price for the construction (after including his own costs and allowance for profit). He is then ready to bid for the building contract.

Estimates. There are many kinds of estimates used in the construction industry. The type of estimate is determined by the manner in which the builder or owner wishes to operate his construction project. If he wishes to relieve himself of the responsibility for material and labor, he requests a *com-*

plete estimate from a general contractor. This estimate is based on the working drawings and specifications as drawn by the architect, and includes the material, labor, and equipment necessary to complete the building and make it ready for occupancy or for sale.

Labor Estimate. Another form of estimate is the labor estimate. A builder may supply all the materials needed for the proper construction of the building but require that subcontractors furnish necessary labor and equipment. For example, a builder may wish to supply all the lumber for framing and sheathing a small home. He asks a contractor to give him an estimate of the labor requirements. (This estimate includes the tools and necessary equipment.) The contractor or his estimator must determine the time and number of men needed in addition to any special equipment.

Quantity take-off. Some organizations prefer to do all work necessary to finish a project. These organizations require a quantity take-off or quantity estimate. (This is basically a *survey* of needed material.) When the quantity of materials is known, the lists are submitted to men who are acquainted with the prices of labor and equipment. In some areas, the supply houses will break down a particular job and list the materials, the cost of materials, and the amounts necessary for a particular project.

The material estimate may be submitted in either of two forms. The first, the most desirable from the builder's standpoint, is a listing of the price of each material on the estimating sheet. These are then added together to obtain a total cost.

The second method is preferred by the supply house. This is the *total* material estimate. For this estimate, the supplier lists the materials and quantities but does not price them separately. The supplier offers the builder a single price for the entire amount of material. (This is also called a *lump sum* material estimate.)

The estimator

The estimator is a person who must assume a great amount of responsibility. He is highly trained or experienced and is familiar with all phases of building construction and general contracting. He is persistent, progressive, observant, and patient. The good estimator will acquire these traits and find that he will also acquire the respect of those with whom he deals in the building industry.

Experience. There is no substitute for experience in the building industry. The contractor should have many years of practical experience in the field which he is estimating, in order to arrive at a competitive and fair bid. Intensive practical experience gives the builder, contractor, or estimator an insight into the construction plans and permits him to calculate the *hidden* items which are not clearly evident on the plans but are necessary to carry the plans to completion. For example, a plastering estimator not only must estimate the area to be plastered, but must be aware of the heights at which the plasterers will be working. An 8 foot high plaster wall does not require special scaffold rigging but a 20 foot high wall (such as a mezzanine room) will require many special set-ups of the same scaffold. Rigging the scaffold for the higher wall will require much more time—perhaps many more hours of labor. Both rooms may have the same total wall and ceiling area, but you can see that different prices will be bid for the rooms. The difference will be in the labor estimate. The estimator calculates the time it will take to readjust the scaffold and bring materials up to the men at the different heights.

Mathematical Ability. It is important that the estimator possess some mathematical ability, especially in computational and tabulating work.

The use of desk calculators and slide rules will speed the work. (Note: The slide rule is not recommended unless you are sure where to place the *decimal points* in the answers. Serious errors can result if you are inexperienced in slide rule operation.) In many offices, the estimator does not do his own tabulating. This is usually done by an apprentice or an individual experienced in the operation of a desk calculator.

Neatness. The estimator should be neat and clean both in his work and personal habits. Experience has shown that a neat, clean individual usually does work of a similar quality. This is a very necessary and desirable trait in estimating because other people will need to read the estimate. There have been cases where the estimator's work was shown to him months later and he was unable to explain the figures and results shown on his tally sheets. This situation can be avoided when the estimator is neat and clean.

Standardization. The estimator must follow a pattern which will be reflected in his completed work. This habit has been the incentive for organizations to compile standard forms for estimating and pricing, and to train all their new personnel in their own particular sys-

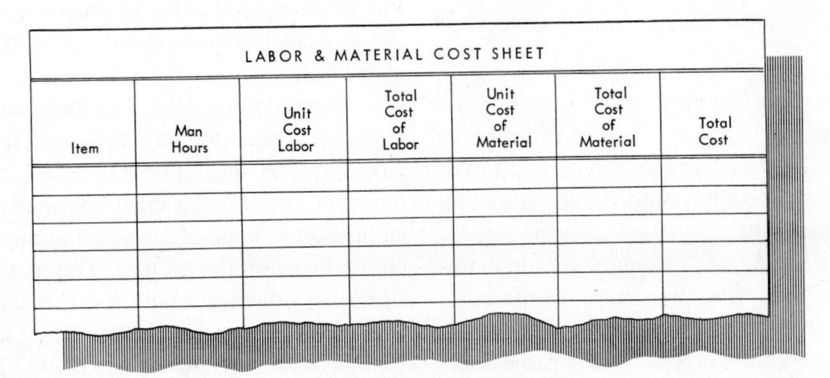

LABOR & MATERIAL COST SHEET

Item	Man Hours	Unit Cost Labor	Total Cost of Labor	Unit Cost of Material	Total Cost of Material	Total Cost

Fig. 1. A sample estimating form.

tem. The estimating forms are often used as actual forms for ordering material and equipment. See Fig. 1.

Knowledge. The estimator must know the various construction materials and their uses. He must also be acquainted with equivalent materials and acceptable substitutes which are available in his area. Obviously the estimator must know the prevailing prices of materials and their equivalents — or how to quickly obtain the prices. On many sets of plans, the architect specifies the materials to be used. If the estimator has a thorough knowledge of the materials, he can save his company much money. Not only can he save money by knowing the available materials (in his area), but by knowing of equivalent materials. This is especially important if a particular brand should change in quantity available (or price) between the time of the estimate and the actual construction. *It should be fully understood that if the builder cannot obtain the specified material, he must ask for permission to substitute another material.* This should be granted (in writing) by the builder, owner, or architect or, in some cases, all the concerned parties.

Keeping Records. A good estimator is a man who knows how to keep records and has the patience to file them over a long period of time. Since estimating consists of quantity take-offs and labor pricing, the estimator should maintain a continuous record of the costs of the various operations during the construction of a particular project. He compiles this record from the daily and weekly progress reports which are sent into the office by the field men. If he has records of similar jobs within a given time span, he is able to determine the average costs for labor, material, and equipment for almost any situation which will arise on a similar job in the future.

Average costs are sometimes used when an estimate is needed quickly. For a particular kind of building, only the *square feet* of floor area or *cubic yards* of volume are necessary in order to determine the average costs. This method seems simple, but remember that it is based upon the average costs of previous projects. Thus accurate records are a necessity for this kind of estimating.

The estimator must be exceedingly thorough. He must know exactly what the working drawings illustrate, and he must be able to understand the specifications. Understanding the specifications is half the battle in the pricing and estimating of a job. Applying the specifications to the job is equally important. If there are items which are not clear in the specifications (or are not to be priced), the estimator lists these on an *exception sheet* or an *omission sheet*.

The estimator knows that the specifications take precedence over the working drawings. For example, the basement or foundation drawing of a small home might show the basement slab poured on the ground. The specifications, however, state that the slab is to be placed on a bed of gravel. The specifications would be followed. If the contractor wanted to pour the slab on the ground, he must obtain permission from the architect and/or owner.

Summary

You can see that accurate estimates are necessary if you wish to make a profit. Estimates are based on quantity take-offs and labor prices; to these are added equipment costs, overhead costs, and an allowance for profit.

The builder who wishes to be an estimator must possess certain traits if he is to be successful. He must be systematic and able to think in an orderly and logical manner. Mathematical ability, neatness, standardization, and complete and accurate records are also necessary.

All these personal traits will have little meaning unless the estimator is able to read and understand specifications and plans. Remember that there is no substitute for experience. You must be familiar with all phases of the construction for which you are estimating.

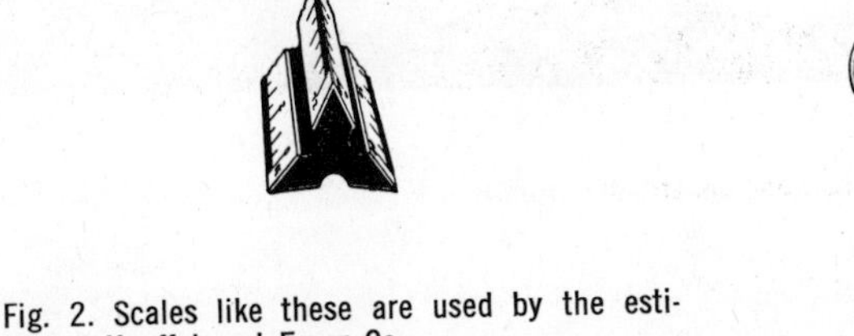

Fig. 2. Scales like these are used by the estimator. Keuffel and Esser Co.

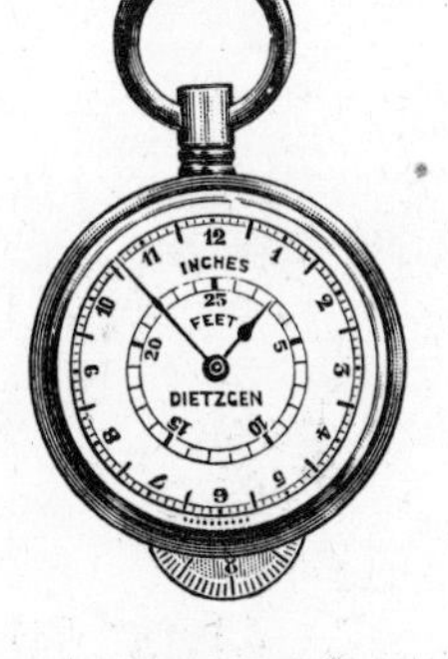

Fig. 3. A map measurer is used to determine distances on maps and plans. Eugene Dietzgen Co.

Fig. 4. Metal tapes are used in the office and in the field. Eugene Dietzgen Co.

A sketch of the finished building constructed from House Plan A.

2

SPECIFICATIONS AND DRAWINGS

A set of plans *consists of* specifications *and* working drawings. *Written specifications supply the contractor with a word description of units of work, types and methods of installation, and all other items that could not, or do not, appear on the drawings. The specifications also explain units that appear which are not fully described by the working drawings.*

Specifications

To avoid misunderstanding and confusion, all the various methods, requirements and materials for the construction of a building are listed on one or more sheets of the specifications. These "specs" are part of the set of plans, and a set of working drawings without specs is not complete.

The specs always accompany the working drawings. At no time should they be separated from the drawings, for constant reference is made from the specs to the drawings and from the drawings to the specs.

The statement, "Work shall be done in a workmanlike manner and in accordance with the working drawings," is usually found in the builder's contract. This means that the contractor must follow both the specs and the drawings in their entirety, as given to him by the architect. He must not make changes for any reason. Lack of materials, lack of "know-how," misinterpretation of drawings or specs does not excuse him from the requirements of this contract. Before he can deviate from the plans, he must obtain a change order or revision to the drawings or specs. *One unauthorized change could upset the design or the strength of a unit of construction.*

The architect makes every effort to cover all items concerned with the construction, but sometimes he unintentionally leaves out some phase or item that on the surface does not appear to be important. Nevertheless, the contractor, once he starts his job, is held responsible for providing the omitted phase or item under the statement, "Work shall be done in a workmanlike manner. . . ."

When preparing his material for bid

purposes the estimator should examine both the working drawings and the specs very carefully. Not only the units of construction but also the *General Conditions of the Specifications* (often shortened to "General Conditions") should be examined carefully; some of the very important dollar items are outlined in this section.

This chapter will examine some phases of the specs and interpret them in layman's language.

Working drawings

The working drawings and specifications are equally important because one cannot estimate without having both. A set of working drawings may have many sheets and consist of the following:

a) *Site Plan and Building Land.* These sheets are usually titled: Site, Sheet No. 1; Site, Sheet No. 2; and so forth for as many sheets as may be necessary.

b) *Architectural.* These sheets are usually titled: A Sheet No. 1, A Sheet No. 2, and so forth for as many sheets as may be necessary.

c) *Structural.* This type of drawing includes the structural steel and the concrete. These are usually titled: S Sheet No. 1, S Sheet No. 2, and so forth for as many sheets as may be necessary.

d) *Details.* These drawings show the details and the blow-ups of units of work that may be on each of the preceding sheets of the drawings. These sheets are titled: Detail Sheet No. 1, Detail Sheet No. 2, and so forth for as many sheets as may be necessary.

e) *Electrical.* These are similarly titled: E-Sheet No. 1, E-Sheet No. 2, and so forth.

f) *Mechanical.* M Sheet No. 1, M Sheet No. 2, and so forth.

g) *Plumbing and Heating.* Pl & H Sheet No. 1, Pl & H Sheet No. 2, and so forth.

h) *Air Conditioning.* A Cond Sheet No. 1, A Cond Sheet No. 2, and so forth.

There may be many more sections to a complete set of drawings. The number depends upon the size and type of building and its location.

Specification analysis

The detailed design of a building is the architect's job. There are meetings between the architect and the owner or developer to determine the needs of the owner or tenant. The architect, after deliberation and conferences, prepares preliminary sketches of the building and submits them to the owner for approval. The owner suggests such changes as he thinks are required and returns the sketches to the architect for the final work on the plans.

In order for a building to be erected from a set of plans, the needs of the structure, the materials that go into the building of the structure, and the methods by which these materials are placed and erected must be clearly specified. This is so that all parties concerned with the building will understand the architect's concept of the structure.

The architect or developer submits both the specifications and working drawings to several contractors for bid purposes. The specifications are considered the law for the builder. If there is a difference between the working drawings and the specifications, the specifications will take precedence over the drawings and govern the materials and methods which will be used for the job.

The specifications are divided into sections which cover the requirements of each trade involved in building the structure. It is understandable that some building specifications will have more sections than others. There is, however, a basic set of specifications which is usually required to carry the average building to completion. This chapter will discuss the most pertinent parts of the specifications to give the reader a good idea of the need for careful analysis of every section, every paragraph, and every line. The misreading of one word, one comma, or one paragraph may distort the entire meaning and requirements of a particular section of the specifications and can lead to a faulty bid or a dangerous deviation from the plans.

General Conditions. The General Conditions section includes *overhead* expenses which do not come under any specific trade. The General Conditions section is included in the specifications because the construction of a building involves many people and many trades. Construction also involves city and county agencies, banks and insurance companies. It is essential that the wishes of the parties involved be stated in writing so that every contractor who is interested in bidding for the job is aware of the requirements. It gives each contractor a fair and equal chance to arrive at the lowest possible bid. If the needs and wishes of the various organizations are not clearly stated in the General Conditions, some contractors might bid too low by avoiding requirements that are not clearly stated. This could cause the contract to be awarded to a contractor who would not receive enough money to carry out the architect's intentions; he may not break even or make a reasonable profit.

Fig. 1 is a typical "General Conditions Summary Sheet" used by an estimator. The most important items in this section of the specifications are discussed in the following sections.

Insurance, Contributions and Bonds. The contractor should not begin his work until he has obtained and paid in full for all insurance required under the contract and until the insurance policies have been approved by the architect, the owner, or both. The policies should be issued by a financially responsible firm and the amount of coverage should be sufficient and adequate to cover any occurrences during the construction of the building. Receipts for complete payment of premiums must be shown to the architect, owner, or both, to indicate that the policy is in force *for the particular work* of the contractor submitting the policy.

There are many types, forms and limits of insurance coverage. It is necessary that the estimator, when preparing a bid, be aware of the fact that the insurance coverage of the various policies will affect the final cost of construction. The estimator must keep

GENERAL CONDITIONS SUMMARY SHEET

JOB ______________________ DATE ______________

Superintendence	$ .00
Engineer & Rod Men	.00
Clerks & Job Stationery	.00
Waterboys	.00
Permits	.00
Security Requirements	.00
Job Site Inspection	.00
Mason Superintendents	.00
Sheds & Office	.00
Toilet Facilities	.00
Heating	.00
Lighting	.00
Telephone	.00
Traveling Expenses	.00
Enclosures	.00
Ladders & Chutes	.00
Cutting & Patching	.00
Signs	.00
Photographs	.00
Surveys	.00
Protection to Adj. Property	.00
Protection to Adj. Utilities	.00
Protection to Existing New Work	.00
Samples	.00
Shop Inspections	.00
Cold Weather Protection	.00
Barricades	.00
Small Tools	.00
Material Hoists	.00
Rental of Equipment	.00
Gasoline & Oil	.00
Trucking	.00
Freight	.00
Scaffolding	.00
Storage Facilities	.00
Fire Extinguishers	.00
Periodic Cleaning & Final	.00
Special Hazard Insurance	.00
Special Overtime Work	.00
Glass Breakage	.00
Bonds	.00
Progress Charts	.00
Glass Cleaning	.00
Special Allowances	.00
Conc. Cylinder Tests	.00
Aluminum Protection	.00
Taxes	.00
Pumping & Baling	.00
Temporary Water	.00
	.00
TOTAL	$ .00

Fig. 1. Sample General Conditions summary sheet. Chicago Estimating Service.

in mind that the costs of these policies vary with the size, location and extent of the job. This being the case, the unit which he would apply to the estimate would necessarily have to vary in form and in percentage. The following types of insurance should be considered by the contractor prior to preparing his bid.

Workmen's Compensation. Each of the United States and territories and the Canadian provinces require this type of insurance protection. Federal laws cover federal employees, private employees in the District of Columbia, and longshoremen and harbor workers. Workmen's Compensation provides employees with certain benefits in the event of injury or death and (in some states) certain diseases.

A number of states have severe penalties for failure to carry this type of insurance; fines, civil damage awards, and even imprisonment of the employer are possible.

Contractors' Public Liability Insurance. The contractor or subcontractor must also carry public liability insurance covering bodily injury or death suffered as a result of any accident occurring from or by reason of, or in the course of operation of, the construction project. This includes accidents occurring by reason of omission or act of the contractor or any of his subcontractors, or by people employed by the contractor or subcontractor. The extent of the coverage is usually determined by the size of the job and/or by the architect or the owner.

Fire and Extended Coverage. The contractor must have and maintain insurance against loss by fire, windstorm, tornado, and so forth, to cover any damages done while the work is in progress. This insurance covers a wide range of items and equipment. Special clauses are usually written into the policy by an experienced insurance agent who has a thorough background in construction coverage liability.

Fire rates vary among areas according to the rating organization having jurisdiction.

As an example of fire rates, the General Rating Rule for Fire Resistive Buildings in the course of construction as stated by the City Division of the New York Insurance Rating Organization is as follows:

1. Base Rate: .17 ($0.17 per hundred dollars)
2. Height: Add .005 for each story of planned height in excess of 25 stories; maximum height charge to be .15.
3. Protection: Above rate may be reduced 60 per cent if the following four items are complied with:

   a) Approved installation of dry line standpipe equipment with one or more properly located siamese connections [also called "Y" connectors] with due regard to accessibility and complete compliance with regulations of the Fire Department.

   b) Elevator ready for use with competent operator at all times, in buildings exceeding 150 feet in height.

c) Approved watchman on premises and so warranted.
d) Acceptable and adequate supply of first aid supplies provided.

For illustrative purposes let us assume a ten story, fire resistive building with a 100 per cent completed value of $1,000,000.

Co-insurance is the dividing of the contractor's insurance among different companies, resulting in a lower insurance rate.

Completed Value Builders Risk rate would be 55 per cent of the 100 per cent co-insurance Builders Risk rate:

1. Base Rate:	.170
2. Height: no charge	
3. Protection: assuming full compliance with the four items listed; less 60 per cent	—.102
	.068
Less 10 per cent for 100 per cent co-insurance	—.006
100 per cent co-insurance Builders Risk Fire rate	.062
Completed Value Builders Risk rate (55 per cent)	.034

Annual Premium = $1,000,000 × .034/100 = $340

Failure to comply with the four items listed under "Protection" would change the rate as follows:

1. Base Rate:	.170
2. Height: no charge	
3. Protection: no credit	
Less 10 per cent for 100 per cent co-insurance	— .017
100 per cent co-insurance Builders Risk Fire rate	.153
Completed Value Builders Risk rate (55 per cent)	.084

Annual Premium = $1,000,000 × .084/100 = $840

The actual premium savings for compliance with the four items listed under Protection is $500, hardly enough to pay for these items, but the added protection they afford against accident is worth much more.

Social Security Payments. The contractor must pay the taxes and assessments for Social Security Insurance and Unemployment and Old Age Benefits. The amount of money to be paid is determined by the wages that the employees of the general contractor receive and also the employees of his subcontractors. The general contractor must usually accept responsibility for these assessments; he usually will release the owner from any responsibility for assessments which might be defaulted by himself or his subcontractors. Some other contributions which must be paid and yet are not required by law would be welfare dues to local unions as well as other local and miscellaneous fees.

To summarize, the general contractor must include in his bid some percentage to cover these and other taxes and assessments.

Performance Bond. When estimating a construction job, the owner and the architect must be assured that the lowest, competent bidder will accept the contract as bid. In order to insure this, the architect or the owner, or both, may require that the subcontractor or contractor post a performance bond. This performance bond can be for the entire cost of the job or a percentage of the job, depending upon the requirements of the architect.

Completion Bond. Many contingencies such as weather, labor, supply of material and so forth make it necessary for the owner to protect himself to the extent that the contractor, once he has started, must guarantee completion of the job. This is usually taken care of by the posting of a completion bond. A completion bond is usually for the cost of the total estimate of the particular job.

Insurance Summary. As an estimator, you are concerned only with the amount of money that it costs your company to secure insurance policies, make contributions and to furnish the various bonds which are required by contract or law. Whether the percentage of insurance is 5 per cent, 8 per cent, or 3 per cent of the total cost of the job, it is necessary that you consider and enter the insurance costs on your estimate.

Job Meetings. It is necessary for efficient operation that the contractor, the architect and the owner meet periodically and discuss items such as payment progress, changes that might be necessary, discrepancies in plans and specifications and other items. It is standard procedure for most large construction companies to hold job meetings to discuss the problems. A stenographer usually takes complete minutes of the meetings, duplicates them, and makes them available to all parties directly concerned with the construction. Meetings are another cost of construction which does not show up anywhere in the building or on the plans, but must be taken into consideration by the estimator since they are a part of the General Conditions of many contracts.

Permits and Inspection Fees. This section requires a contractor or subcontractor to pay fees for a license to practice his trade. He must obtain all necessary permits required in the locality where the construction takes place.

Line and Grade. The section of the specifications known as "Line and Grade" requires engineering services under which the contractors, at their own expense, establish lines and grades required for the installation of their work.

The contractor must lay out his own work and be responsible for all lines, elevations and measurements for the work. He must verify the figures shown on the drawings before laying out the work, and he is responsible for any error resulting from his failure to exercise this precaution. He also must verify lines and grades furnished him by the subcontractor for general construction and is held responsible for any errors resulting from failure to exercise such precaution.

Rubbish Removal. The following statement is quoted from an actual specification: "Each contractor shall clean and gather daily all debris from the construction site and place the debris in containers or appropriate cans which are provided by the contractor at the entrance of the building unit or units, and have same removed by a bonafide rubbish removal contractor."

This item may seem insignificant, but it isn't if you are estimating a one hundred unit apartment building. When you consider the refrigerator cartons, the packings around the kitchen cabinets, and the cardboard containers and the plastic coverings for the bathtubs and showers, this rubbish situation becomes a large problem. Although this cost compared to the total construction cost does not seem large, we wish to emphasize that no contractor should willfully overlook or give away money.

Rubbish removal in most areas is governed by city and state codes. A careful study of codes, plans and building site can prevent rubbish removal from becoming a very costly problem.

As an example, it may be necessary to construct a rubbish chute. The cost of erecting a suitable rubbish chute for a multiple unit dwelling increases in relation to the height of the building. It can be related to the cost of building the forms for a very long, large concrete column. A rubbish chute is as costly to construct and in many cases more costly because it must be anchored securely to the building at intervals and at floors and must be self-supporting. The average column in a building only runs from floor to floor or from floor to mezzanine or balcony. The breakout or rupturing factor for the column is much less than for a rubbish chute. If a rubbish chute collapses it could mean the loss of life to many construction men on the site. Since it must be made safe, anchored securely, and built firmly, it will be expensive. The removal of dirt and rubbish from a building site and structures must be entered into the General Conditions of the contract to make sure that your costs are complete.

Window Cleaning. This item, sometimes neglected by the specifications, is a very important and costly item to the contractor. If the estimator omits this unit it will cost the contractor money he did not plan to spend. Windows are marked with chalk, whitewash, or other methods so that all trades will know there is a pane of glass in the sash and thereby avoid breakage. When the building is complete, and before the turnover to the owner, it must be presentable. The contractor must clean each and every window in the building. An estimator should include this cost even though it may be omitted from the specifications.

Winter Protection. The contractor must protect his work from the elements. This includes snow, sleet, rain, windstorms, cyclones, tornadoes, hail and flood. The elements must be taken into consideration for the entire time the building will be under construction. The most costly item is winter protection. Winter protection includes the rigging of canvas or plastic around the structure and heating the inside by some temporary means, such as salamanders (cans with open fires). The heating and the canvas or plastic are necessary if the various trades are to work during inclement weather. Heating is also necessary for the proper curing of concrete. The duration of the job must be considered for winter protection. If the job is to be protected for one or more winters, the cost on the General Conditions must be entered and taken into serious consideration.

Watchmen. On most large jobs it is necessary for the contractor to provide adequate watchmen for general patrolling of the site, both day and night, including Saturdays, Sundays and holidays. The watchman is necessary to prevent the theft of materials. He is also necessary to prevent trespassing on the construction site which can lead to bodily injury, for which the general contractor could be held responsible. A watchman is another costly item which must be entered into the General Condition costs to give you a complete picture of the job.

Inspection and Testing. This item is an important section of the specifications. Under it the contractor must furnish facilities and assistance for inspection, examination and tests that the building inspectors may require. He must also secure for the building inspectors free access to factories and plants in which any materials are being manufactured or prepared, and to all parts of the construction. The contractors may be required to give the inspectors advance notice of preparation, manufacture, or shipment of any materials. This section becomes very important when concrete is used; innumerable slump tests and samples may be required.

Field Office. This item is no longer the well-constructed shanty that took several carpenters possibly a week to construct. Today, the field office is usually a beautiful mobile home which contains not only office furniture, but also has toilet facilities, water, electricity, calculators, typewriters and all the necessities of a modern office. This cost may seem trivial, for this mobile home can be used from job to job, but what one fails to take into consideration is the two, three or four months wait between jobs. The cost must then be prorated (divided proportionately) and can become large. Therefore, in figuring your estimate, always take care of this item.

Project Sign. This item consists of a sign naming the general contractor, the architect and the project. Not only must the sign be erected, the words lettered with paint and protected by a surface that will withstand the elements, but it must also be kept in a good state of repair throughout the life of the project. We recommend that this item be taken into consideration when estimating.

Site Fence. A fence is a costly thing to erect. Not only must you use a surveyor to get the proper line, but you must bring a fence to security. Security means vertical, horizontal and sometimes diagonal support so that a storm or childish horseplay cannot do any damage. As an example, we quote from a specification which states:

> "The entire area must be enclosed by a fence 7′-0″ high above the respective grades as indicated on drawings which are attached to this specification. Should there be

any trees, the contractor shall also include the cost of maintaining and providing fences for such trees. These are covered by additional drawings. If there is any existing fence at the site, such fencing may be used to an extent. If the fence is built along the curb, free access from the street to hydrants, fire and police alarm box and light standards must be maintained."

The above paragraph should give the reader an idea of what he may run into when constructing a fence around a proposed construction site. The structure might be a block square and the fence, therefore, four blocks long. Provisions must also be made for entry of trucks, workmen, and so forth. This fence must be constructed and possibly painted to withstand the elements.

Temporary Light and Power. The contractor must provide, on the construction site, light and power for the various trades and subcontractors under the jurisdiction of his contract. The light and power must meet with approval of the county, township or city in which the building is to be erected. The installation of proper size meters and proper height electrical poles can become a costly item. This cost is generally determined by the length of live wire which must be brought in from existing lines to the central point of the construction site. Temporary light and power is just another one of the many items which the estimator must price in relation to the size and duration of his project.

Temporary Plumbing. The contractor must provide temporary heating and toilet facilities.

Temporary heat is a very costly item. It includes such things as heating and ventilating, radiators, piping and other units required to heat the structure while under construction. This is the contractor's responsibility and therefore becomes a building cost. The large dollar item consists of labor and fuel. The estimator should give strict attention to the specifications of this unit.

Temporary toilet facilities range from the simple "outhouse" to elaborate, multiple unit water closet and lavatory installations, depending on the size, location, and type of construction.

Local and state laws, as well as union regulations, must be considered in estimating this unit of cost. In "high-rise" construction it may be necessary to provide at least one water closet for every 30 persons, located no more than four stories above or below the place of work. These toilet facilities must be sheltered from view, from weather, and from falling objects.

The estimator must check laws and union regulations carefully; even if this item is omitted from the specifications the contractor might still be required to pay the cost of temporary heat and toilet facilities.

Progress Records and Schedules. The General Conditions covered so far will normally be found in the specifications for large construction projects. This is by no means a complete list. For example, some jobs may call for progress photographs. This item entails the hiring of a competent commercial photographer who must periodically submit photographs to the owner and the architect to illustrate the good construction practices being used.

Progress Schedules. These are called for on most large jobs. The making of such a schedule requires the services of a competent man who knows the times required for each phase of the work and each trade.

Progress Records. This is simply a record of the work which has been completed. It is usually based on daily work reports.

Temporary plumbing, discussed in the previous section, often requires daily work reports which must be submitted to inspectors. The inspectors may go over these reports and not be satisfied with the fuel consumption or attendance of engineers and require the contractor to put on additional men or other items. Additional costs such as this can be discovered while the job is under construction.

There are, of course, many more items which the estimator must consider. Even if these items do not appear in the specifications, the contractor and estimator will have to include their costs in the estimate if it is to be accurate and complete, and result in a profit for the contractor.

Sample Specifications

Specifications for a single resident dwelling may be quite short and still be adequate, but for larger and more complicated structures they are often as thick as the book you are now reading. The following specifications are a small section from a set of specifications.

SECTION 11A - STRUCTURAL STEEL

1. SCOPE OF WORK

The work covered by this Section of the Specifications consists of furnishing all plant, labor, equipment, appliances, and materials, and in performing all operations in connection with the furnishing, fabricating and erecting of all STRUCTURAL STEEL, and all other items related thereto, complete, in strict accordance with this Section of the Specifications and the applicable Drawings, and subject to the General and Supplementary Conditions as hereinbefore stated, and generally including, but not limited to the following:

a. All structural ASTM A-36 steel shown on the Drawings and including all structural steel shapes, plates, stiffeners, bearing plates, anchors, bolts rivets, connections and framing on roof and openings, outriggers as indicated on Drawings, etc., as required to complete the work of this Section.

b. Columns, bases, caps, beams and anchor bolts.

c. Lintels shown on the structural Drawings.

d. Shop and field painting of all structural steel.

e. Steel decking for the canopy.

f. The cutting, punching, and drilling required for the work of other contractors shall be performed on the shop, based upon information this Contractor will receive from them, prior to the approval of Shop Drawings. Cutting, etc., for the General Contractor's trades shall be performed by this Contractor in the field and/or in the shop.

NOTE: Steel joists are specified under another Section of the Specifications.

2. SHOP DRAWINGS

a. Shop Drawings shall be prepared and submitted in accordance with the General Conditions. The length and fillet dimensions of each weld shall be shown, and each structural steel member shall be numbered for erection identification. Fabrication shall not be started until Shop Drawings have been finally approved.

b. Submit one transparency (sepia) and two prints. Sepia to be returned for distribution. Furnish Architect's field office with one print of approved drawings. The use of copies of Contract Drawings for submission as Shop Drawings will not be permitted.

3. INSPECTION AND TESTS

a. Mill Reports shall be submitted for all structural steel used.

b. This Contractor shall allow in his bid the sum of THREE ($3.00) DOLLARS per ton, to cover the Field Inspection by a Professional Engineer of all structural steel, and any other structural steel members. The Professional Engineer will be selected by the Architect, but will be employed and paid by this Contractor.

c. The Architect or his Structural Engineer reserve the right to reject materials and workmanship at any time prior to final acceptance of the work in place, if they do not conform to plans and specifications.

4. ADDITIONAL STEEL

a. This Contractor shall include an allowance for furnishing and installing two tons of steel in addition to that now indicated on the Structural Steel Drawings, or as required under this Section of the Specifications. Steel shall be of type, size, length, and design as will be determined by the Architect's Structural Engineer. In the event all this additional structural steel is not required or used, an adjustment will be made and the Owner credited therewith. The amount allowed for this two tons of additional structural steel shall be entered on the Form of Bid in the space provided for same.

5. SHOP AND FIELD CONNECTIONS

a. All shop connections shall be fully riveted or welded or shall be made with approved rivet bolts. Field connections shall be suitable bolted or welded. Bolts shall be 3/4" round with shanks of the proper lengths to provide full grip. Heavy washers shall be provided under the nuts so that threads will not bear on the members. Thread shall be upset after nuts are drawn up tight. All field riveting shall be performed with a pneumatic hammer in conformance with AI SC Specifications. All field connections for fascia angles or curtain-wall angles shall have slotted holes to allow full adjustment in the field.

6. WELDING

a. Welding shall be performed by the electric arc process in conformance with the latest revised code of the American Welding Society. All surfaces to be welded shall be free of paint, dirt, rust and moisture. Surface conditions shall be approved by the Architect prior to actual welding. Welding electrodes shall be heavily coated and not over 5/16" thick, exclusive of the coating. All electric current required for welding shall be furnished by this Contractor.

b. Only welders who are skilled in the type of welding required by this contract shall be employed. When so directed by the Architect, each welder employed shall prepare test samples of his work with materials supplied by this Contractor, to the satisfaction of the Architect.

7. FIELD ERECTION

a. All structural steel and iron shall be erected true and in its designed location. All members shall be plumb or level as shown. Temporary bracing shall be installed where necessary, and such bracing and shoring shall remain in place as long as may be required for safety of the structure. Leveling plates 1/4" thick shall be provided and placed on 3/4" grout under all column base plates. Bases shall be fabricated as an integral part of the column. Loose base plates required for steel beams shall be set in same manner as leveling plate. Structural steel members will be considered plumb or level where the error does not exceed 1 in 500.

8. LINTELS

a. A masonry bearing shall be provided at each end of lintels of not less than 6". An approved adjustable type of hanger shall be used where hangers are required for lintels. In addition to the prime coat, all parts that will be inaccessible after erection shall be given 2 additional coats of asphaltic base paint before erection.

9. PIPE OR TUBE COLUMNS

a. Pipe or tube columns shall be ASTM A-36 steel and shall be rammed full of concrete at the factory as shown on the Drawings, or remain hollow as shown on the Drawings, and shall be complete with welded on base and cap plates or as detailed. Pipe or tube columns shall be of an approved type, similar and equal to those manufactured by the Lally Column Company. For the concrete column fill, use 1:1 1/2:3 mix with the coarse aggregate well graded up to 1/2" in its greatest dimensions and consisting of blue trap rock or washed gravel. Where pipe or tube columns are exposed, they shall be belt wheeled or sanded to a perfect, smooth surface, free from rust, scale or dirt.

10. STEEL DECKING

a. Steel deck shall be type D as manufactured by Fenestra Steel Products or approved equal. Steel shall conform to latest revision of ASTM A-245. Deck panels shall be galvanized, 24" wide, ribs 3" deep spaced 12" on center, interlocking type.

b. Where required furnish closures, roof sump recesses (to be welded in place by this Contractor) and any other accessories which must be attached directly to deck panels to provide a finished surface for the application of insulation and roofing.

c. Caulking material shall be as recommended by the manufacturer. If design of side joints will not make effective vapor seal when caulked, polyethylene sheet of approved thickness installed with approved adhesive shall be installed.

d. Place and adjust deck panels in initial bay; check coverage and alignment to avoid fanning. Anchor deck panels by welding directly through the bottom of rib to every structural support at minimum spacing of 12" o.c. Weld the female half of each side joint. Weld side joints between supports at 3'-0" intervals.

e. Touch up top surface of field weld area and all damaged places with paint as specified for shop painting immediately upon completion of erection.

11. PAINTING

a. All steel work shall be thoroughly cleaned of all rust, scale and dirt and shall be given a shop coat of red lead and linseed oil. Shop coats shall be composed as follows: For every 100 lbs, of red lead paste, use 1 gallon of raw linseed oil, 1 gallon of boiled linseed oil, 1/2 gallon of turpentine, and 2 pints of an approved dryer. All bolts, anchors, separators and other accessories which will remain permanently in the work shall be dip-coated in the above solution. All surfaces that will be inaccessible after assembly shall be given 2 shop coats of paint prior to assembly.

b. Immediately after erection of the steel, a field coat of paint shall be applied. Before applying this field coat, all marred surfaces of the shop coat shall be touched up with the shop coat paint. This shall include touch up of all erection marks and all field rivets. This field coat shall be composed of the following:

> For every one hundred (100) lbs. of paste red lead, use 2-5/8 gallons of linseed oil (1/2 boiled, 1/2 raw) and twelve (12) ounces lamp black.

c. No painting shall be performed on surfaces that are damp or frosty or when the outside temperature is below 55 Deg. F., unless otherwise approved by the Architect.

HOUSE PLAN A

The set of plans on the following pages are for a single-unit residence. The plans were originally drawn to 1/48 actual size but were further reduced to fit this book. These sample plans are referred to in the following chapters.

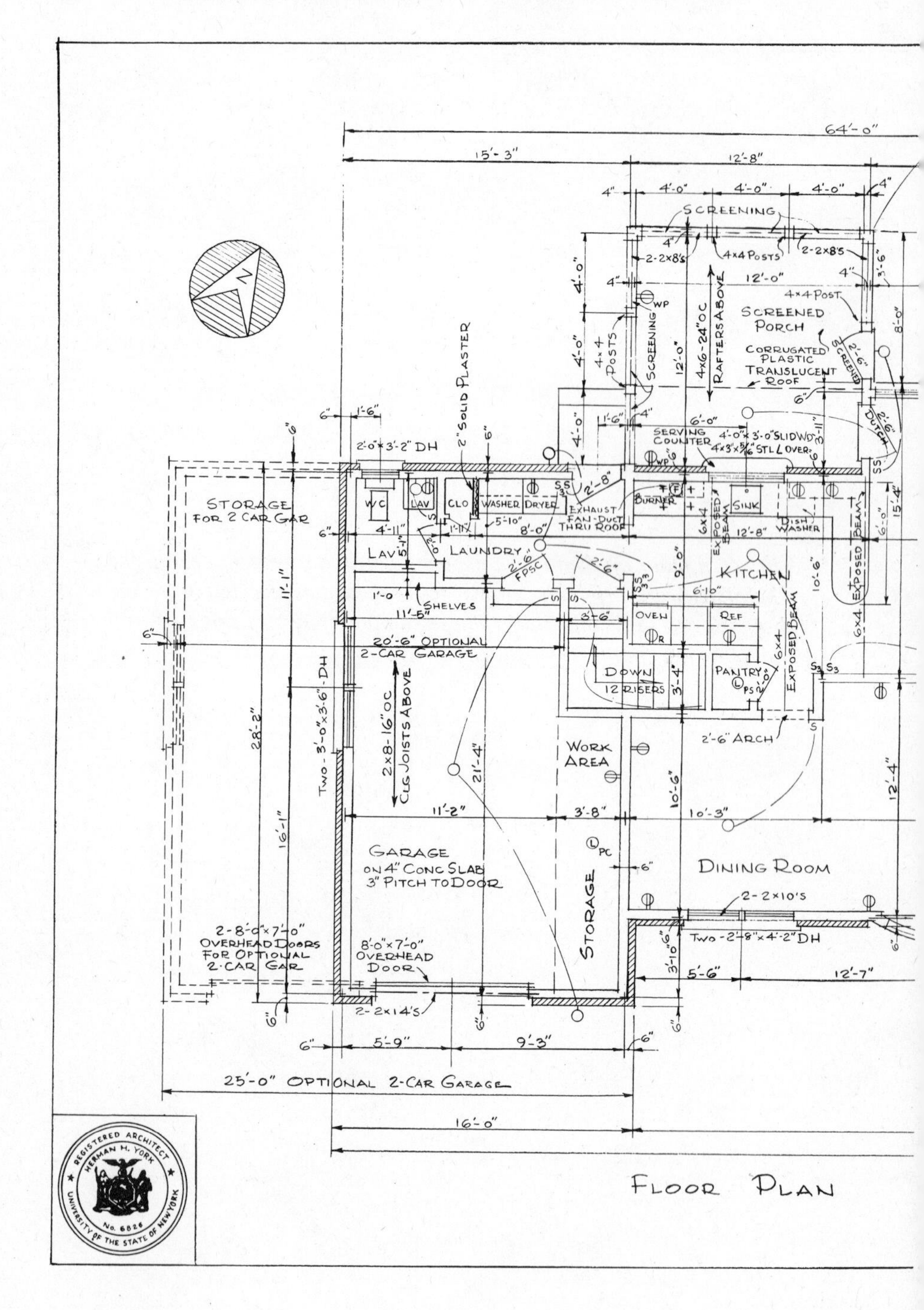
SCREENED PORCH
CORRUGATED PLASTIC TRANSLUCENT ROOF
SERVING COUNTER
KITCHEN
STORAGE FOR 2 CAR GAR
LAUNDRY
WC
LAV
CLO
WASHER
DRYER
OVEN
REF
PANTRY
DOWN 12 RISERS
WORK AREA
GARAGE ON 4" CONC SLAB 3" PITCH TO DOOR
STORAGE
DINING ROOM
20'-6" OPTIONAL 2-CAR GARAGE
2x8-16" OC CLG JOISTS ABOVE
2-8'-0"x7'-0" OVERHEAD DOORS FOR OPTIONAL 2-CAR GAR
8'-0"x7'-0" OVERHEAD DOOR
25'-0" OPTIONAL 2-CAR GARAGE
64'-0"
REGISTERED ARCHITECT HERMAN H. YORK No. 6826 UNIVERSITY OF THE STATE OF NEW YORK
FLOOR PLAN

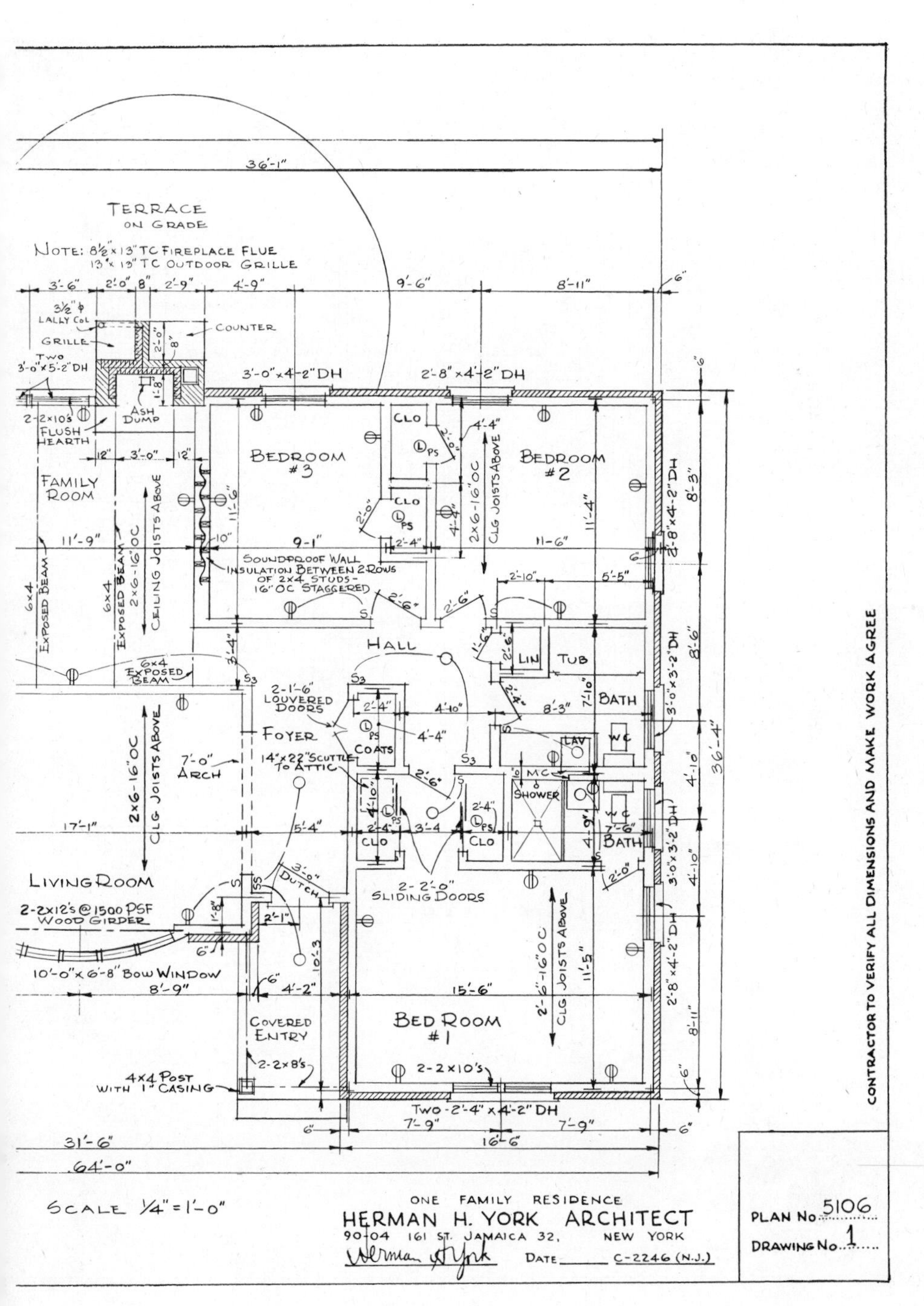
36'-1"
TERRACE ON GRADE
NOTE: 8½"x13" TC FIREPLACE FLUE
13"x13" TC OUTDOOR GRILLE
3½" ϕ LALLY COL
GRILLE
COUNTER
TWO 3'-0"x5'-2" DH
3'-0"x4'-2" DH
2'-8"x4'-2" DH
ASH DUMP
2-2x10's FLUSH HEARTH
FAMILY ROOM
BEDROOM #3
BEDROOM #2
CLO
SOUNDPROOF WALL INSULATION BETWEEN 2 ROWS OF 2x4 STUDS-16" OC STAGGERED
CEILING JOISTS ABOVE
6x4 EXPOSED BEAM
HALL
LIN
TUB
BATH
2-1'-6" LOUVERED DOORS
FOYER
COATS
14"x22" SCUTTLE TO ATTIC
7'-0" ARCH
CLG JOISTS ABOVE
SHOWER
LAV
WC
LIVING ROOM
2-2x12's @ 1500 PSF WOOD GIRDER
2-2'-0" SLIDING DOORS
DUTCH
10'-0"x6'-8" BOW WINDOW
COVERED ENTRY
BED ROOM #1
2-2x10's
2-2x8's
4x4 POST WITH 1" CASING
TWO-2'-4"x4'-2" DH
31'-6"
64'-0"
16'-6"
36'-4"
CONTRACTOR TO VERIFY ALL DIMENSIONS AND MAKE WORK AGREE
SCALE ¼"=1'-0"
ONE FAMILY RESIDENCE
HERMAN H. YORK ARCHITECT
90-04 161 ST. JAMAICA 32, NEW YORK
DATE
C-2246 (N.J.)
PLAN No. 5106
DRAWING No. 1

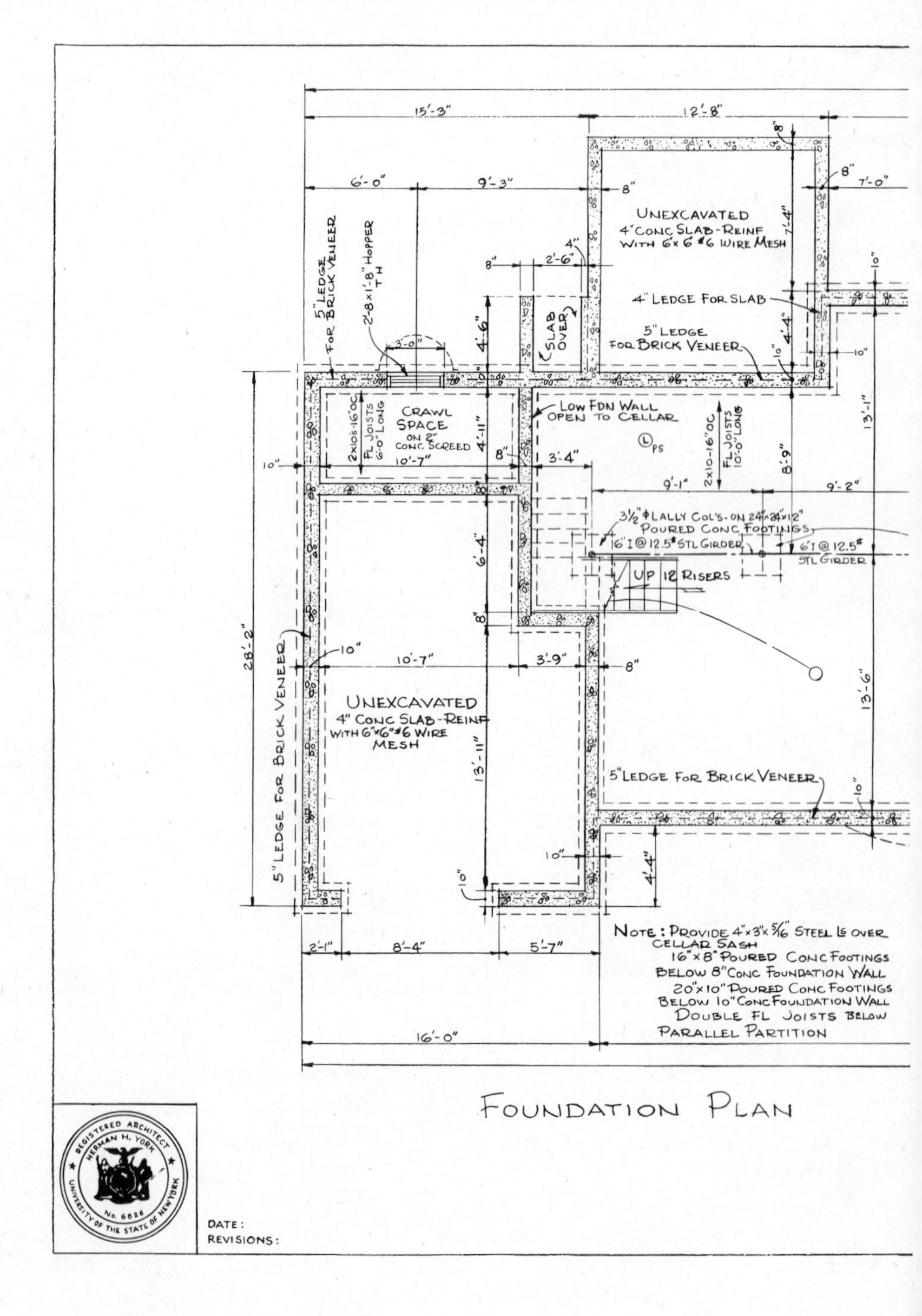
15'-3"
12'-8"
6'-0"
9'-3"
7'-0"
8"
UNEXCAVATED
4" CONC SLAB-REINF
WITH 6x6 #6 WIRE MESH
5" LEDGE FOR BRICK VENEER
2'-8x1'-8" HOPPER T H
4"
2'-6"
SLAB OVER
4" LEDGE FOR SLAB
5" LEDGE FOR BRICK VENEER
3'-0"
4'-6"
7'-4"
4'-4"
10"
13'-1"
CRAWL SPACE
ON 2" CONC SCREED
2x10's-16"OC
FL JOISTS 6'-0" LONG
10'-7"
4'-11"
LOW FDN WALL OPEN TO CELLAR
2x10-16"OC
FL JOISTS 10'-0" LONG
8'-9"
3'-4"
9'-1"
9'-2"
3 1/2" ϕ LALLY COL'S-ON 24"x24"x12" POURED CONC FOOTINGS
16' I @ 12.5# STL GIRDER
6' I @ 12.5# STL GIRDER
UP 12 RISERS
28'-2"
6'-4"
10'-7"
3'-9"
13'-6"
5" LEDGE FOR BRICK VENEER
UNEXCAVATED
4" CONC SLAB-REINF
WITH 6"x6" #6 WIRE MESH
13'-11"
5" LEDGE FOR BRICK VENEER
4'-4"
2'-1"
8'-4"
5'-7"
16'-0"
NOTE: PROVIDE 4"x3"x5/16" STEEL Ls OVER CELLAR SASH
16"x8" POURED CONC FOOTINGS BELOW 8" CONC FOUNDATION WALL
20"x10" POURED CONC FOOTINGS BELOW 10" CONC FOUNDATION WALL
DOUBLE FL JOISTS BELOW PARALLEL PARTITION
FOUNDATION PLAN
REGISTERED ARCHITECT
HERMAN H. YORK
No. 6826
UNIVERSITY OF THE STATE OF NEW YORK
DATE:
REVISIONS:

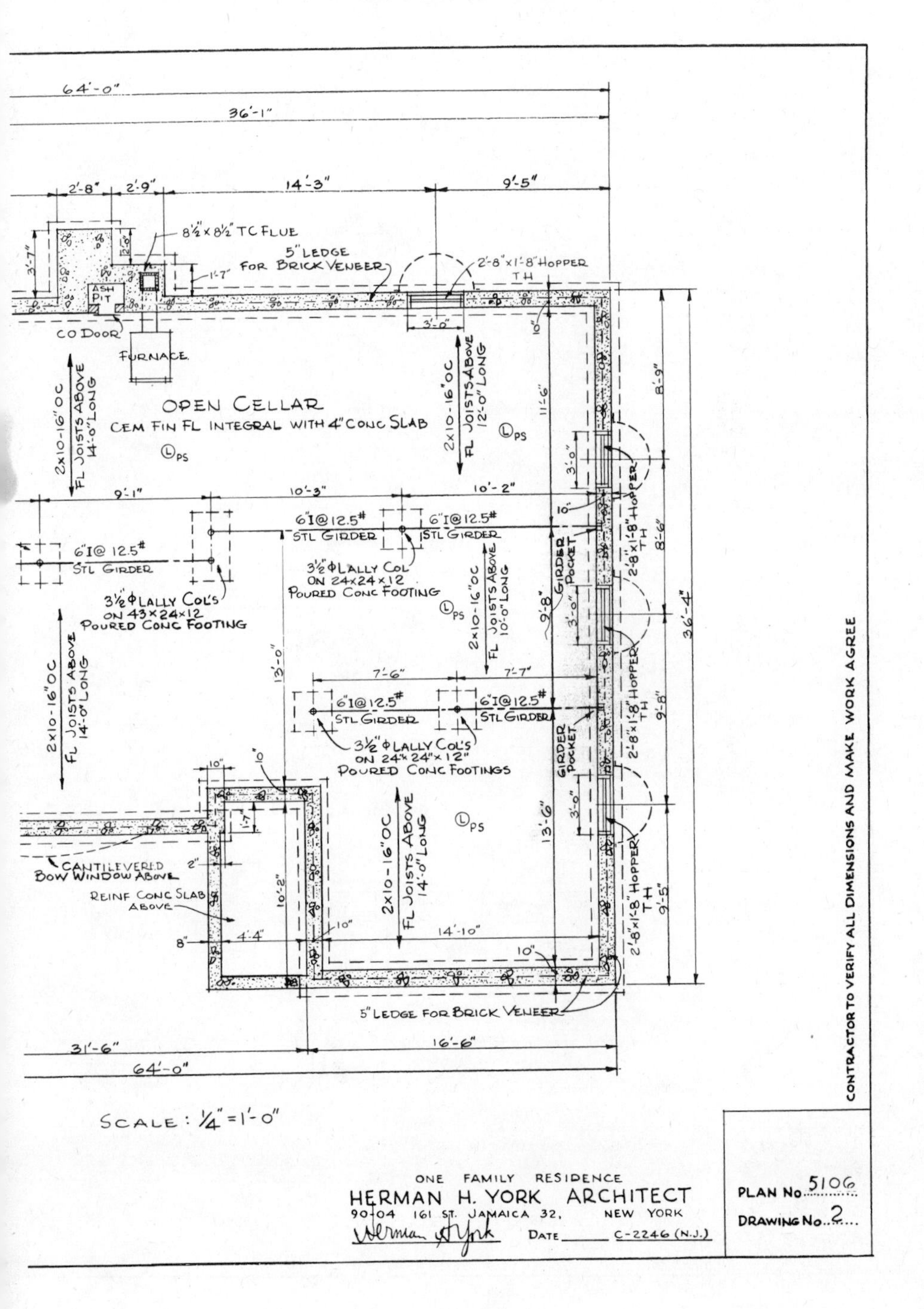
64'-0"
36'-1"
2'-8"
2'-9"
14'-3"
9'-5"
8½" x 8½" TC FLUE
5" LEDGE FOR BRICK VENEER
2'-8" x 1'-8" HOPPER TH
ASH PIT
CO DOOR
FURNACE
OPEN CELLAR
CEM FIN FL INTEGRAL WITH 4" CONC SLAB
2x10-16" OC FL JOISTS ABOVE 14'-0" LONG
2x10-16" OC FL JOISTS ABOVE 12'-0" LONG
6"I@12.5# STL GIRDER
3½" ϕ LALLY COL'S ON 43x24x12 POURED CONC FOOTING
3½" ϕ LALLY COL ON 24x24x12 POURED CONC FOOTING
2x10-16" OC FL JOISTS ABOVE 10'-0" LONG
GIRDER POCKET
3½" ϕ LALLY COL'S ON 24" x 24" x 12" POURED CONC FOOTINGS
CANTILEVERED BOW WINDOW ABOVE
REINF CONC SLAB ABOVE
5" LEDGE FOR BRICK VENEER
31'-6"
16'-6"
64'-0"
36'-4"
SCALE: ¼" = 1'-0"
CONTRACTOR TO VERIFY ALL DIMENSIONS AND MAKE WORK AGREE
ONE FAMILY RESIDENCE
HERMAN H. YORK ARCHITECT
90-04 161 ST. JAMAICA 32, NEW YORK
DATE
C-2246 (N.J.)
PLAN No. 5106
DRAWING No. 2

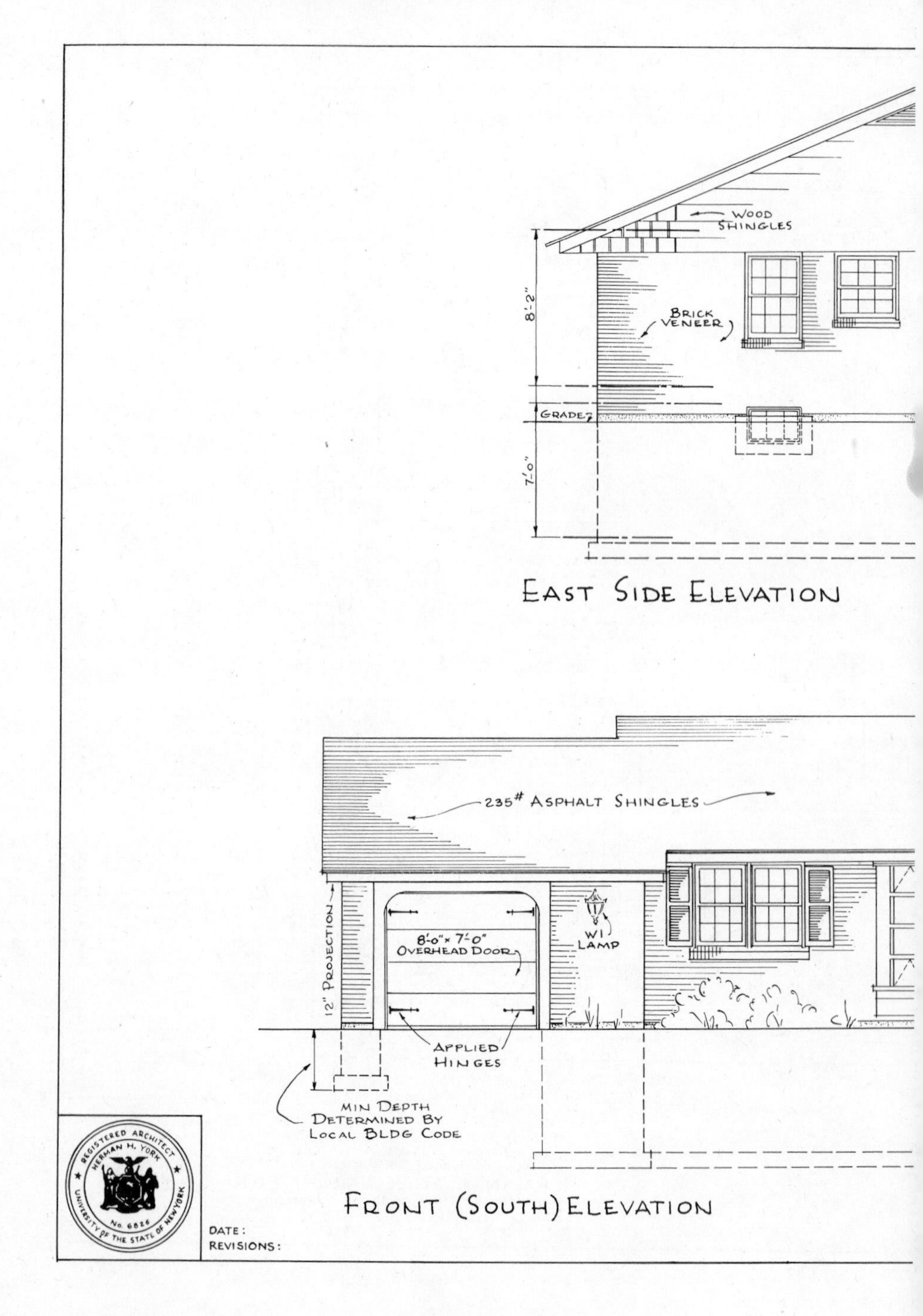
WOOD SHINGLES
8'-2"
BRICK VENEER
GRADE
7'-0"
EAST SIDE ELEVATION
235# ASPHALT SHINGLES
12" PROJECTION
8'-0" × 7'-0" OVERHEAD DOOR
WI LAMP
APPLIED HINGES
MIN DEPTH DETERMINED BY LOCAL BLDG CODE
REGISTERED ARCHITECT HERMAN H. YORK No. 6826 UNIVERSITY OF THE STATE OF NEW YORK
DATE:
REVISIONS:
FRONT (SOUTH) ELEVATION

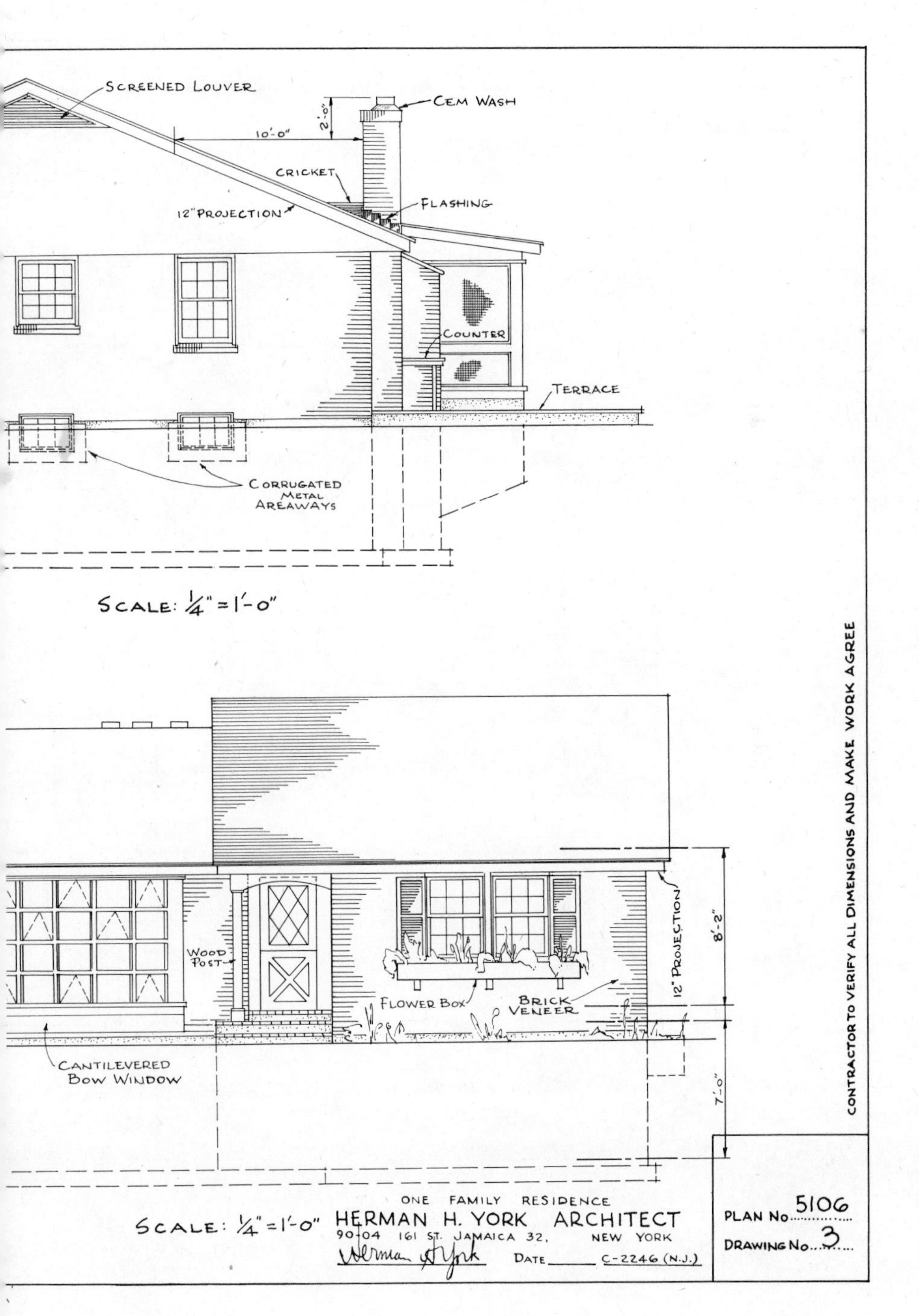
SCREENED LOUVER
CEM WASH
2'-0"
10'-0"
CRICKET
12" PROJECTION
FLASHING
COUNTER
TERRACE
CORRUGATED METAL AREAWAYS
SCALE: 1/4" = 1'-0"
WOOD POST
FLOWER BOX
BRICK VENEER
12" PROJECTION
8'-2"
7'-0"
CANTILEVERED BOW WINDOW
CONTRACTOR TO VERIFY ALL DIMENSIONS AND MAKE WORK AGREE
SCALE: 1/4" = 1'-0"
ONE FAMILY RESIDENCE
HERMAN H. YORK ARCHITECT
90-04 161 ST. JAMAICA 32, NEW YORK
DATE
C-2246 (N.J.)
PLAN No. 5106
DRAWING No. 3

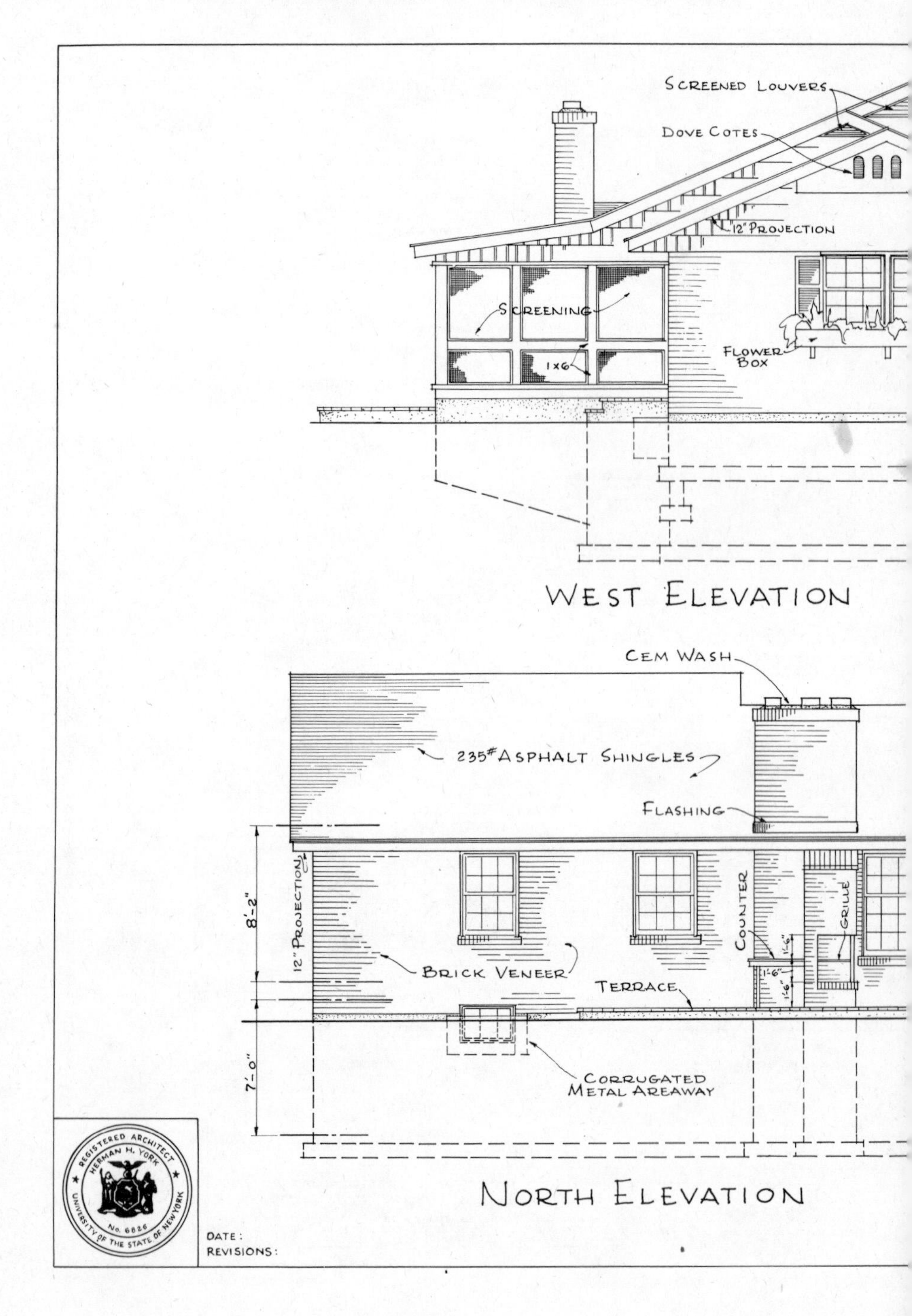
SCREENED LOUVERS
DOVE COTES
12" PROJECTION
SCREENING
1x6
FLOWER BOX
WEST ELEVATION
CEM WASH
235# ASPHALT SHINGLES
FLASHING
8'-2"
12" PROJECTION
BRICK VENEER
COUNTER
GRILLE
1'-6"
TERRACE
7'-0"
CORRUGATED METAL AREAWAY
NORTH ELEVATION
REGISTERED ARCHITECT
HERMAN H. YORK
No. 6826
UNIVERSITY OF THE STATE OF NEW YORK
DATE:
REVISIONS:

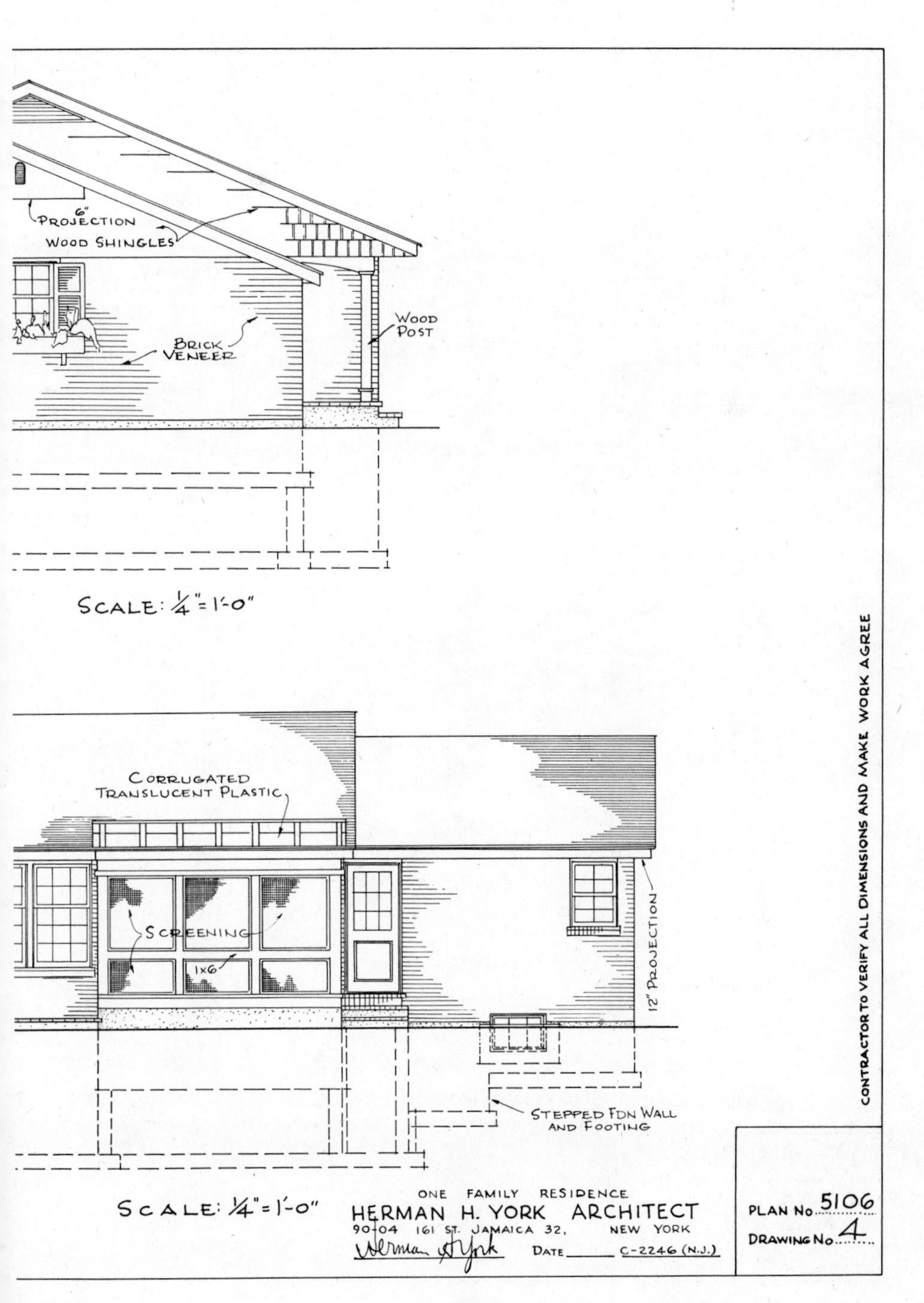
6" PROJECTION
WOOD SHINGLES
BRICK VENEER
WOOD POST
SCALE: 1/4" = 1'-0"
CORRUGATED TRANSLUCENT PLASTIC
SCREENING
1x6
12" PROJECTION
STEPPED FDN WALL AND FOOTING
SCALE: 1/4" = 1'-0"
ONE FAMILY RESIDENCE
HERMAN H. YORK ARCHITECT
90-04 161 ST. JAMAICA 32, NEW YORK
DATE C-2246 (N.J.)
PLAN No. 5106
DRAWING No. 4
CONTRACTOR TO VERIFY ALL DIMENSIONS AND MAKE WORK AGREE

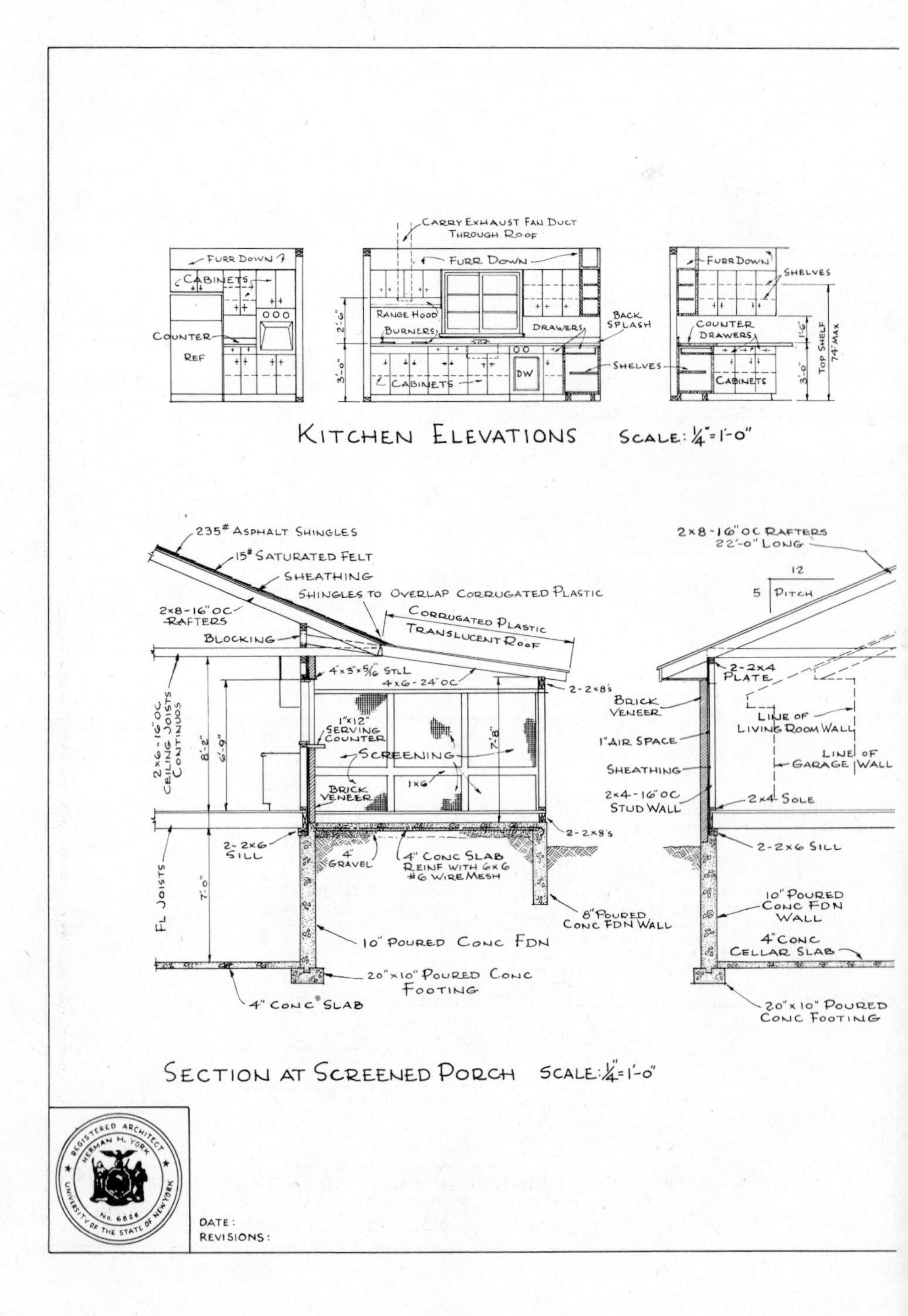

Carry Exhaust Fan Duct Through Roof
Furr Down
Cabinets
Counter
Ref
Range Hood
Burners
Drawers
Back Splash
Shelves
DW
Shelves
Counter Drawers
Top Shelf 7'4" Max
Kitchen Elevations Scale: 1/4"=1'-0"
235# Asphalt Shingles
15# Saturated Felt
Sheathing
Shingles to Overlap Corrugated Plastic
Corrugated Plastic Translucent Roof
2×8-16" OC Rafters
Blocking
4×3"×5/16 Stl
4×6-24" OC
2-2×8's
2×6-16" OC Ceiling Joists Continuos
1"×12" Serving Counter
Screening
1×6
Brick Veneer
2-2×6 Sill
4" Gravel
4" Conc Slab Reinf with 6×6 #6 Wire Mesh
8" Poured Conc Fdn Wall
FL Joists
10" Poured Conc Fdn
20"×10" Poured Conc Footing
4" Conc Slab
2×8-16" OC Rafters 22'-0" Long
12
5 Pitch
2-2×4 Plate
Brick Veneer
1" Air Space
Sheathing
2×4-16" OC Stud Wall
Line of Living Room Wall
Line of Garage Wall
2×4 Sole
2-2×6 Sill
10" Poured Conc Fdn Wall
4" Conc Cellar Slab
20"×10" Poured Conc Footing
Section at Screened Porch Scale: 1/4"=1'-0"
Registered Architect Herman H. York
University of the State of New York No. 6626
Date:
Revisions:

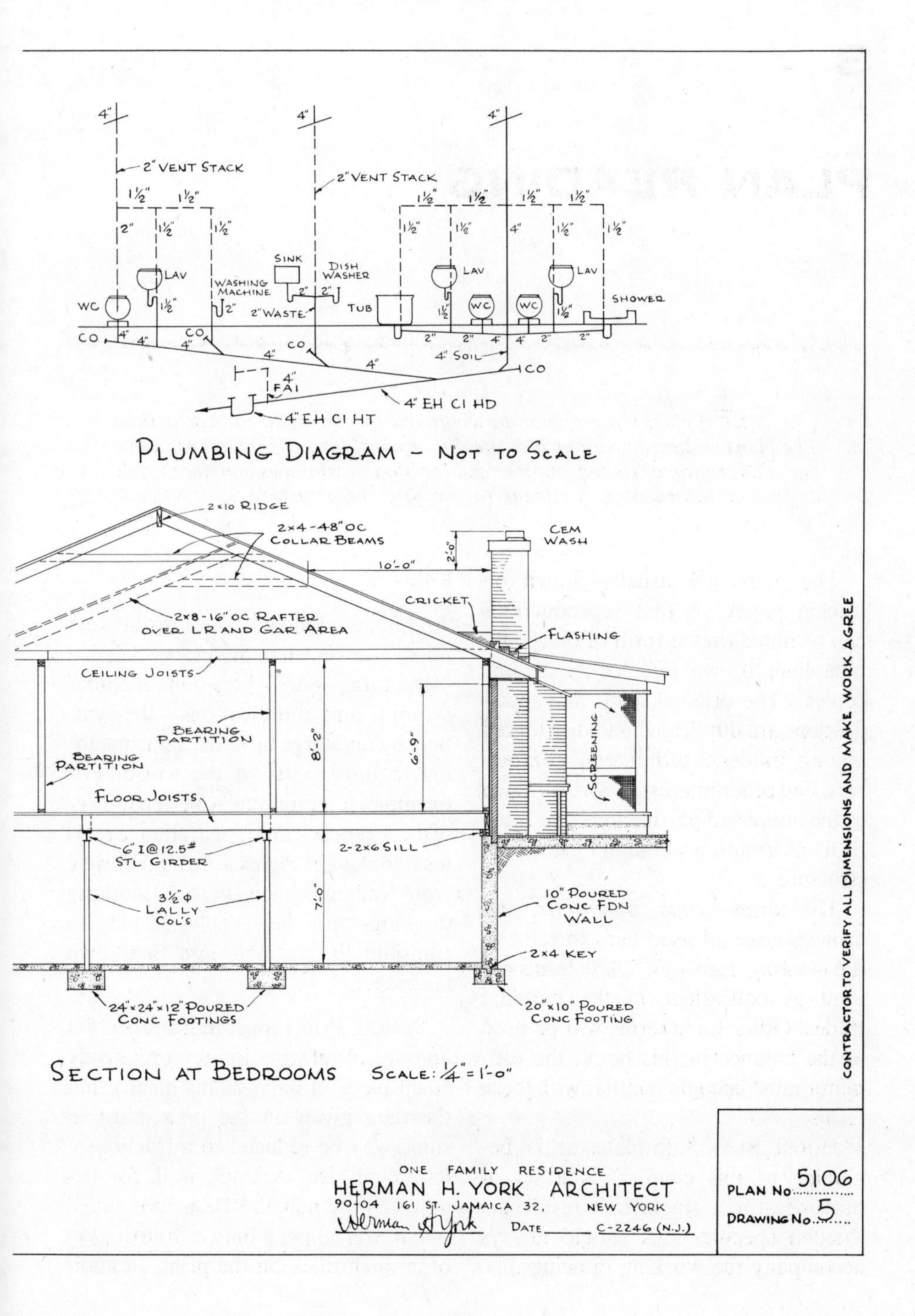
2" VENT STACK
2" VENT STACK
WC
LAV
WASHING MACHINE
SINK
DISH WASHER
2" WASTE
TUB
LAV
WC
WC
LAV
SHOWER
CO
4" SOIL
FAI
4" EH CI HD
4" EH CI HT
PLUMBING DIAGRAM - NOT TO SCALE
2x10 RIDGE
2x4-48" OC COLLAR BEAMS
CEM WASH
10'-0"
CRICKET
2x8-16" OC RAFTER OVER LR AND GAR AREA
FLASHING
CEILING JOISTS
BEARING PARTITION
BEARING PARTITION
8'-2"
6'-9"
SCREENING
FLOOR JOISTS
6" I @ 12.5# STL GIRDER
2-2x6 SILL
3½" φ LALLY COL'S
7'-0"
10" POURED CONC FDN WALL
2x4 KEY
24"x24"x12" POURED CONC FOOTINGS
20"x10" POURED CONC FOOTING
SECTION AT BEDROOMS SCALE: ¼"=1'-0"
CONTRACTOR TO VERIFY ALL DIMENSIONS AND MAKE WORK AGREE
ONE FAMILY RESIDENCE
HERMAN H. YORK ARCHITECT
90-04 161 ST. JAMAICA 32, NEW YORK
DATE
C-2246 (N.J.)
PLAN No. 5106
DRAWING No. 5

3

PLAN READING

A set of plans is a group of drawings and specifications for a structure. The plans and specifications indicate dimensions, kinds of materials, number and location of rooms, number and location of windows and doors, and any other information pertinent to creating the structure.

The plans are usually drawn on tracing paper so that reproductions can be made (in the form of blueprint, blue line, brown line, or black line copies). The original plans and specifications are duplicated and distributed among bidders, estimators, contractors, and other interested parties. Each of the interested parties will have identical information about the proposed structure.

The terms *prints, blueprints,* and *drawings* are all used here to refer to the *working drawings*. These terms are used as equivalent in the building trades. Other trade terms will be used in the balance of this book; the estimator must become familiar with these terms.

HOUSE PLAN A (5 plates at the beginning of this chapter) is a set of drawings for a single unit residence. Written specifications should always accompany the working drawings.

Prints

Architects, builders, and estimators make use of what might be called a "structural code." This code includes symbols and abbreviations—the symbols by and large have the same meaning in most parts of the world. For example, if a group of American, Swedish, French, and Australian architects looked at Plates 1 through 5 they would all recognize these as working drawings and they would be able to translate the symbols into their own languages.

Symbols. Prints must indicate a great amount of information on a relatively small piece of paper. This means that the data given on the print must in some way be reduced to a fraction of its actual size. A brick wall, for example, may actually be a foot thick, but it will appear only a fourth (¼) of an inch thick on the print. In addi-

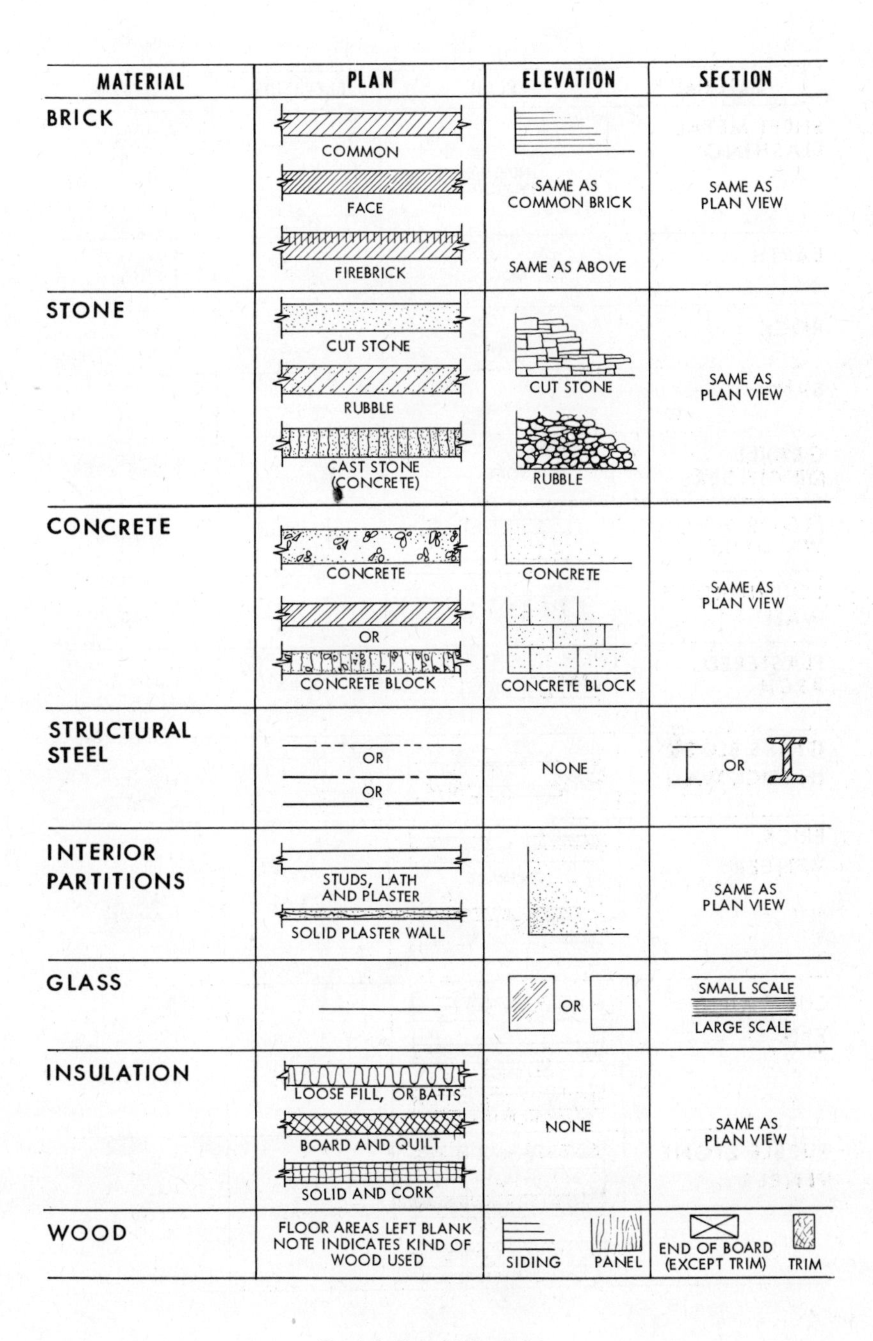

MATERIAL	PLAN	ELEVATION	SECTION
BRICK	COMMON FACE FIREBRICK	SAME AS COMMON BRICK SAME AS ABOVE	SAME AS PLAN VIEW
STONE	CUT STONE RUBBLE CAST STONE (CONCRETE)	CUT STONE RUBBLE	SAME AS PLAN VIEW
CONCRETE	CONCRETE OR CONCRETE BLOCK	CONCRETE CONCRETE BLOCK	SAME AS PLAN VIEW
STRUCTURAL STEEL	OR OR	NONE	OR
INTERIOR PARTITIONS	STUDS, LATH AND PLASTER SOLID PLASTER WALL		SAME AS PLAN VIEW
GLASS		OR	SMALL SCALE LARGE SCALE
INSULATION	LOOSE FILL, OR BATTS BOARD AND QUILT SOLID AND CORK	NONE	SAME AS PLAN VIEW
WOOD	FLOOR AREAS LEFT BLANK NOTE INDICATES KIND OF WOOD USED	SIDING PANEL	END OF BOARD (EXCEPT TRIM) TRIM

Fig. 1-A. Material Symbols.

MATERIAL	PLAN	ELEVATION	SECTION
SHEET METAL FLASHING	INDICATE BY NOTE		HEAVY LINE SHAPED TO CONFORM
EARTH	NONE	NONE	
ROCK	NONE	NONE	
SAND	NONE	NONE	
GRAVEL OR CINDERS	NONE	NONE	
FLOOR AND WALL TILE			
SOUNDPROOF WALL		NONE	NONE
PLASTERED ARCH		DESIGN VARIES	SAME AS ELEVATION VIEW
GLASS BLOCK IN BRICK WALL			SAME AS ELEVATION VIEW
BRICK VENEER	OR ON FRAME	SAME AS BRICK	SAME AS PLAN VIEW
CUT STONE VENEER	OR ON BRICK ON CONCRETE BLOCK	SAME AS CUT STONE	SAME AS PLAN VIEW
RUBBLE STONE VENEER	OR ON FRAME ON BRICK ON CONCRETE BLOCK	SAME AS RUBBLE	SAME AS PLAN VIEW

Fig. 1-B. Material Symbols.

ABBREVIATIONS

Term	Abbreviation
Access Door	AD
Access Panel	AP
Acoustic	ACST
Aggregate	AGGR
Alternating Current	AC
Aluminum	AL
Anchor Bolt	AB
Angle	∠
Apartment	APT
Asbestos	ASB
Asbestos Board	AB
At	@
Barrel	BBL
Basement	BSMT
Bathroom	B
Bath Tub	BT
Beam	BM
Bedroom	BR
Blueprint	BP
Bolts	BT
Boundary	BDY
Brass	BRS
Brick	BRK
Broom Closet	BC
Building	BLDG
Building Line	BL
Cabinet	CAB
Casing	CSG
Cast Iron	CI
Catch Basin	CB
Ceiling	CLG
Cellar	CEL
Cement	CEM
Center	CTR
Center Line	CL, ℄
Center Matched	CM
Center-to-Center	C to C
Ceramic	CER
Cesspool	CP
Channel	CHAN
Cinder Block	CIN BL
Circuit Breaker	CIR BKR
Clean Out	CO
Clear	CLR
Clear Glass	CL GL
Closet	CLO, CL, C
Cold Air	CA
Cold Water	CW
Collar Beam	COL B
Column	COL
Concrete	CONC
Concrete Block	CONC B
Conduit	CND
Copper	COP
Counter	CTR
Cubic Feet	CU FT
Cubic Yards	CU YDS, ₵
Cut Out	CO
Dampproofing	DP
Detail	DET
Diagram	DIAG
Diameter	DIA, ϕ
Dimension	DIM
Dinette	DT
Dining Alcove	DA
Dining Room	DR
Direct Current	DC
Dishwasher	DW
Ditto	DO, "
Door	DR
Double-Acting Door	DAD
Double-Hung Window	DHW, DH
Double Strength Glass	DSG
Down	DN, D
Downspout	DS
Drain	D, DR
Drain Board	DB
Drawing	DWG
Dressed & Matched	D & M
Drip Cap	DC
Dryer	D
Electric Panel	EP
Electric	ELEC
Elevation	ELEV, EL
End-to-End	E to E
Entrance	ENT
Excavate	EXC
Expansion Joint	EXP JT
Exterior	EXT
Finish	FIN
Finished Floor	FIN FL
Firebrick	FBRK
Fireplace	FP
Fireproof	FPRF
Fireproof Solid Core	FPSC
Flashing	FL
Floor	FL
Flooring	FLG
Fluorescent	FLUOR
Flush	FL
Foot or Feet	FT,
Footing	FTG
Foundation	FDN
Frame	FR
Fresh Air Intake	FAI
Full Size	FS
Furring	FURR
Games Room	GR
Galvanized	GALV
Galvanized Iron	GI
Garage	GAR
Gas	G
Gage	GA
Glass	GL
Glass Block	GL BL
Grade	GR

ABBREVIATIONS

Grade Line	GL	Porch	P
Gypsum	GYP	Pound	LB, #
Hall	H	Pounds per Square Foot	PSF
Hardware	HDW	Powder Room	PR
Height	HGT, H, HT	Precast	PRCST
High Point	H PT	Prefabricated	PREFAB
Hot Air	HA	Pull Chain	PC, P
Hot Water Tank	HWT	Pull Switch	PS
I Beam	I	Radiator	RAD
Inches	IN., "	Radius	R, r
Inside Diameter	ID	Recessed	REC
Insulation	INS	Refrigerator	REF
Interior	INT	Reinforced	REINF
Iron	I	Revision	REV
Jamb	JB	Riser	R
Joist Space	JS	Rivet	RIV
Kick Plate	KP	Roof	RF
Kitchen	K	Roof Drain	RD
Kitchen Cabinet	KC	Roofing	RFG
Kilowatt	KW	Room	RM, R
Knocked Down	KD	Rubber	RUB
Landing	LDG	Rubber Tile	R TILE
Lath	LTH	Scale	SC
Laundry	LAU	Screen	SCR
Lavatory	LAV, L	Section	SECT
Leader	L	Sewer	SEW
Length	L, LG, lgth	Sheathing	SHTHG
Length Overall	LOA	Sheet	SH
Level	LEV	Shelving	SHELV
Light	LT	Shiplap	SHLP
Limestone	LS	Shower	SH
Line	L	Siding	SDG
Linen Closet	LIN, L CL	Single Strength Glass	SSG
Lining	LN, Lng	Sink	SK, S
Linoleum	Lino	Sliding	SLID
Living Room	LR	Soil Pipe	SP
Louver	LV	Specifications	SPEC
Low Point	LP	Square	SQ, sq.
Maximum	MAX	Square Inch	SQ IN , sq. in.
Medicine Cabinet	MC	Square Foot	SQ FT , sq. ft.
Metal	MET, M	Stairs	ST
Minimum	MIN	Standard	STD
Miscellaneous	MISC	Steel	STL
Mixture	MIX	Stone	STN
Molding	MLDG	Steel Sash	SS
Mortar	MOR	Storage	STG
Movable Partition	M PART	Switch	SW, S
Mullion	MULL	Telephone	TEL
Number	NO., #	Terra Cotta	TC
On Center	OC	Thermostat	THERMO
Opening	OPNG	Thick or Thickness	THK, T
Outlet	OUT	Threshold	TH
Outside Diameter	OD	Thousand	M
Panel	PNL	Through	THRU
Partition	PTN	Tongue & Groove	T & G
Perpendicular	PERP	Top Hinged	TH
Plaster	PLAS, PL	Tread	TR, T
Plate	PL	Unexcavated	UNEXC
Plumbing	PLMB	Utility Room	UR

ABBREVIATIONS

Vent	V
Ventilating or Ventilation	VENT
Vertical	VERT
Vinyl Tile	V TILE
Volt	V, v
Volume	VOL, V
Washing Machine	WM
Washroom	WR
Water Closet	WC
Water Heater	WH
Waterproofing	WP
Watts	W, w
Weather Stripping	WS
White Pine	WP
Width	W, WTH
Window	WDW
Window Radiator	WR
Wire Glass	W GL
Wood	WD
Wrought Iron	WI
Yard	YD, yd.
Yellow Pine	YP

tion to indicating thickness (which is done by scale drawing—explained in a later section) the print must also indicate the material in the wall. In this case, the material is brick. No draftsman would want to draw a likeness of each individual brick and no firm of architects would find it economical to do so. To facilitate drawing and make them less expensive to produce—a standard symbol is used. Figs. 1 through 4 show some of the symbols used in working drawings. The estimator should learn to recognize these symbols quickly and accurately. (*Warning:* Some localities and trades use other symbols. Be sure to check the symbols given here against local custom. The symbols presented here, however, are widely used and recognized.)

Abbreviations. Like symbols, abbreviations are used to save time and to conserve space. Again as with symbols, there are non-standard abbreviations in use, but their meanings usually become clear as you study the prints on which they appear.

Plan Views. Every set of plans includes a plan view of each floor of the house. (Plan views are also known as *floor plans;* the meaning is the same.) Fig.

PLUMBING SYMBOLS

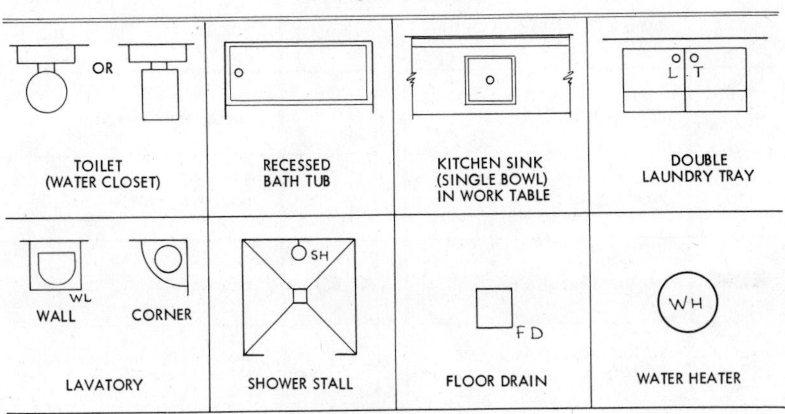

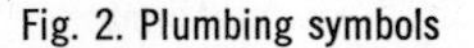
Fig. 2. Plumbing symbols

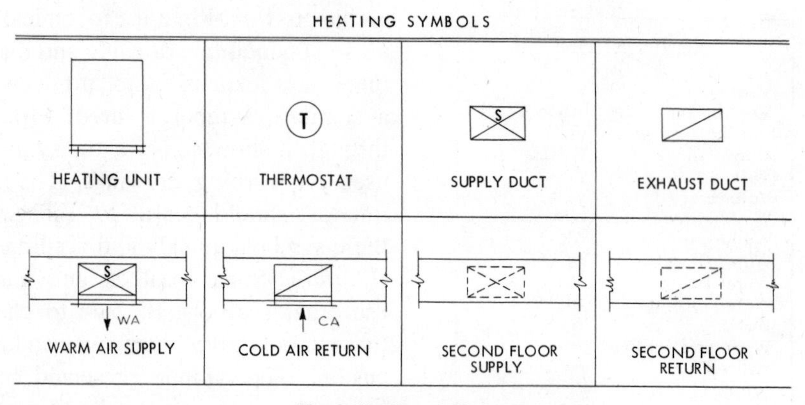

Fig. 3. Heating symbols.

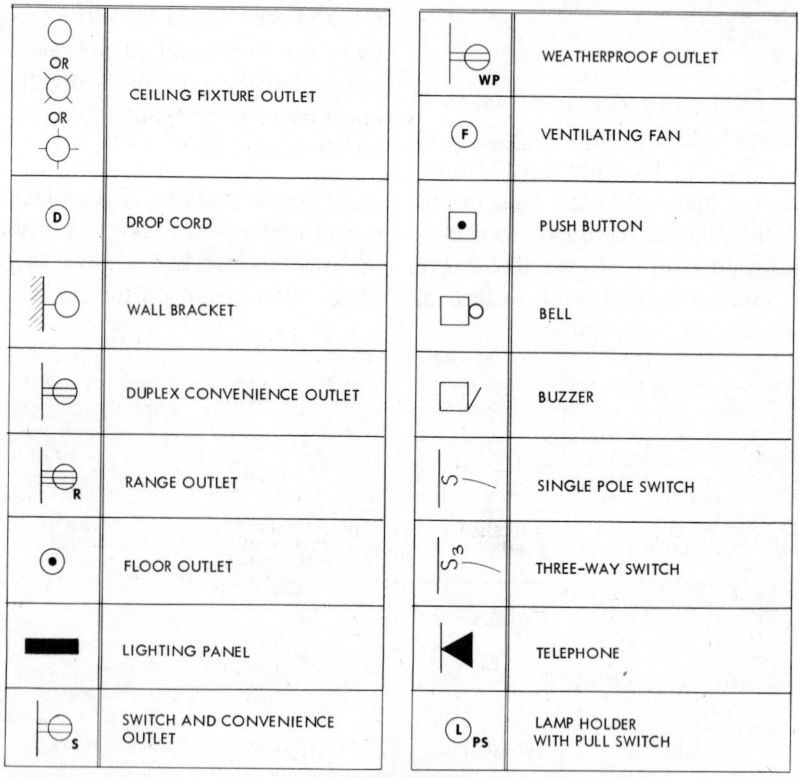

Fig. 4. Electrical symbols.

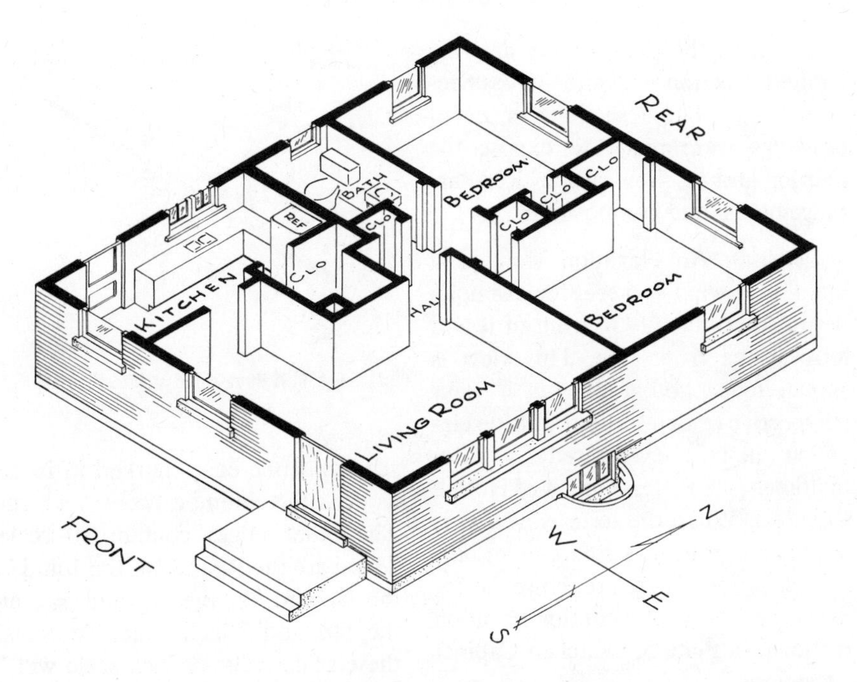

Fig. 5. Cutaway isometric view of small house.

5 shows a cutaway isometric view of a small house. The house appears to have been sawed through from one side to the other at the level of the first floor windows. The top has been left out of the drawing. To understand a plan view visualize the heavy black lines of Fig. 5 directly below your eyes so that only the heavy lines are visible. You now have a plan view. Plate 1 is such a view.

Sections. When you cut through the house in Fig. 5, you are *sectioning* the structure as well as revealing its interior plan. Plan views, to a large extent, represent simplified sectional views. Another example of sectioning might involve a watermelon. The outer surface (rind) gives no indication of the interior. By cutting the melon in half you can see the interior structure —red meat and black seeds. The cut surface of the melon can be called a melon·because it shows the makeup or construction of the melon, including the skin thickness, distribution of the seeds, and the apportionment of red meat.

Sections can be taken from either a plan view or from an elevation view. Plan view sections are called *cross sections* or simply *sections*. Sections taken from elevation views are called *longitudinal sections* or *sectional elevations*.

Section drawings (or details) are made to illustrate structural details not shown in the plan or elevation drawings. Look at Plate 5 of House Plan

A and note the parts of the drawing entitled "Section . . ." and "Cross Section. . . ." These sections or detail drawings were made to expose the interior and to show proportions, arrangements, and composition.

Elevations. An elevation view of a building is what you see, for example, if you stand directly in front of it and look at the front side. This view is similar to a photograph, but it lacks perspective or depth. Four outside elevation views are required for most buildings (see Plates 3 and 4 of HOUSE PLAN A). When the term *elevation* is used alone, it usually refers to an *outside elevation*. There are, however, *interior elevations*. An interior elevation is shown in Plate 5, "Kitchen Cabinet Elevations."

Scale

Drawings must be made smaller than the actual size of the building they represent. Working drawings are made to a predetermined ratio of the actual dimensions. This practice is known as drawing to scale. In order to scale it is necessary to adopt some unit of measurement so that the ratio of drawing dimensions is constant. (The ratio should be stated on the plans.)

A common twelve inch ruler is a scale that provides a unit of measure. The twelve inches are marked and divided into halves, quarters, eighths, and sixteenths. Architects, however, use an *Architect's Scale* which may be either flat or three-sided. Fig. 6 shows a three-sided scale. The three-sided

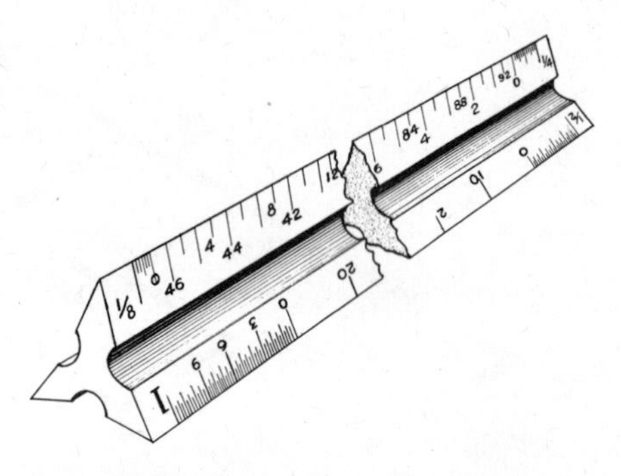

Fig. 6. A three-sided architect's scale.

scale has one edge marked in inches, as does the common twelve inch rule. The other edges contain 10 scales. These are the ⅛ and ¼, the 1 and ½, the ¾ and ⅜, the 3/16 and 3/32, and the 1½ and 3-inch scale. To explain these scales, the ¼ inch scale will be used.

Note the edge marked "¼". This edge is read from right to left. It is divided into spaces each ¼-inch apart. At one end, one of the ¼-inch spaces has been divided into halves, quarters, and twelfths. One of the ¼-inch spaces represents 1 foot, or 12 inches. One of the halves, in that space divided into 12 equal parts, represents 6 inches, and so forth.

Suppose in drawing a plan view you are confronted with the problem of drawing a line 8 feet long according to the ¼-inch scale. To do this, you would just count off 8 spaces on the scale. As another example, suppose you must draw a line representing a dimension of 6′-6″, using the same ¼-inch scale. First count off 6 of the ¼-inch spaces; then add 6 spaces from that space which is divided into

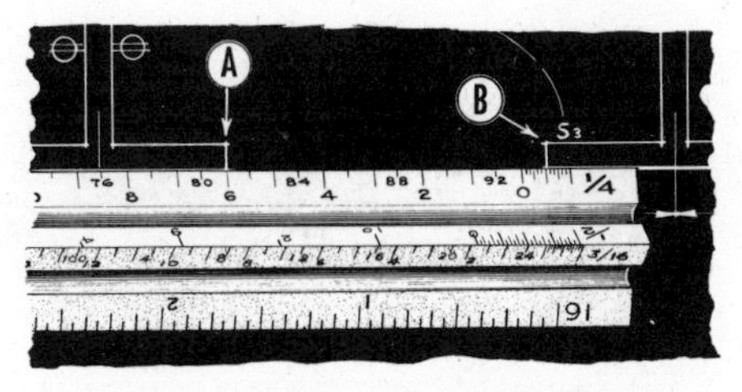

Fig. 7. Scaling a working drawing.

12 equal parts. (This is the distance from A to B in Fig. 7.) In drawing to the ¼-inch scale, therefore, you are simply substituting quarter inches for feet.

The other scales operate in the same manner as the ¼-inch scale. Every part of a drawing must be drawn accurately according to whatever scale is being used, unless that portion of the drawing is labeled "not to scale." This procedure enables the designer to make his drawings in exact ratio to the proposed building.

Generally, drawings are made using the ¼ inch scale. However, in designing large buildings, a smaller scale is sometimes used. Conversely, small details are often drawn to a large scale.

Structural conventions

Dimensions. Dimensions are the all-important guide for the estimator and builder. The prints should show dimensions for all rooms, windows and door locations, heights, and so forth. Fig. 8 illustrates typical dimension lines. This can be seen in Plates 1 through 5.

Look at Plate 1 and locate the 16′-6″ dimension in the lower right corner. If this dimension had not been shown, you could find it by using simple arithmetic. Looking at the other dimensions, you see that the total length of the house front is 64′-0″. Two partial lengths are given as 16′-0″ and 31′-6″. Subtracting the sum of these two partial lengths from the total length, you can find the missing dimension:

$$\begin{array}{rl} 64'\text{-}0'' & \\ -\underline{47'\text{-}6''} & \quad (16'\text{-}0'' + 31'\text{-}6'') \\ 16'\text{-}6'' & \end{array}$$

Note that 64′-0″ = 63′-12″. This makes it easy to subtract 47′-6″. Converting to decimal fractions makes the calculations even easier. (This is discussed in the following chapter.)

Missing dimensions can be found by scaling, that is, by using a set of dividers to measure the desired dimension on the prints and then placing the dividers along a printed scale. Another method of finding the missing dimensions is by using a ruler (preferably an *Architect's Scale*) and measuring the dimensions directly from the prints. Direct measuring should be avoided, however, because of inevitable inaccuracies in drawing and because the prints can shrink.

Windows and Doors. Fig. 9 shows some of the commonly used types of windows. Fig. 10 illustrates how these windows would be shown on drawings. Fig. 11 shows how doors are indicated on drawings. These illustrations are self-explanatory and should be studied carefully. Now look at Plates 1 through 5 of HOUSE PLAN A to identify the window and door types used in this building.

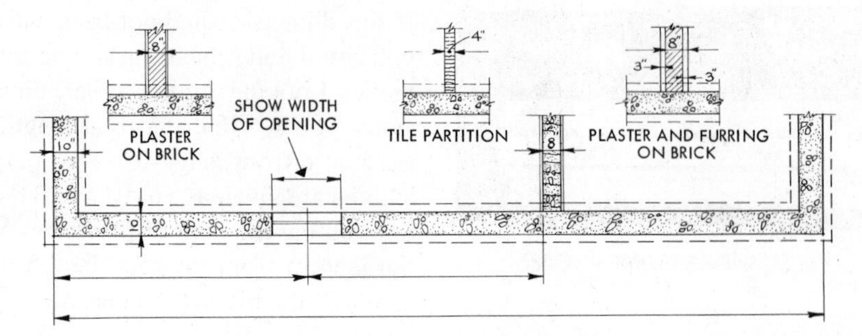

CONCRETE FOUNDATION

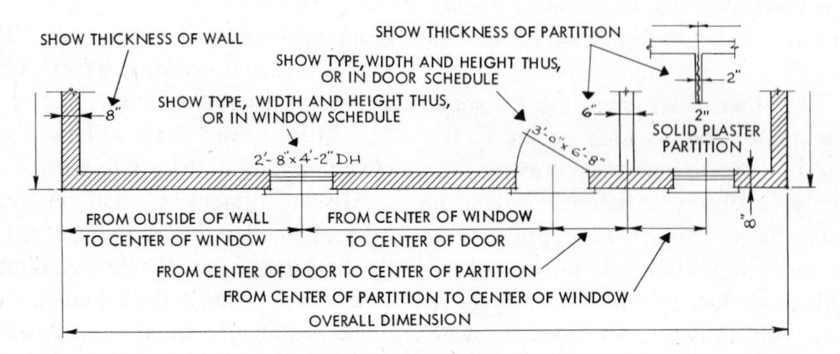

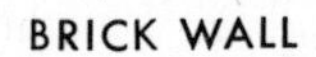
BRICK WALL

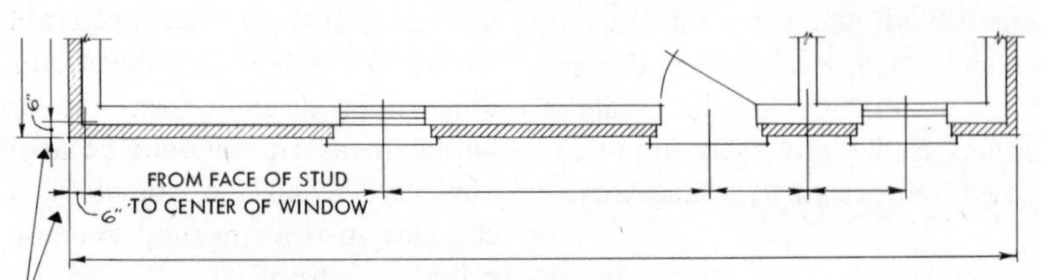

BRICK VENEER

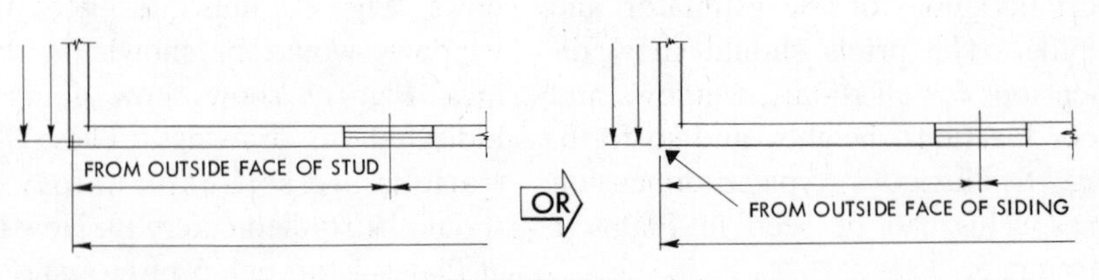

FRAME WALL

Fig. 8. Typical dimension lines used on drawings.

(A)

(B)

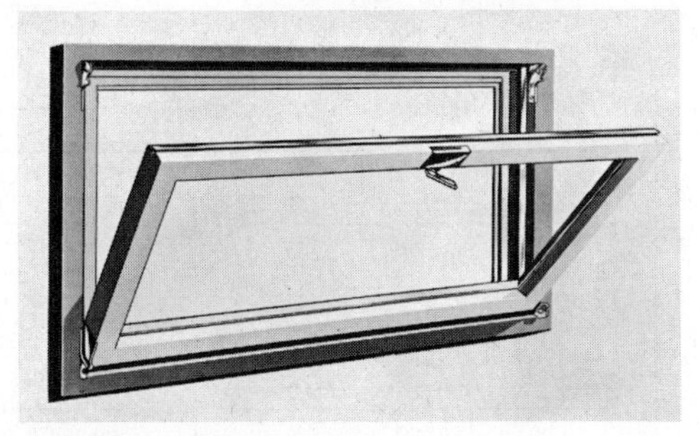

(C)

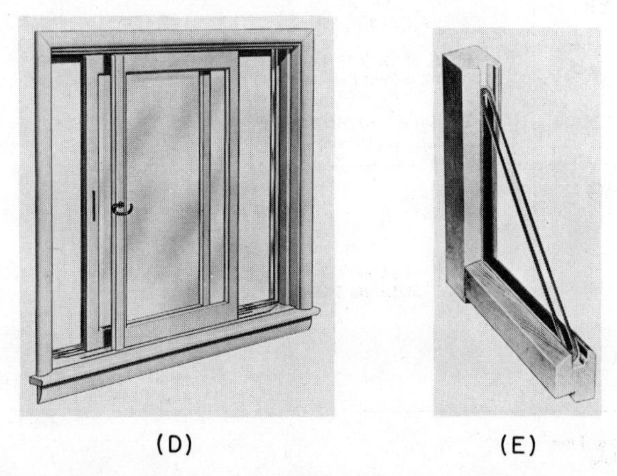

(D) (E)

Fig. 9. Windows commonly used for residential construction: (A) double hung, (B) casement, (C) hopper, (D) sliding sash, (E) insulating glass.

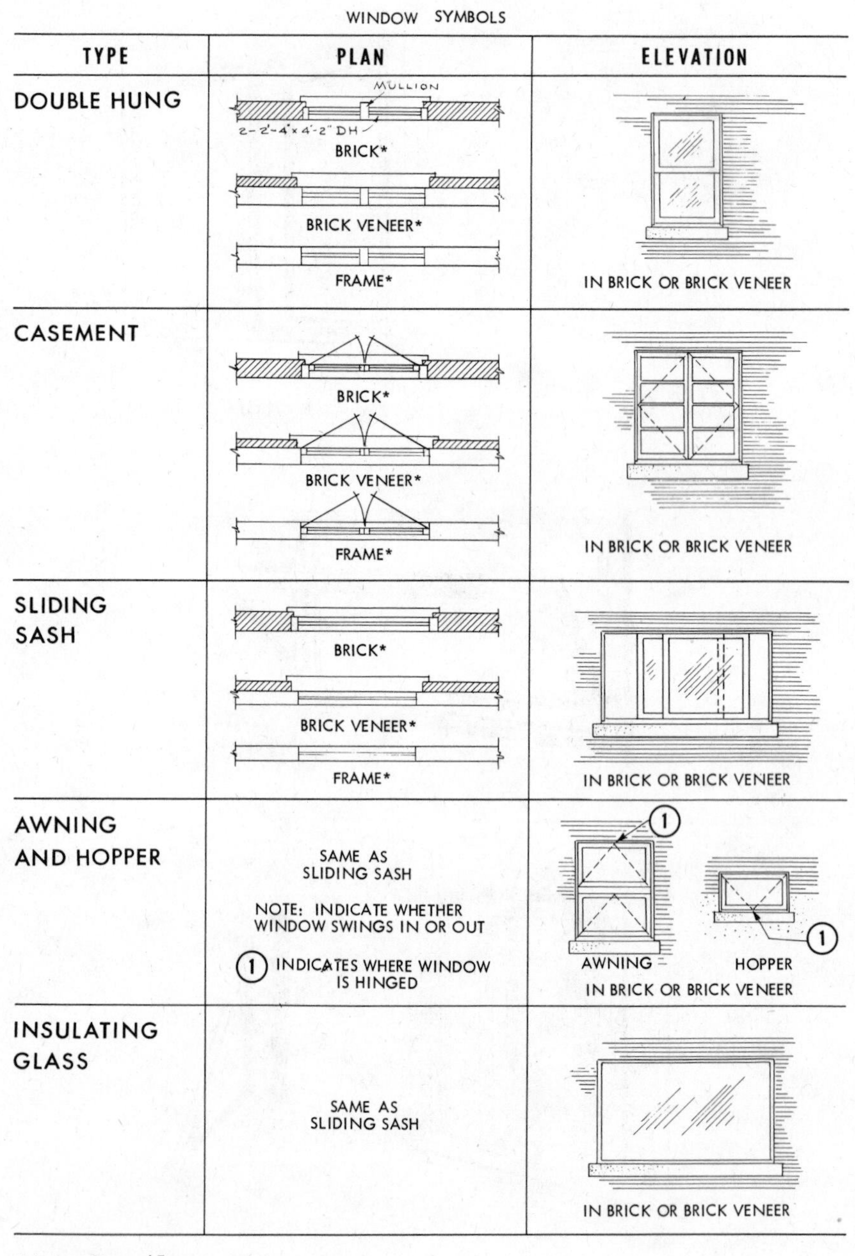

*Number, Size and Type Indicated by Note

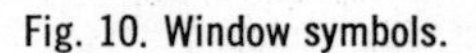
Fig. 10. Window symbols.

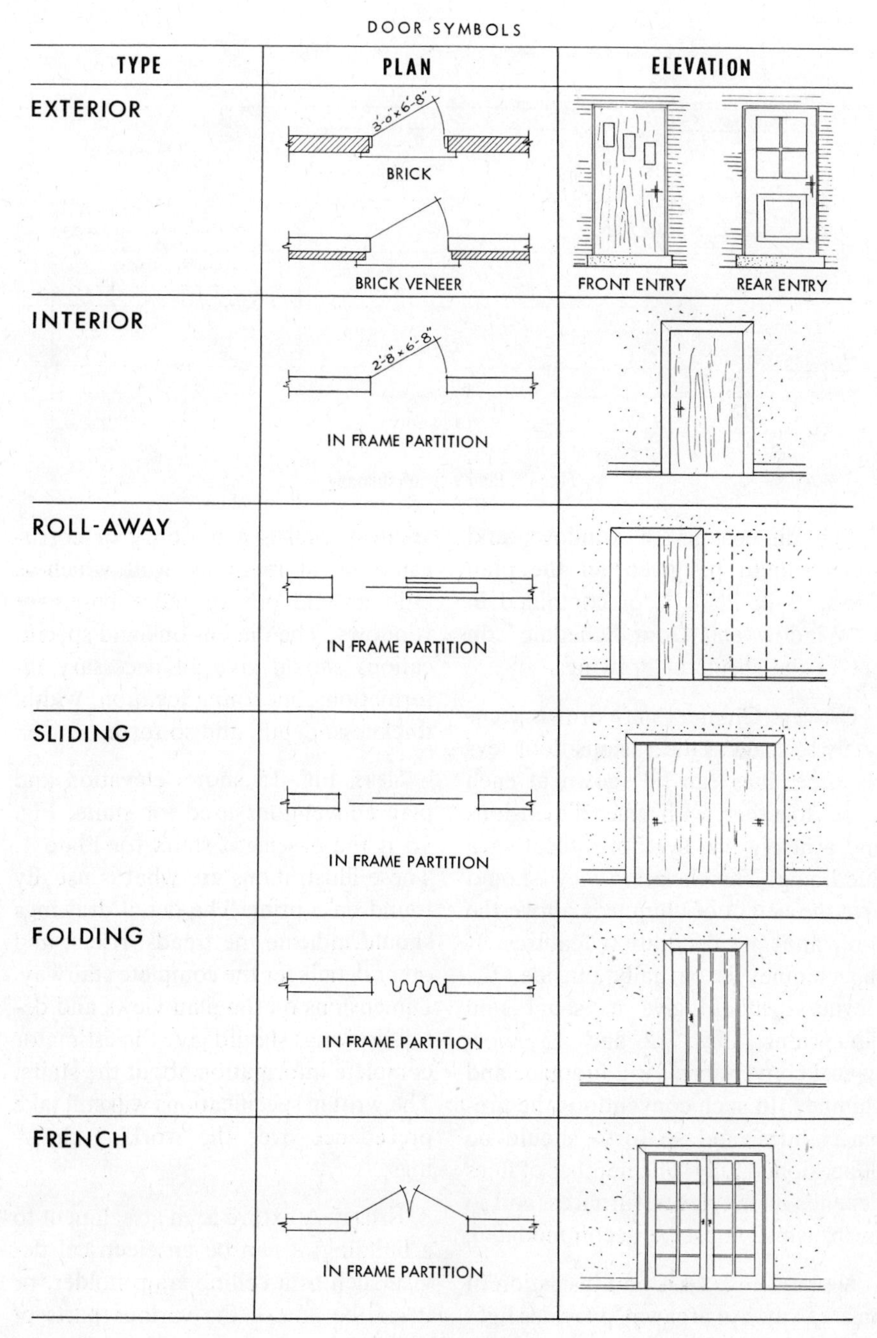

Fig. 11. Door symbols.

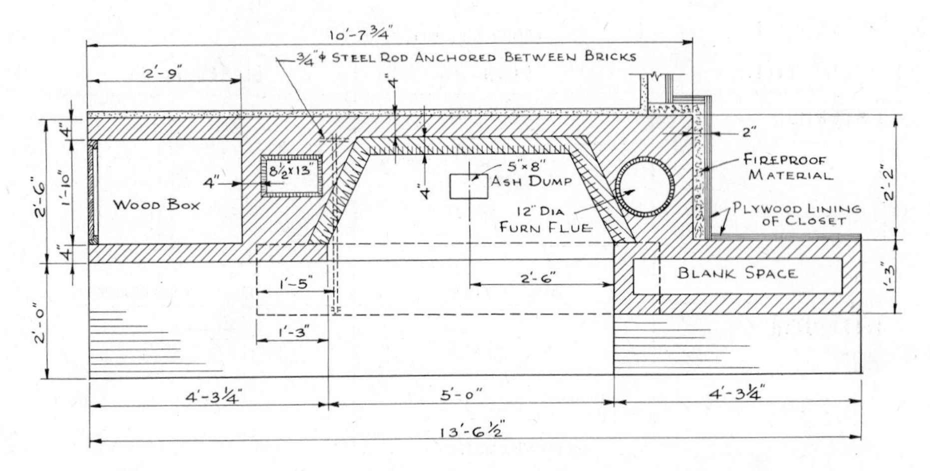

Fig. 12. Plan view of chimney.

The dimensions of windows and doors should be given on the plan views, as in Plate 1, or contained in a "Window and Door Schedule" on one of the sheets of drawings.

Chimneys. Chimneys are drawn accurately to scale at the various floor levels. The flues can be shown at each level, together with flue dimensions and a heavy line symbol indicating a flue lining. The elevation view should give the height of chimneys above the roof, and any decoration features. If the chimney is partially outside, the elevation should give its shape and dimensions. Figs. 12 and 13 show typical conventions for a fireplace and chimney. In such conventions the fireplace, flues, and so forth should be dimensioned fully. The number of flues depends on fireplaces, furnaces, and so forth, which must be accommodated.

Areaways. Fig. 14 is an illustration of an areaway. An areaway provides light and ventilation for basement areas. An areaway wall is a masonry or corrugated metal retaining wall which is built around one or more basement windows. The dimensions and specifications should give all necessary information concerning location, width, thickness, depth, and so forth.

Stairs. Fig. 15 shows elevation and plan conventions used for stairs. Fig. 16 is the basement stairs for Plate 1. These illustrations are what is usually found on a print. The detail drawings should indicate the treads, risers, and other details for the complete stairway. Dimensions on the plan views and detail drawings should give the estimator complete information about the stairs. The written specifications will still take precedence over the working drawings.

Fixtures. A fixture is an attachment to a building; it can be an electrical device such as a ceiling lamp holder, or it can be any of the various parts of the plumbing system.

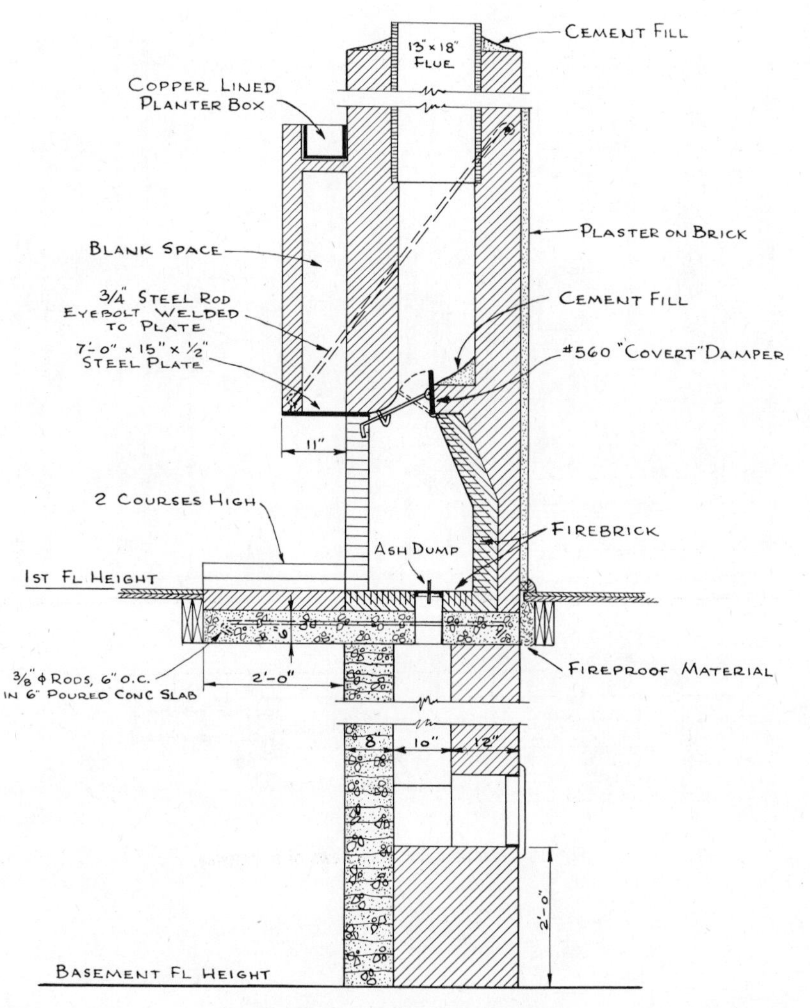

Fig. 13. Elevation section view of chimney.

Fig. 14. Areaway with corrugated metal retaining wall.

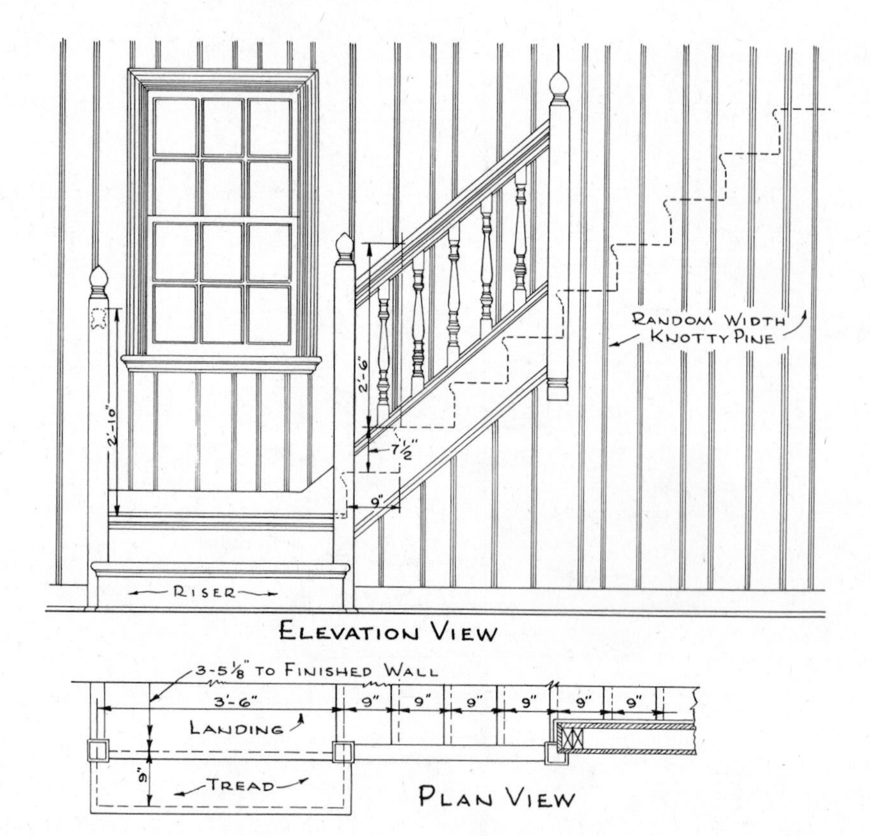

Fig. 15. Elevation and plan views of a stairway.

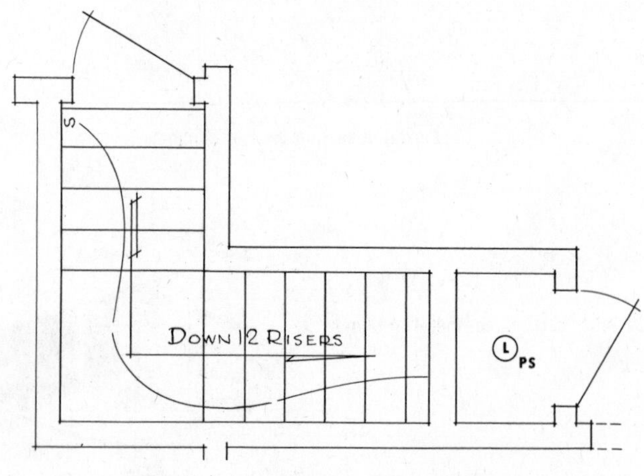

Fig. 16. Basement stairs of House Plan A.

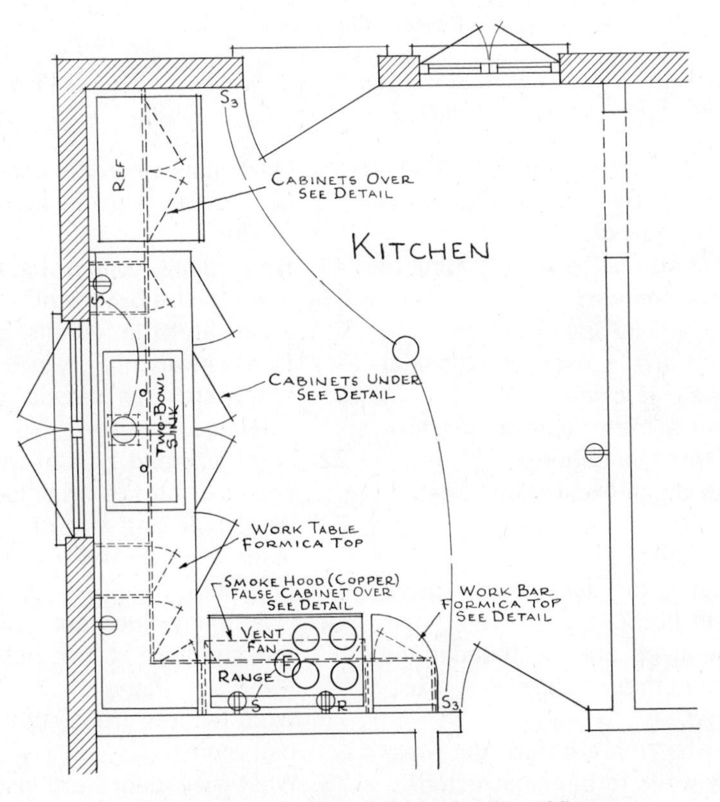

Fig. 17. Floor plan of a kitchen, showing locations of fixtures.

Bathroom fixtures include lavatories, toilets (also called water closets), bathtubs, showers, and medical cabinets and tile floors. These conventions must be drawn accurately to scale. The written specifications tell the style, quantity, and color.

Kitchen fixtures such as an electric range or gas stove, refrigerator, and table and cabinets can all be indicated by rectangles drawn to scale. These are placed in their proper positions and sometimes named on the drawings as in Fig. 17. The written specifications should give the style, quantity, color, and so forth.

Laundry fixtures can be indicated by rectangles or circles representing washtubs and washing machines. Ironing boards, washing machines, and dryers can also be represented by rectangles.

Sewers and Drains. Sewer pipes and water pipes are usually represented by small circles or by dot-and-dash lines. Catch basins (cisterns) are indicated by large circles. Downspouts (also called "leaders") are represented by small circles on plan views and by parallel lines on elevations.

Remember that the plans and specifications must be used *together*.

Review Questions

The following questions refer to plates 1 through 5 of House Plan A, *at the beginning of this chapter.*

1. What are the interior dimensions of the foundation for the screened porch?
2. What size is the door leading to the basement?
3. Is any steel specified?
4. Is all of the basement floor at the same level?
5. What are the main dimensions of the foundations?
6. How much excavation is necessary?
7. How are footings indicated?
8. What is the depth of the basement floor?
9. How deep are the foundations under those areas not excavated?
10. Of what material are the areaway walls to be constructed?
11. What size and kind of basement windows are indicated?
12. How many fireplaces are to be constructed?
13. How many steps are necessary between the basement and the first floor?
14. How wide are the basement stairs?
15. Are lintels necessary over the basement windows?
16. Of what material are the garage walls to be constructed?
17. What are some of the fireplace details?
18. How many electrical baseboard outlets are there in the Family Room?
19. How many ceiling lights are there in the basement?
20. What are the details of the front entrance?
21. What are the details of the screened porch?
22. What size and type of windows are used on the first floor?
23. What type and size of opening leads from the foyer to the living room?
24. How many flues are visible in the chimney at the basement level?
25. What fixtures are located in the bathrooms?
26. What size doors are used for the bedrooms?
27. What size closet doors are used?
28. What type of louvers are shown?
29. What are the details of a typical wall section?
30. What type of outside wall construction is used?
31. What type of door is used for the garage?
32. How is the top of the chimney finished?
33. What is the roof pitch?
34. What type of shingles are to be used?
35. Are shutters indicated?

4

SQUARE AND CUBIC MEASURE

As you begin to study this chapter you might ask yourself, "Is it necessary to know all these figures and shapes and their areas and volumes"? For accurate quantity take-offs and estimates, the answer is "Yes," as you will realize when you work the problems. You will be reviewing mensuration, *which is the branch of mathematics dealing with* length, area, *and* volume. *Problems at the end of each section will illustrate the relevant types of calculations.*

Remember that all units must be the same when you do the calculations. For example, you cannot multiply feet by inches or yards by feet. When the dimensions are given in feet, the area is in square feet *and the volume is in* cubic feet.

Decimal fractions are used whenever possible. (See Table 1 for the decimal equivalents.) For example, instead of using 2'6" or 2½', it is easier to use 2.5'. By converting to decimals, the calculations are made easier and you can also use slide rules or adding and calculating machines. Using the slide rule or these machines helps to reduce errors and can save much time and effort.

Square measure

A *plane* is a flat surface bounded by straight or curved lines and having no thickness. All building sites may be considered plane surfaces. The figure is a *polygon* if bounded by straight lines. The sum of the sides of a polygon, that is, the distance around, is the *perimeter*. The perimeter of a circle is called the *circumference*. Typical plane figures are illustrated in Fig. 1.

The main point to remember about plane figures is that they are considered as having area only. (That is, they have no depth.) The surface of a roof, floor, or sidewalk is a plane figure when regarded solely in terms of its boundary lines. Area can therefore be thought of as describing the extent of

TABLE 1. DECIMAL EQUIVALENTS OF INCHES IN FEET

1 INCH	= 0.08 FEET	7 INCHES	= 0.58 FEET
2 INCHES	= 0.17 "	8 "	= 0.67 "
3 "	= 0.25 "	9 "	= 0.75 "
4 "	= 0.33 "	10 "	= 0.83 "
5 "	= 0.42 "	11 "	= 0.92 "
6 "	= 0.5 "	12 "	= 1.0 "

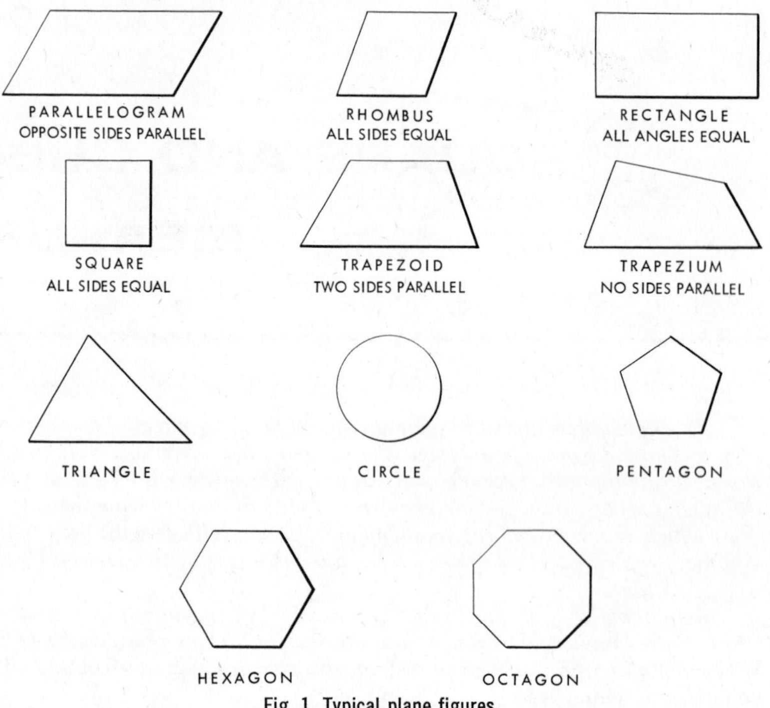

Fig. 1. Typical plane figures.

a plane figure. When you say that a farm contains 160 acres you are thinking only in terms of the surface area occupied by the farm. You are not considering depth of the soil. The same idea applies to all plane figures.

Areas are calculated in terms of square feet, square yards, acres, square miles, and so forth. See Table II. Generally, the estimator is concerned with square feet and square yards.

TABLE II. SQUARE OR SURFACE MEASURE

144 SQUARE INCHES	= 1 SQUARE FOOT
9 SQUARE FEET	= 1 SQUARE YARD
30 1/4 SQUARE YARDS	= 1 SQUARE ROD
160 SQUARE RODS	= 1 ACRE
43,560 SQUARE FEET	= 1 ACRE
4,840 SQUARE YARDS	= 1 ACRE
640 ACRES	= 1 SQUARE MILE = 1 SECTION
36 SECTIONS	= 1 TOWNSHIP

Fig. 2 illustrates square measure. The part bounded by *A, B, C,* and *D* represents one square foot (*sq. ft.*). It is a *square* foot because each of its four sides is one foot long. You know that 1 foot equals 12 inches, so in the square *ABCD* the sides are each divided into 12 equal parts, or "inches." If lines are drawn horizontally and vertically, you can see that the square *ABCD* is divided into 144 smaller squares, each equal to 1 square inch (*sq. in.*). The area is the length times the width, or 12 inches × 12 inches equals 144 square inches; thus 144 square inches equals 1 square foot.

The larger square in Fig. 2, *AEFG,* represents one square yard (*sq. yd.*). It

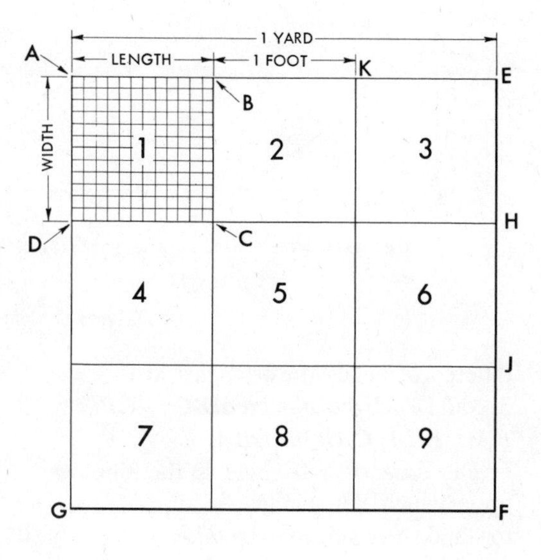

Fig. 2. A square yard contains 9 square feet. A square foot contains 144 square inches.

is a square yard because all sides are one yard, or three feet long. The distances *AB, BK, KE, EH, HJ,* and *JF* are each one foot long. Since the square marked 1 equals one square foot, a square yard contains nine square feet.

Rectangles and Squares. *The area of a rectangle or square is equal to the product of its length and width.*

$$A = a \times b$$

where A = area
a = length
b = width

NOTE: Since the sides of a *square* are equal, the formula becomes

$$A = a \times a = a^2*$$

Example 1. Calculate the area of a rectangular floor whose length is 13′-6″ and whose width is 10′-8″.

*The small figure "2" means "squared"; e.g. $3^2 = 3 \times 3 = 9$.

Solution: First convert to decimals (use Table I).

13′-6″ = 13.5′ 10′-8″ = 10.667′

(Round off to two places in the computation.)

$$\text{Area, } A = a \times b = 13.5' \times 10.67' = 144 \text{ sq. ft.}$$

Example 2. How many *square yards* are there in the floor of Example 1?

Solution: 9 sq. ft. = 1 sq. yd.

$$\text{Area, } A = \frac{144}{9} = 16 \text{ sq. yd.}$$

Triangles. A *triangle* is a polygon enclosed by three straight lines called sides, Fig. 3. Many building elements have a triangular shape. Two examples are the gabled roof and most shed roofs (elevation views).

Triangles are named by reading the letters at each angle just like a rectangle or other polygon. The order in which the triangles are read makes no

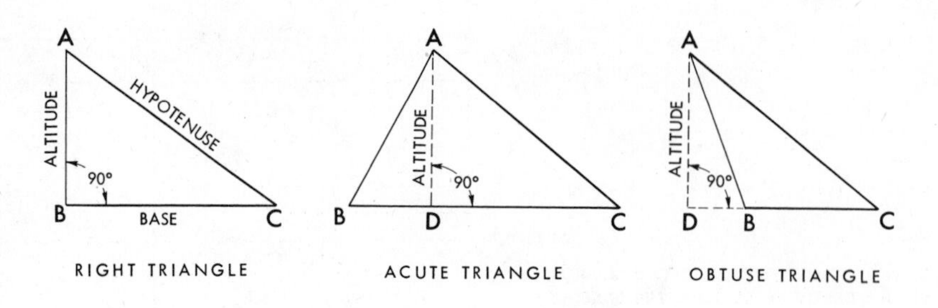

Fig. 3. Types of triangles.

difference; hence the triangles in Fig. 3 could each be named *ABC, ACB, BAC, BCA, CAB,* or *CBA*.

The *base* of a triangle is the side upon which the triangle is supposed to stand. Any side may be taken as the base.

The *altitude* of a triangle is the perpendicular drawn from the vertex to the base. It may be necessary to extend the base of an "obtuse" triangle (Fig. 3) so that the perpendicular will meet it.

A *right-angled* triangle, often called a *right triangle,* is one that has a right angle. The longest side (opposite the right angle) is called the *hypotenuse,* and the other sides are often referred to as *legs*. In a right triangle if two angles are equal then two legs will be equal. One leg may be considered as the base and the other leg as the altitude of the triangle.

Right Triangle Law. Pythagoras, a Greek mathematician and philosopher (about 580 B.C.), formulated the *Pythagorean Theorem,* also known as the *Right Triangle Law:*

The square of the hypotenuse of a right triangle is equal to the sum of the squares of the other two sides.

$$c^2 = a^2 + b^2$$

where c = hypotenuse
a = one side
b = other side

It is always possible to find the length of one side of a right triangle if the other two sides are known.

$$a^2 = c^2 - b^2$$
$$b^2 = c^2 - a^2$$

Example 3. A tool shed has a roof with a rise of 6 feet and a run of 10 feet. What is the length of the rafter? (That is, in arithmetical terms, what is the hypotenuse in a triangle with one leg of 6 feet and the other 10 feet.)

Solution: Substitute in the formula,

$$c^2 = a^2 + b^2$$
$$c^2 = 6^2 + 10^2$$
$$= 36 + 100$$

$$c = \sqrt{136}$$
$$= 11.66 \text{ ft., or } 11'\text{-}8''$$

NOTE: A table of square roots is included in the Appendix.

Area of Triangles. *The area of a triangle is equal to one-half the product of the base and the altitude.* (See Fig. 3.)

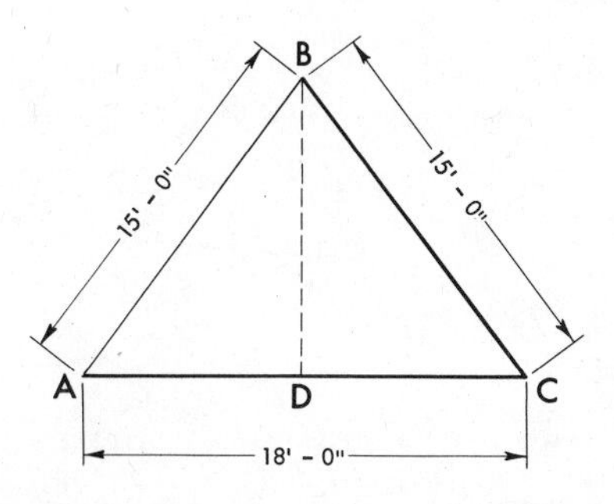

Fig. 4. An isoceles triangle has two sides of equal length.

$$A = \frac{1}{2} \times b \times h$$

where b = base
h = altitude

Example 4. Given that the base of a triangle is 10 feet and the altitude is 8 feet. Find the area.

Solution: Substitute in the formula,

$$A = \frac{1}{2} \times b \times h$$
$$A = \frac{1}{2} \times 10 \times 8$$
$$= 40 \text{ sq. ft.}$$

Example 5. Refer to Fig. 4. The base of the triangle is 18 feet and each side is 15 feet. Find the area.

Solution: First find the altitude, *BD*. The line *BD* is a perpendicular, and in an *isoceles* triangle (two sides equal) it divides the triangle into equal right triangles; thus triangle *ABD* equals triangle *BCD,* and *AD* equals *DC*.

$$AD = DC = 9 \text{ ft.}$$
$$a^2 = c^2 - b^2$$
$$(BD)^2 = (AB)^2 - (AD)^2$$
$$= 15^2 - 9^2$$
$$= 225 - 81$$
$$= 144 \text{ sq. ft.}$$
$$BD = \sqrt{144} = 12 \text{ ft.}$$

The area is now calculated:

$$A = \frac{1}{2} \times b \times h$$
$$= \frac{1}{2} \times 18 \times 12$$
$$= 108 \text{ sq. ft.}$$

Circles. The circle, Fig. 5, is a common shape in the building trades. Structural columns, pipes, and fixtures involve the circle.

A diameter of a circle is a straight line drawn through the center of the circle, ending at two points on the *circumference (perimeter)* of the circle.

A radius of a circle is a straight line joining the center and the circumference. All radii of the same circle are equal and their length is always one-half that of the diameter.

An arc is any part of the circumference of a circle.

A tangent to a circle is a straight line touching the circumference only at one point. A radius drawn to this point forms a right angle with the tangent.

A chord is a straight line joining the extremities of an arc. When a number

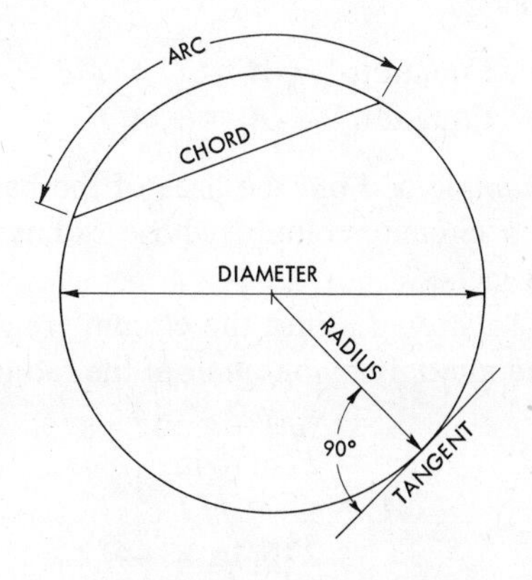

Fig. 5. The various parts of a circle.

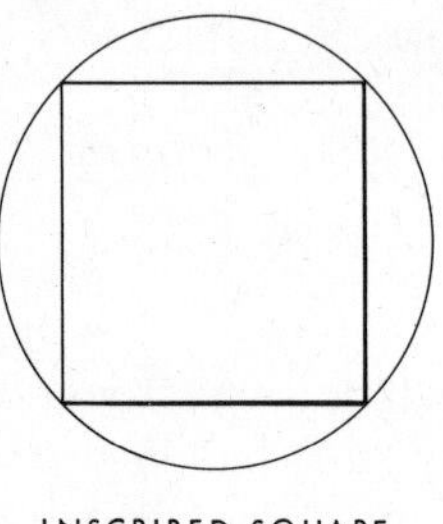

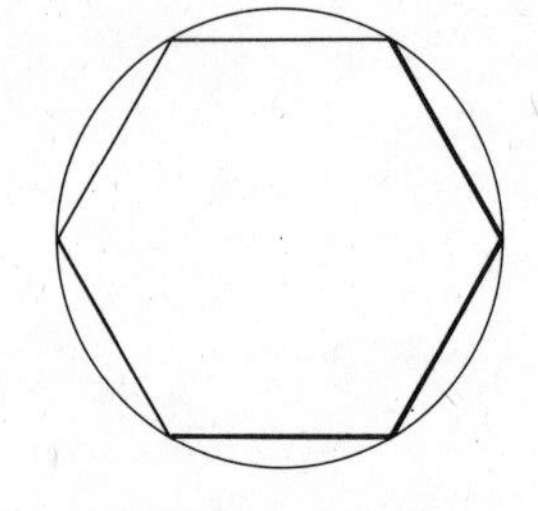

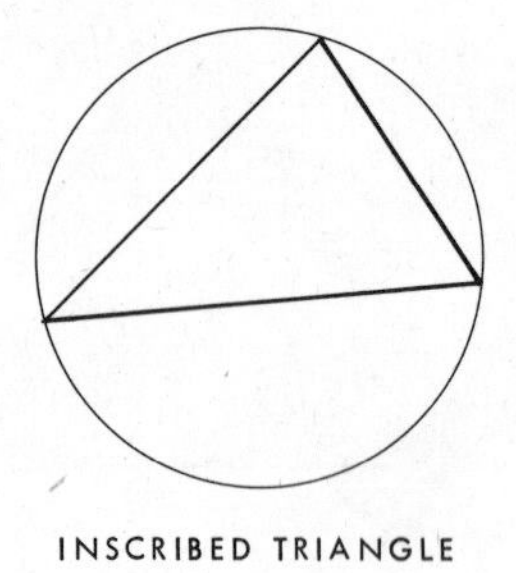

Fig. 6. Inscribed polygons.

of chords form the sides of a polygon, the polygon is said to be inscribed in the circle, Fig. 6.

The circumference of a circle is equal to π times the diameter, or π times twice the radius.

$$C = \pi \times d$$
$$C = \pi \times 2r$$

where C = circumference
d = diameter
r = radius
π = 3.1416 or 22/7

The area of a circle is equal to the circumference multiplied by one-half the radius, or π times the radius squared.

Equation 1—$A = C \times r/2$
Equation 2—$A = \pi \times r^2$

Example 6. Find the area of the base of a circular column whose radius is 44 inches.

Solution 1: Find the circumference and multiply by one-half of the radius.

$$C = \pi \times 2 \times 44$$
$$= 276.46 \text{ in.}$$
$$A = C \times r/2$$
$$= 276.46 \times 44/2$$
$$= 6{,}082.1 \text{ sq. in.}$$

Solution 2: This solution uses the second formula for circle area.

$$A = 3.1416 \times 44^2$$
$$= 3.1416 \times 1{,}936$$
$$= 6{,}082.1 \text{ sq. in.}$$

Area in *square feet:*

$$\frac{6{,}082.1}{144} = 42.2 \text{ sq. ft.}$$

Trapezoids. *The area of a trapezoid is equal to the product of the altitude and one-half the sum of the bases.* (The trapezoid is shown in Fig. 1.)

$$A = h \times \frac{b_1 + b_2}{2}$$

where h = altitude
b_1 = lower base
b_2 = upper base

Example 7. Find the area of a lot shaped like a trapezoid whose bases are 80 feet and 60 feet and whose altitude is 30 feet.

Solution: Substitute in the formula.

$$A = 30 \times \frac{80 + 60}{2}$$
$$= 30 \times 70$$
$$= 2{,}100 \text{ sq. ft.}$$

Hexagons. The estimator will occasionally find it necessary to calculate the area of a pentagon, hexagon, or more complex polygon when doing quantity take-offs. There are formulas for rapidly determining these areas, but they apply only to *regular* polygons, that is, polygons whose sides are *equilateral* (equal length) and whose angles are *equiangular* (equal angles). It is possible, however, to find the area of a polygon by dividing it into triangles and adding together the areas of the triangles.

Triangles are used to find the area of the regular hexagon in Fig. 7. Diagonals BE, AD, and CF divide the hexagon into six equal triangles. Line OL is the *apothem*. It is perpendicular to FE and is the altitude of triangle FOE. In a regular hexagon, the sides of the triangles are equal; $OF = FE = OE$.

The area of triangle FOE is one-half FE times OL, or

Area of $FOE = \frac{1}{2} \times FE \times OL$.

Since there are six triangles in the hexagon, the area of the hexagon is six times the area of triangle FOE, or

$$A = 6\left(\tfrac{1}{2} \times FE \times OL\right) = 3 \times FE \times OL.$$

The area of a regular hexagon is equal to three times the product of the apothem and one side.

$$A = 3 \times s \times R$$

where s = one side of hexagon
R = apothem

Example 8. The perimeter of a regular hexagon is 36 feet. What is the length of the apothem? (Refer to Fig. 7.)

Solution: Each side of a regular

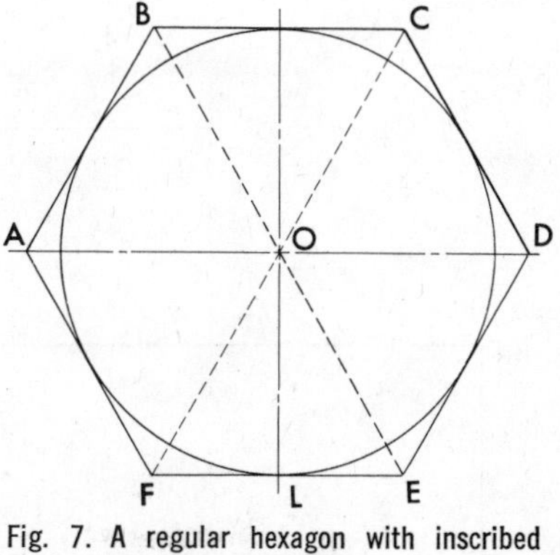

Fig. 7. A regular hexagon with inscribed circle.

hexagon is one-sixth of the perimeter, or 6 feet. OF and OE are therefore 6 feet. The apothem, OL, divides the base of triangle FOL into two equal parts. $FL = LE = 3$ feet. The apothem is one leg of the right triangle FOL and is found by using the Right Triangle Law.

$$\begin{aligned} OL^2 &= OF^2 - FL^2 \\ &= 6^2 - 3^2 \\ &= 27 \\ OL &= \sqrt{27} = 5.19 \text{ ft.} \end{aligned}$$

Example 9. What is the area of a regular hexagon whose sides are 2 yards in length and whose apothem is 1.7 yards?

Solution: Substitute directly into the formula for area.

$$\begin{aligned} A &= 3 \times s \times R \\ &= 3 \times 2 \times 1.7 \\ &= 10.2 \text{ sq. yd.} \end{aligned}$$

Irregular Shapes. To find the area of an irregular shape, the usual procedure is to divide the area into smaller areas of common shapes, calculate the areas and then add them together. This was done in the previous section; the area

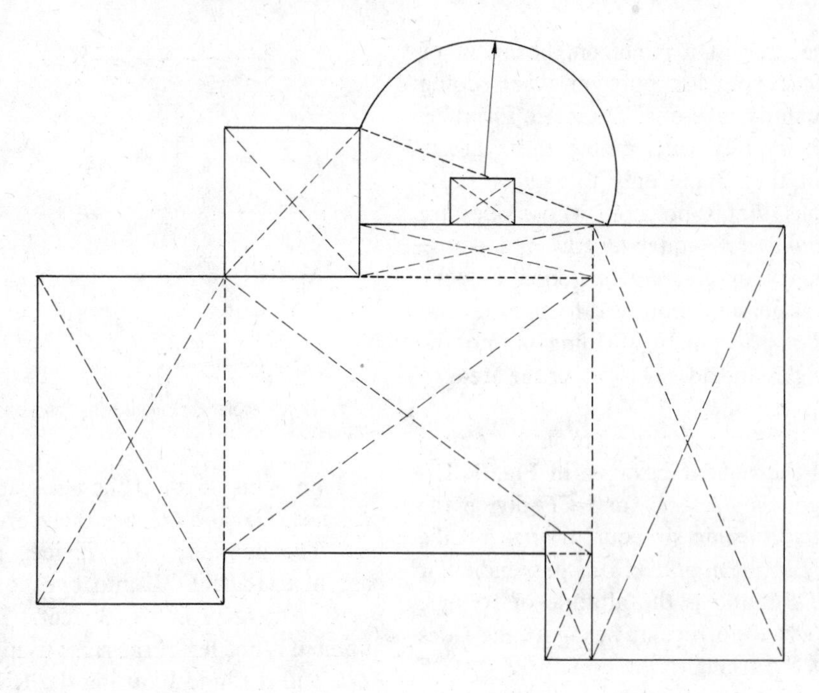

Fig. 8. The area of House Plan A is found by dividing the area into convenient shapes.

of a hexagon was found by dividing it into triangles.

Fig. 8 is the outline of the house shown in the plans at the beginning of Chapter 3. It is an irregular figure which was divided (as indicated) into common shapes: rectangles, a triangle, and a circle. The area of each is calculated separately and then added together to obtain the total area.

Fig. 9 is an irregularly-shaped figure which is bounded on three sides by straight lines. The widths of the figure at five points are measured and indicated as shown. A typical method of calculating the area is as follows.

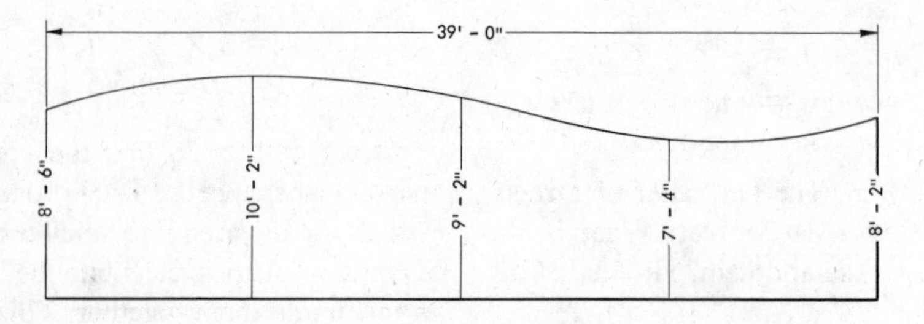

Fig. 9. The area of an irregular shape can be accurately calculated.

TABLE III. AREA FORMULAS

PARALLELOGRAMS

$A = a \times b$

$A = b \times h$

(for squares)

$A = b^2$ or $A = h^2$

TRIANGLES

$A = (b \times h)/2$

Right Triangle Law:

$c^2 = a^2 + b^2$

CIRCLES

$C = \pi \times d$

$C = \pi \times 2r$

$A = C \times r/2$

$A = \pi \times r^2$

TRAPEZOIDS

$A = h \times (b_1 + b_2)/2$

HEXAGONS

$A = 3 \times s \times R$

A = area
a = one side or leg
b = second side or base
h = altitude
c = hypotenuse
b_1 = lower base
b_2 = upper base
s = length of 1 side
r = radius
R = apothem
C = circumference
d = diameter
π = 3.14 or 22/7

The sum of the widths is 43.33 feet. (Use Table I to convert inches to decimal fractions.) The mean or *average width* is $\frac{43.33}{5}$ or 8.67 feet. The area is the length times the average width.

$$A = 39 \times 8.67$$
$$= 338 \text{ sq. ft.}$$

NOTE: Greater accuracy is obtained by using more widths when initially determining the average width.

Table III summarizes the formulas for circumference, length, and area.

Cubic measure

A solid is a body, such as a sack of cement, a building, or a cabinet. Fig. 10 illustrates other, less specific solids. All calculations relating to solids are based on those shown in Fig. 10.

Plane figures are considered in terms of the area they cover. Another consideration exists for solids, namely, *depth* (or *height* or *thickness*). To visualize the difference between a plane figure and a solid, consider a sheet of paper as a plane surface and a book as a solid. The book is made of many sheets of paper in layers. For the paper alone, only length and width are considered; but the book has thickness and that must also be considered. The comparison is immediately apparent if you imagine a sheet of paper so thin as to have no discernible thickness. A plane surface lacks *volume.*

Cubic, or cubage, means volume and gives the size of a body in terms of its bulk. The volume of the sand or earth used to cover a lot to a depth of two feet is specified in terms of cubic measure because it has depth or thickness as well as length and width. The same is true if the lot is *excavated* to a depth of two feet. Thus a foundation has length, width, and thickness and contains so many *cubic* yards of concrete. A room has length, width, and height and therefore has a certain cubage or volume. The same is true for a sand or gravel truck. Its capacity is specified in terms of cubic yards or other units of volume.

Fig. 11 shows a large *cube.* A cube is a solid whose six sides are all equal in length and width. The cube represents one cubic foot (*cu. ft.*) in volume because each side (such as *AB, AN,* and *AC*) is equal to one foot. If you keep in mind that the volume of a cube is length times width times thickness you will easily see that 1 ft. × 1 ft. ×

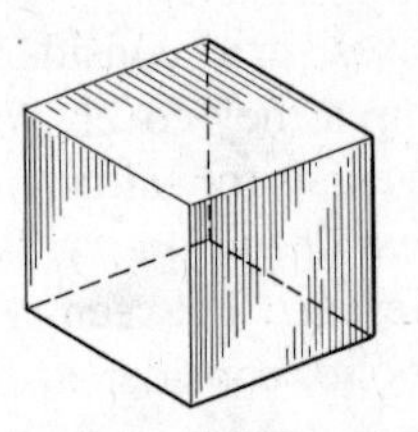

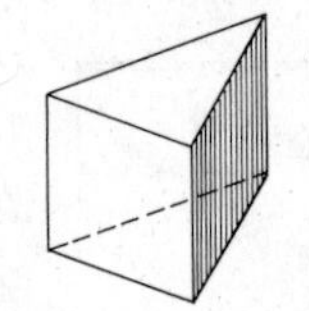

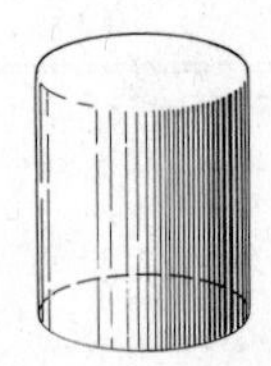

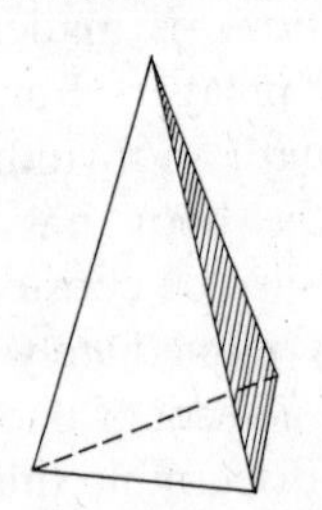

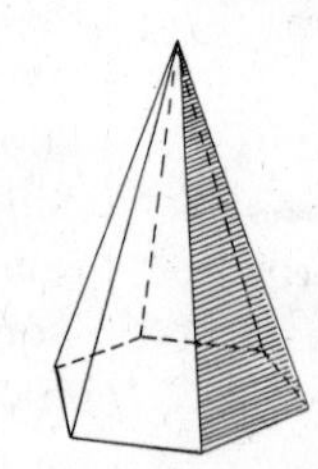

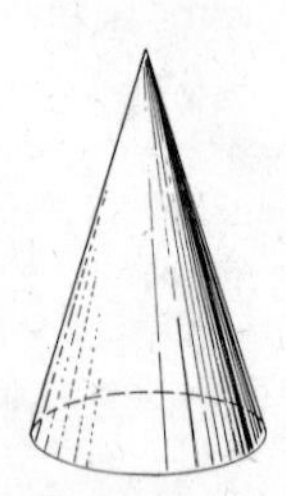

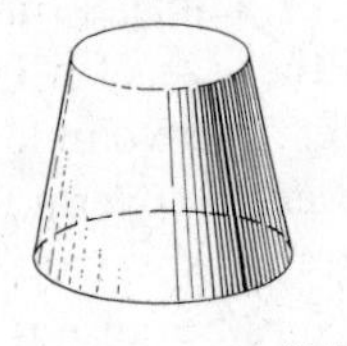

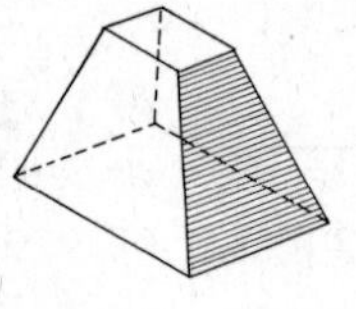

Fig. 10. Typical solids.

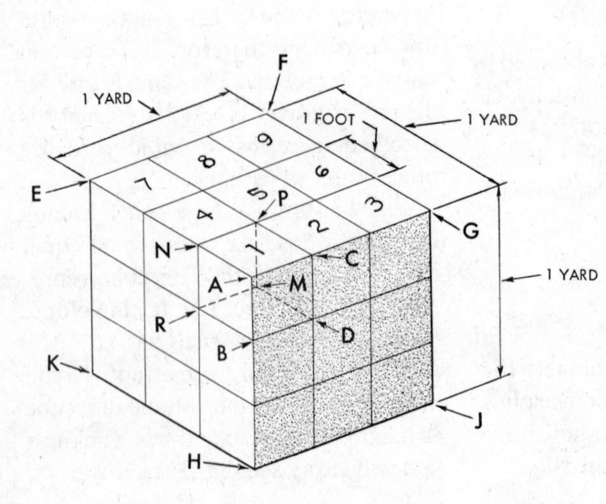

Fig. 11. A cubic yard contains 27 cubic feet.

1 ft. = 1 *cubic* ft. (*Thickness* can be called *height* or *depth.*)

Lay off horizontal and vertical lines one inch apart on all six faces of *ABCDNPMR*. Cut or saw this cube into 12 equal layers or slices as illustrated in Fig. 12. Note slice *CDMP* which is numbered 1. If the small squares in this slice were counted their sum would be 12 × 12, or 144 square inches. Since each of these slices is one inch deep, the volume of slice number 1 is 144 × 1, or 144 cubic inches. There are 12 equal slices, so the total volume is 12 × 144, or 1,728 cubic inches (*cu. in.*). This illustrates the fact that 1 cubic foot contains 1,728 cubic inches.

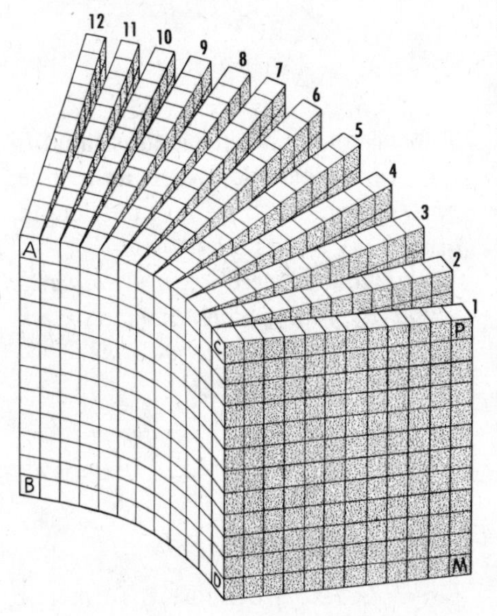

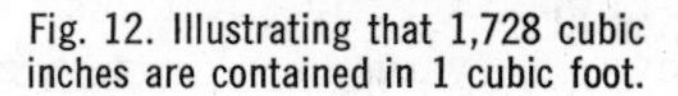

Fig. 12. Illustrating that 1,728 cubic inches are contained in 1 cubic foot.

Refer again to Fig. 11. Note that the large square *AEFG* contains nine smaller squares. The large square is also three layers high. Each layer is one square yard and therefore contains nine square feet. Since each layer is one foot deep, the three layers are three feet deep, and each layer contains nine cubic feet. The three layers contain 3 × 9, or 27 cubic feet, which is the number of cubic feet in a cubic yard (*cu. yd.*). Each side of the cube is one yard long.) Table IV lists some of the units of cubic measure.

Prisms. A *prism* is a solid whose bases or ends (*top* and *bottom*) are equal in area, similar in shape, and parallel plane surfaces. The *lateral faces* or sides of a prism are parallelograms. Fig. 13 illustrates some typical prisms.

The lateral area of a prism, S_l, *is the combined area of all its lateral faces.*

TABLE IV. CUBIC MEASURE

1,728 CUBIC INCHES	= 1 CUBIC FOOT
9 CUBIC FEET	= 1 CUBIC YARD
128 CUBIC FEET	= 1 CORD

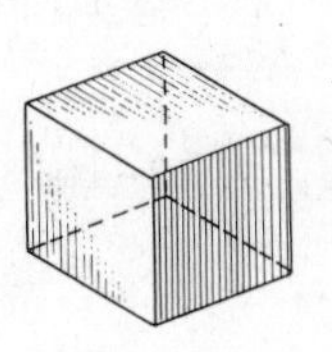

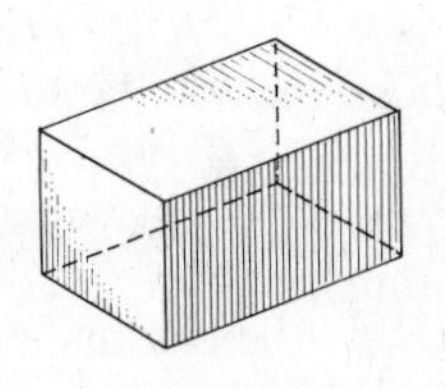

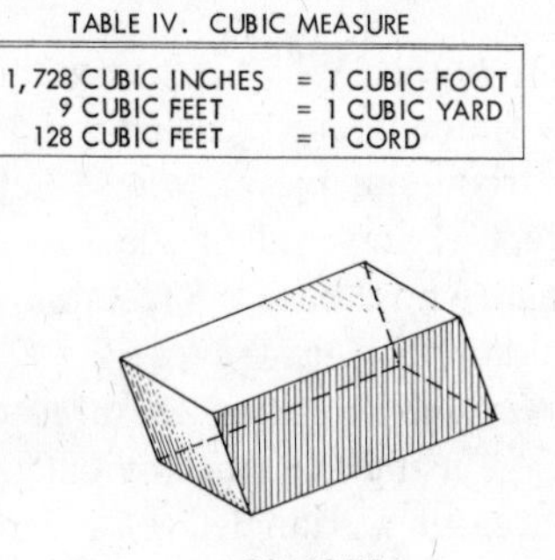

Fig. 13. Typical prisms.

The total area of a prism is the combined area of the lateral faces and the bases.

$$A = S_l + A_B$$

where A = total area
S_l = lateral area
A_B = area of bases

NOTE: The total area of a cube is six times the area of one side, or
$A = 6a^2$ where a = length of side

The volume, V, of any prism is found by multiplying the area of a base by the altitude.

$$V = A_b \times h$$

where A_b = base area

Example 10. Calculate the volume of a room whose dimensions are 12 feet long by 12 feet wide and whose ceiling height is 10 feet. Convert your answer to cubic yards.

Solution: The volume is found by direct substitution in the formula.

$$V = 12 \times 12 \times 10$$
$$= 1{,}440 \text{ cu. ft.}$$
$$\frac{1{,}440}{27} = 53.33 \text{ cu. yds.}$$

Example 11. What is the volume of an A-frame house, which has the shape of a triangular prism, whose height is 20 feet? Each end of the prism is a triangle whose base is 8 feet and whose altitude is 5 feet.

Solution: First find the area of the base by using the rule for calculating the area of a triangle, $A = \frac{1}{2} \times b \times h$.

$$A = \tfrac{1}{2} \times 8 \times 5$$
$$= 20 \text{ sq. ft.}$$

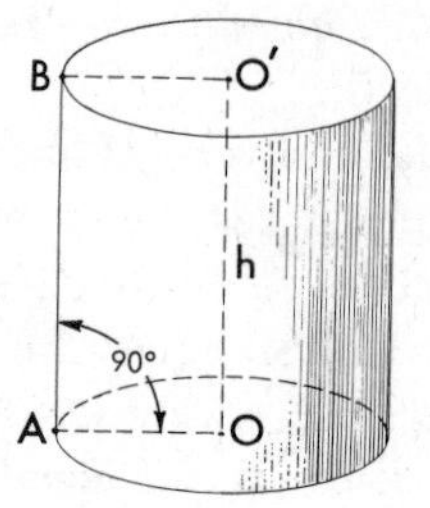

Fig. 14. A right circular cylinder.

The volume is the area of the base times the altitude.

$$V = 20 \times 20$$
$$= 400 \text{ cu. ft.}$$

Cylinders. A *cylinder* is illustrated in Fig. 14. A cylinder, such as a pipe, post, or column, may have any length, h, and any radius, OA.

The lateral area of a cylinder is found by multiplying the circumference of the base by the altitude.

$$S_l = C \times h$$

The volume of a cylinder is found by multiplying the area of its base by its altitude.

$$V = A_b \times h$$

Example 12. A cylindrical form is 6 feet in diameter and 8 feet long. How many cubic yards of concrete are required to fill it?

Solution: The base area is found by using the formula $A = \pi \times r^2$

$$A_b = 3.1416 \times 3^2$$
$$= 28.27 \text{ sq. ft.}$$

The volume is $V = 28.27 \times 8.0$
$$= 226.16 \text{ cu. ft.}$$
$$\frac{226.16}{27} = 8.38 \text{ cu. yds.}$$

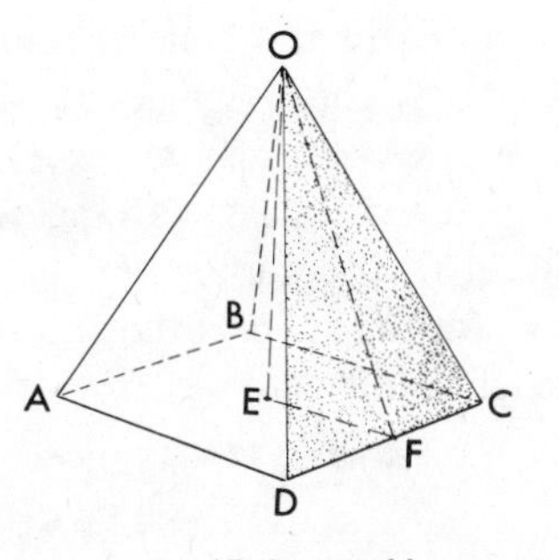

Fig. 15. A pyramid.

Pyramids. A *pyramid,* Fig. 15, is a solid whose base is a polygon and whose sides are triangles. The triangles meet at a common point to form the vertex, O. The altitude is OE and the slant height is OF.

The *slant height* of a regular pyramid (base a regular polygon) is a line drawn on a side from the vertex, and is perpendicular to a side of the base, OF in Fig. 15. In other words, it is the altitude of one of the triangles which form the sides.

The *lateral edges* of a pyramid are the intersections of the triangular sides.

The lateral area of a pyramid is equal to the perimeter of the base multiplied by one-half the slant height.

$$S_l = P_b \times L/2$$

where $P_b =$ perimeter of base
$L =$ slant height

The volume of a pyramid is one-third its base area times its altitude.

$$V = \tfrac{1}{3} \times A_b \times h$$

Example 13. Find the volume of a regular hexagonal pyramid whose altitude is 12 feet. One side of the base is 6 feet. Refer to Figs. 7 and 10.

Solution: First find the area of the base. Refer to the section on "Hexagons." The apothem, R, of the hexagon bisects one side of the hexagon; half of one side is 3 feet. The apothem is now found by the Right Triangle Law:

$$\begin{aligned} R^2 &= 6^2 - 3^2 \\ &= 27 \\ R &= \sqrt{27} = 5.196 \text{ ft.} \end{aligned}$$

Substituting in the formula for area of a hexagon,

$$\begin{aligned} A &= 3 \times s \times R \\ &= 3 \times 6 \times 5.196 \\ &= 93.528 \text{ sq. ft.} \end{aligned}$$

The volume of the pyramid is found by substituting in the volume formula:

$$\begin{aligned} V &= \tfrac{1}{3} \times A_b \times h \\ &= \tfrac{1}{3} \times 93.528 \times 12 \\ &= 374.112 \text{ cu. ft.} \end{aligned}$$

Cones. A *cone* is a solid whose base is a circle and whose lateral surface tapers to a point—the vertex or top, Figs. 10 and 16. A cone may be considered a pyramid with numerous sides or faces, each side so small that the

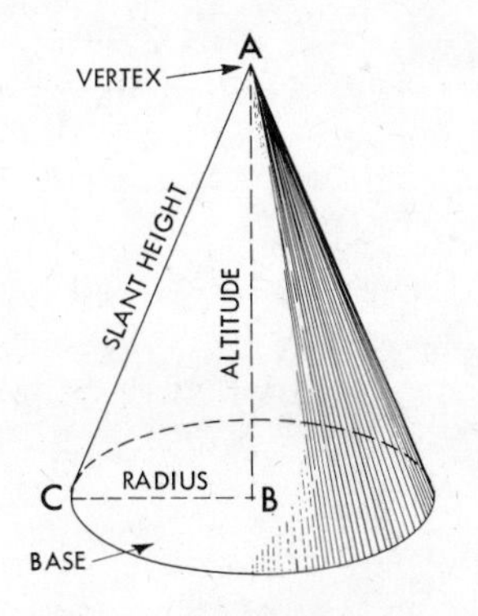

Fig. 16. A right circular cone.

surface or lateral area has no edges and appears smooth.

The lateral area of a cone is the area of the tapering side.

The rules for finding areas and volumes of cones are the same as those for areas and volumes of pyramids.

The lateral area of a cone is the circumference of the base multiplied by one-half the slant height.

$$S_l = C \times L/2$$

The volume of a cone is one-third the product of its base area and altitude.

$$V = 1/3 \times A_B \times h$$

Example 14. Both the circumference of the base and the slant height of a cone are 25 inches. Find the volume.

Solution: Finding the volume requires both altitude and base area. Neither are given, so they must be calculated. (See Fig. 16.) First find the radius, then calculate the base area.

$$C = \pi \times d$$
$$d = C/\pi$$
$$= 25/3.1416$$
$$= 7.95 \text{ in.}$$
$$r = 3.98 \text{ in.}$$
$$A_B = \pi \times r^2$$
$$= 3.14 \times 3.98^2$$
$$= 49.76 \text{ sq. in.}$$

The altitude of the cone is found by using the Right Triangle Law. The slant height, AC, is 25 inches. This is the hypotenuse of right triangle ABC in Fig. 16. The radius, BC, is 3.98 inches. Altitude AB is the second leg of the triangle.

$$AB^2 = 25^2 - 3.98^2$$
$$= 625 - 15.84$$
$$= 609.16$$
$$AB = 24.7 \text{ inches}$$

Substituting in the formula for volume,

$$V = 1/3 \times 49.76 \times 24.7$$
$$= 409.8 \text{ cu. in.}$$

Frustums. Suppose that the top of a pyramid or cone is cut off parallel to its base. When the top part is removed it leaves a *frustum,* Fig. 17. Concrete footings often have this shape. The *altitude, MN,* and the slant height, OL, are indicated.

The lateral area of a frustum of a right regular pyramid is one-half the sum of the perimeters of the two bases times the slant height.

$$S_l = 1/2 \times (P_1 + P_2) \times L$$

where P_1 = perimeter of one base
P_2 = Perimeter of other base

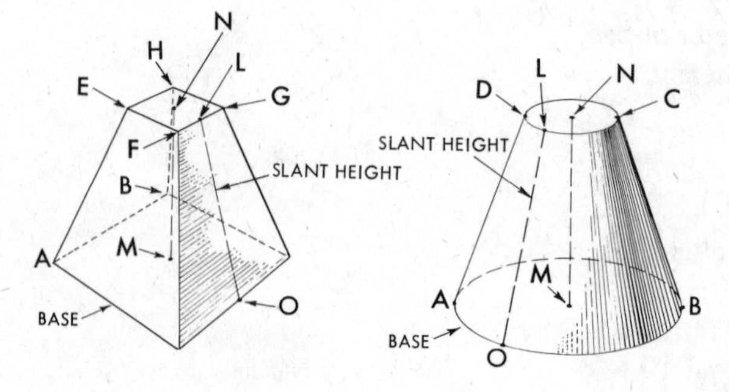

Fig. 17. Frustums of a pyramid and cone.

Since a cone may be considered as a pyramid with numerous sides so small that the surface appears smooth, a similar rule is used for finding the lateral area of a frustum of a cone.

The lateral area of a frustum of a right regular cone is one-half the sum of the circumferences of the bases multiplied by the slant height.

$$A_L = \tfrac{1}{2} \times (C_1 + C_2) \times L$$

where C_1 = circumference of one base

C_2 = circumference of other base

The total area of a frustum is the sum of the lateral area and the two bases.

$$A = S_l + A_B$$

The volume of a frustum is the sum of the areas of the two bases added to the square root of the product of the two bases and the result multiplied by one-third of the altitude.

$$V = (A_b + A_t \times \sqrt{A_b \times A_t}) \times \frac{h}{3}$$

Example 15. What is the volume of the frustum of a square pyramid whose altitude is 15 feet? The sides of the lower and upper bases are 8 feet and 2 feet long respectively.

Solution: First find the areas of the bases.

$$A_b = 8^2 = 64 \text{ sq. ft.}$$
$$A_t = 2^2 = 4 \text{ sq. ft.}$$
$$A_b \times A_t = 64 \times 4 = 256$$
$$\sqrt{A_b \times A_t} = \sqrt{256} = 16 \text{ sq. ft.}$$

The volume is now found by substituting in the formula:

$$V = (64 + 4 + 16) \times \frac{15}{3}$$
$$= 84 \times 5$$
$$= 420 \text{ cu. ft.}$$

Spheres. A *sphere* is a solid bounded by a curved surface every point of which is equally distant from the center. In other words, it is a perfectly round ball, Fig. 18.

A plane is *tangent* to a sphere when it touches the sphere at only one point, as plane *PNQ* (seen edgewise) in Fig. 18, which touches the sphere at *N*.

When a plane cuts through a sphere, the section is a circle such as plane *ACBD*. If the plane cuts through the center of the sphere, the resulting section is a *great circle* such as *ACBD, ANBM,* and *NCMD*. A *small circle, LFE,* is formed when the plane does not pass through the center.

The *circumference* of a sphere is the circumference of a great circle.

The surface area of a sphere is π times the diameter squared.

$$S = \pi \times d^2$$

where S = surface area

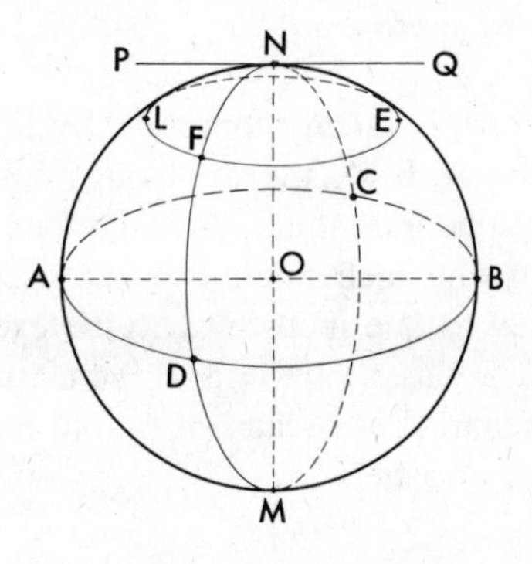

Fig. 18. A sphere.

Since $d = 2r$,

$$S = \pi \times 4r^2$$

The volume of a sphere equals the surface area multiplied by one-third of the radius.

$$V = S \times \frac{r}{3}$$

Since $S = 4r^2 \times \pi$, you can substitute for S in the formula:

$$V = 4 \times r^2 \times \pi \times \frac{r}{3}$$

$$= \frac{4}{3} \times \pi \times r^3$$

Example 16. Find the volume of a sphere whose radius is 5 feet.

Solution: Substitute directly into the formula for volume.

$$V = \frac{4}{3} \times \pi \times 5^3$$

$$= \frac{4}{3} \times 3.1416 \times 125$$

$$= 523.6 \text{ cu. ft.}$$

Table V contains formulas for the areas and volumes of the most common solids.

TABLE V. VOLUME FORMULAS

PRISMS

$A = S_1 + A_B$

$A = 6 \times a^2$ (for squares)

$V = A_b \times h$

CYLINDERS

$S_1 = C \times h$

$V = A_b \times h$

PYRAMIDS

$S_1 = P_b \times L/2$

$V = A_b \times h/3$

FRUSTUMS

$S_1 = (P_b + P_t) \times L/2$ (right pyramid)

$S_1 = (C_b + C_t) \times L/2$ (right circular cone)

$A = S_1 + A_B$

$V = (A_b + A_t + \sqrt{A_b \times A_t}) \times h/3$

SPHERES

$S = \pi \times d^2$

$S = \pi \times 4r^2$

$V = \pi \times 4r^3/3$

A_1 = lateral surface area
A = total area
A_B = total base area
A_b = base area
A_t = top area
P_b = perimeter of base
P_t = perimeter of top
L = slant height
C_b = circumference of base
C_t = circumference of top
S = total surface area

Board measure

Lumber is usually computed by *Board Measure,* B.M., the unit being a square foot one inch thick. Any number less than one inch thick is usually computed as one inch thick. (One exception to this is plywood; it is measured in square feet because it is sold in the form of panels.)

Steel Square Method. The back of a blade of a typical *steel square* is shown in Fig. 19. On the back of this blade is the Board Measure, where eight parallel lines along the length of the blade are shown and divided at every inch by cross-lines. Under 12, on the outer edge of the blade, are found the various lengths of the boards, as 8, 9, 10, 11, 13, and so forth. For example, take a board 14 feet long and 9 inches wide. To find the number of board feet, look under 12, and find 14; then follow this space to the cross-line under 9, the width of the board; here is found 10 feet 6 inches, indicating the number of board feet of the board.

Calculation Method. The usual method of calculating the B.M. of lumber is to multiply the length in feet by the width and thickness in inches, and di-

Fig. 19. A steel square. The back of the blade is used for board measure calculations.

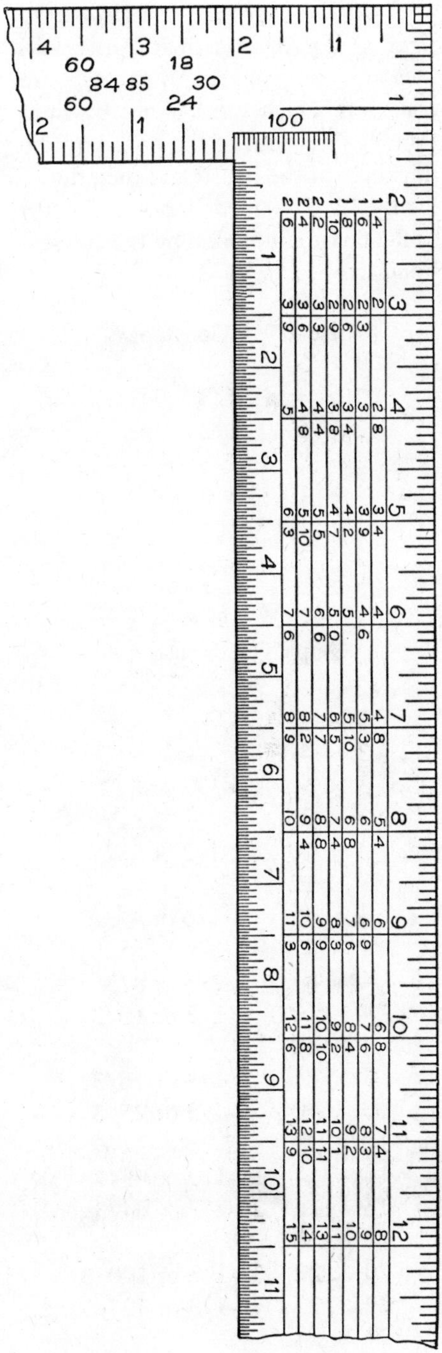

vide the product by 12. This is a slow process and is a violation of our rule about multiplying feet by inches (discussed at the beginning of this chapter). The following system, as shown by examples, is recommended as a quicker method of calculation.

Example	No. of Pieces	Size	Length	B.M.
1.	1	2″ × 8″	30′	40
2.	1	4″ × 10″	18′	60
3.	1	10″ × 10″	36′	300
4.	1	20″ × 20″	60′	2000

1.

2 × 8 equals 16; 16 divided by 12 equals 16/12 or 4/3. When this is multiplied by the length (30′) the answer is 40, the Board-Measure.

$$\text{B.M.} = \frac{2 \times 8}{12} \times 30 = 40$$

2.

4 × 10 equals 40; 40 divided by 12 equals 10/3; multiplying 18 by 10/3 is 60.

$$\text{B.M.} = \frac{4 \times 10}{12} \times 18 = 60$$

3.

$$\text{B.M.} = \frac{10 \times 10}{12} \times 36 = 300$$

4.

$$\text{B.M.} = \frac{20 \times 20}{12} \times 60 = 2{,}000$$

The following is a list of *standard sizes* of boards and their *conversion factors*. The conversion factors are constants which were found by multiplying the width of the board by the thickness of the board and then dividing by 12. The *linear footage* (length) of the board multiplied by the conversion factor is the B.M.

Size	Constant
1 × 3	1/4
1 × 4	1/3
1 × 6	1/2
1 × 8	2/3
1 × 10	5/6
1 × 12	1
2 × 3	1/2
2 × 4	2/3
2 × 8	1-1/3 or 4/3
2 × 10	1-2/3 or 5/3
2 × 12	2
3 × 3	3/4
3 × 4	1
3 × 6	1-1/2 or 3/2
3 × 8	2
3 × 10	2-1/2 or 5/2
3 × 12	3
4 × 4	1-1/3 or 4/3
4 × 6	2
4 × 8	2-2/3 or 8/3
4 × 10	3-1/3 or 10/3
4 × 12	4
8 × 8	5-1/3 or 16/3
10 × 10	8-1/3 or 25/3
12 × 12	12
14 × 14	16-1/3 or 49/3
16 × 16	21-1/3 or 64/3
18 × 18	27
20 × 20	33-1/3 or 100/3
22 × 22	40-1/3 or 121/3
24 × 24	48

A Variation in Method. A convenient method for computing B.M. is as follows:

For all 12 ft. lengths multiply width by thickness.

For all 14 ft. lengths multiply width by thickness and add 1/6 of the resulting total.

For all 16 ft. lengths multiply width by thickness and add 1/3 of the resulting total.

For all 18 ft. lengths multiply width by thickness and add 1/2 of the resulting total.

For all 20 ft. lengths multiply width by thickness and add 2/3 of the resulting total.

For all 22 ft. lengths multiply width by thickness and add 5/6 of the resulting total.

For all 24 ft. lengths multiply width by thickness and double the resulting total.

Some objection may be taken to the use of 2/3 and 5/6, but often you may be able to substitute 1/6, 1/3, or 1/2 as in the following examples:

1) You need 10 pieces of 1 × 18 boards, each 22 feet long. In order to avoid the awkward fraction 5/6, you reverse the 18 and the 22. In this problem consider that you need 10 pieces of 1 × 22's, each 18 feet long. In this way you may use the formula for 18′ lengths, with its more convenient fraction of 1/2, without affecting the solution.
2) Similarly, 16 pieces of 1 × 22's each 20 feet long may be considered as 20 pieces of 1 × 22's, each 16 feet long.

The above system is very convenient when calculating lumber from 12 to 24 feet long, particularly where odd widths and thicknesses occur frequently.

Converting Board-Measure to Lineal Feet. Simply reverse the multiplier used to bring lineal feet to Board-Measure; in other words, multiply board feet by 12 and divide by thickness and width.

Example 17. How many lineal feet are there in 1,000 feet Board-Measure of 2 × 8?

Solution:

$$\frac{12 \times 1{,}000}{2 \times 8} = 750 \text{ lineal feet}$$

Example 18. Car orders frequently call for a specified amount of sizes containing special lengths. Before proceeding to load, it is necessary to find the number of pieces required. Find the number of pieces in the following order:

1,000 ft. B.M. 2 × 4—14
1,000 ft. B.M. 2 × 4—16
1,000 ft. B.M. 2 × 4—20

Solution: Change the Board-Measure to lineal feet as shown in Example 1; then divide the length into lineal feet. The result is the number of pieces.

$$\frac{12 \times 1{,}000}{2 \times 4} = 1{,}500 \text{ lineal feet}$$

$$\frac{1{,}500}{14} = 107 \text{ pcs.} \qquad \frac{1{,}500}{16} = 94 \text{ pcs.}$$

$$\frac{1{,}500}{20} = 75 \text{ pcs.}$$

107 pcs.	2 × 4—14 containing	998 ft. 8 in. B.M.
94 pcs.	2 × 4—16 containing	1,002 ft. 8 in. B.M.
75 pcs.	2 × 4—20 containing	1,000 ft. B.M.
276		3,001 ft. 4 in. B.M.

Review Questions

1. One leg of a right triangle is 1′ long and the other leg is 3′ long. What is the length of the third side? (Answer: 3.16′)
2. A right triangle has a hypotenuse 7″ long and a leg which is 2″ long. Find the length of the third side. (Answer: 6.71″)
3. A stairway 35′ long reaches the top of a wall 28′ high. How far is the foot of the stairs from the base of the wall? (Answer: 21.0′)
4. Refer to the obtuse triangle in Fig. 3. Let the base BC equal 9′ and the side AC equal 15′. A perpendicular, AD, intersects line BC at D. The distance from B to D is 3′. Find the area of triangle ABC. (Answer: 40.5 sq. ft.)
5. Find the circumference of a drain pipe which is 3.5″ in diameter. (Answer: 11.0″)
6. Find the area (in sq. yds.) of the base of a silo with a radius

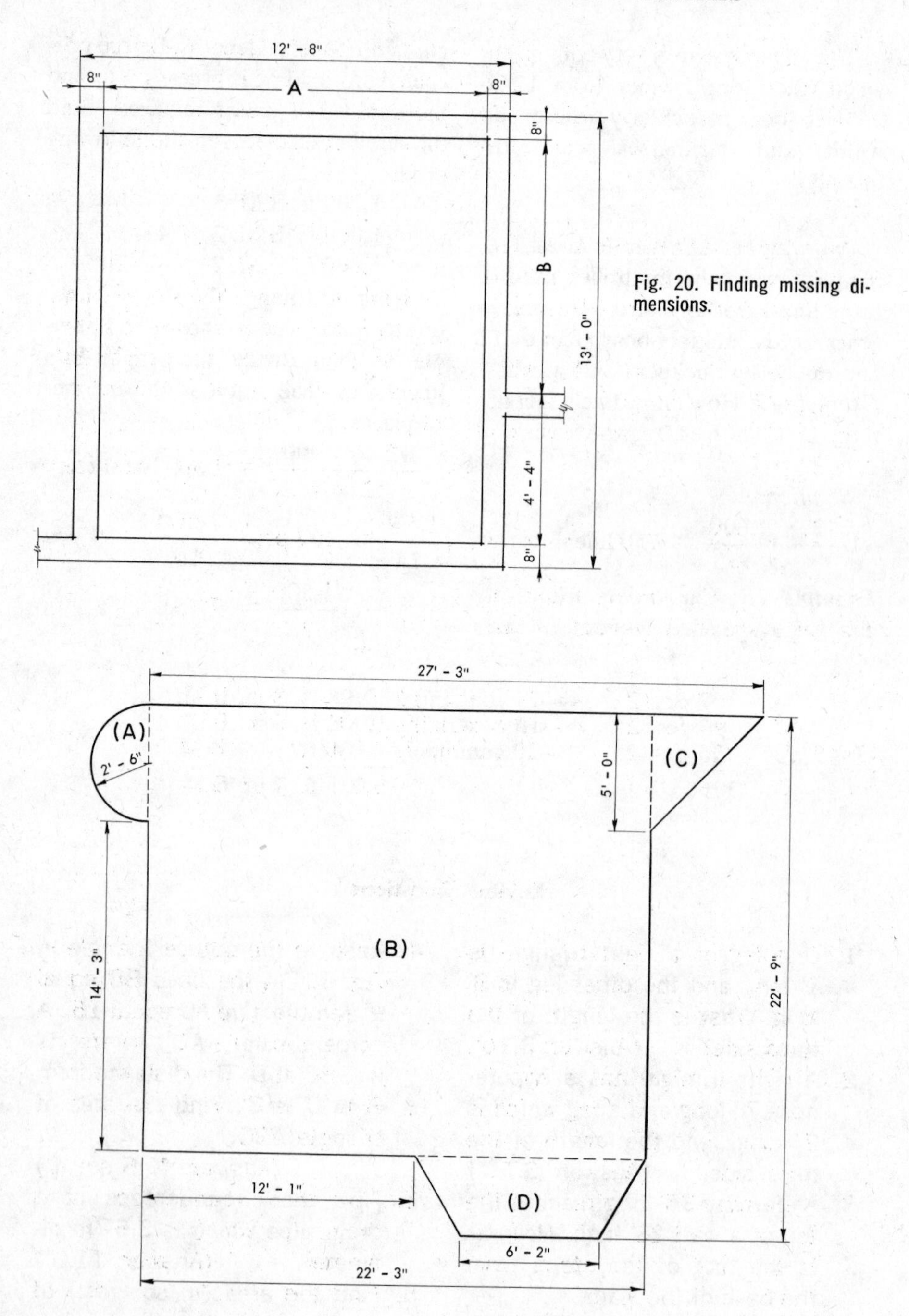

Fig. 20. Finding missing dimensions.

Fig. 21. Finding the area of an irregular figure.

of 10′. (Answer: 34.91 sq. yd.)

7. How many square feet are there in a patio which is in the shape of a half-circle, and whose radius is 75″?
(Answer: 61.36 sq. ft.)
8. Find the area of a trapezoid whose bases are 12.25′ and 16.25′ and whose altitude is 12.0′. (Answer: 171.0 sq. ft.)
9. What is the volume of a rectangular prism whose height is 9′? The dimensions of the base are 3′-6″ × 2′-0″.
(Answer: 63.0 cu. ft.)
10. A circular excavation is to be 20 yards in diameter and 4 yards deep. What is the volume of the excavation?
(Answer: 251.33 cu. yds.)
11. What is the volume of a regular triangular pyramid whose altitude is 20′? The sides of the base are each 15′ long.
(Answer: 649.5 cu. ft.)
12. Find the volume of a cone whose slant height is 20′ and whose base is 22′ in circumference.
(Answer: 253 cu. ft.)
13. What is the volume of the frustum of a cone, the radiuses of the bases being 4′ and 2′, and the altitude 10′?
(Answer: 293.22 cu. ft.)
14. Find the surface area and volume of a spherical storage tank whose diameter is 24′.
(Answer: S = 1,809.56 sq. ft.,
V = 7,238.25 cu. ft.)
15. Find the number of board feet contained in a board 10′ long and 7″ wide. Use the steel square in Fig. 19.
(Answer: 5′-10″)
16. Find the B.M. of the following pieces of lumber. Each is a single piece:
(a) width = 4″, thickness = 2″, length = 24″
(Answer: 16)
(b) width = 8″, length = 30′-0″, thickness = 5″
(Answer: 100)
(c) thickness = 8″, width = 8″, length = 60′-0″
(Answer: 320)
17. Find the missing dimensions in Fig 20. (Answers: A = 11′-4″,
B = 7′-4″)
18. Find the area of the irregular figure shown in Fig. 21. (Break the area into simple shapes.)
(Ans.: A = 9.82 sq. ft.,
B = 428.31 sq. ft.,
C = 12.5 sq. ft.,
D = 28.6 sq. ft.,
Total area = 479.23 sq. ft.)

5

SURVEYING AND EXCAVATION

A set of plans for which an estimate is being made usually includes a plot or location plan on which a surveyor has previously noted the exact lot position in addition to the elevations. ("Elevation" as it is used in this chapter refers to the distance above sea level or above an arbitrary elevation of 0 or 100.) The estimator must have a general understanding of surveying principles in order to compute excavation yardage accurately in all cases.

If a lot is nearly level the calculation of excavation yardage is quite simple. Only the elevation under the basement floor and the surface elevation need be known in order to compute the volume of the excavation. Or, if the lot is level and it is known that the elevation must be 6′-4″ deep the computation is equally simple, as will be shown later in this chapter. If a lot is sloping or irregular in surface the calculation of excavation yardage is much more complicated, and it becomes necessary to have more detailed elevation data.

This chapter will acquaint you with general principles relating to excavation calculations and with surveying operations and principles required for the excavation calculations.

Surveying

Topographic Maps. Frequently, topographic maps are made as a preliminary step to construction work and are called plot, survey, or location plans. These maps (as they are technically called) should show all details which will be of interest to the architect, the estimator, the contractor, and the owner. Such details include trees, elevations, terraces, slopes, building location, lot lines, street and sidewalk lines, sewer lines, water lines, and utility lines. The estimator is primarily interested in the location of the house, the elevations, and the character of the surface of the ground. In fact, if a set of plans does not contain such a map and if the lot is irregular in elevation it will be necessary for the estimator to have a map made or to visit the site and make such a map himself.

Contours. A contour is a line joining points which have the same elevation. An elevation is a known distance above or below a given level, such as sea level or an arbitrary elevation of 0 + 00 or 100 + 00, the latter being used to avoid negative numbers. Every town and city has one or more permanent points whose elevation has been established and from which elevations on a given lot may be determined. In the larger cities these points of known elevation are situated at convenient locations in different parts of the city. Sometimes they are marked on buildings and sometimes by iron rods fixed in concrete bases. The exact number of feet is determined and marked on a *bench mark* or master marker. At various points in the city, other markers give the elevation at their location. Elevations vary because of slopes or other irregularities of the surface. In Chicago, which is practically level, the variation is slight, while at Davenport, Iowa, located among the rolling hills of the Mississippi River Valley, elevation reading differ sharply.

Some cities have what is known as a *city datum* from which all elevations or levels can be calculated.

The simplest way to understand the nature and variability of contours is to picture them as elevation marks on a cone. This is illustrated in Fig. 1. Suppose a right circular cone, Fig. 1-*A*, is 6 feet high and is placed in a

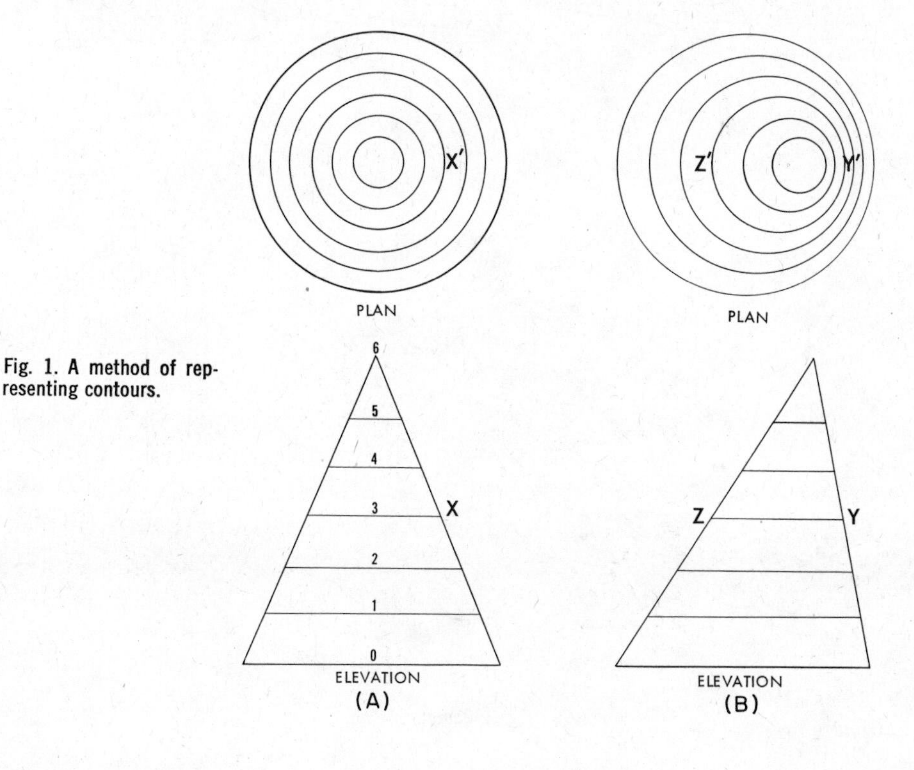

Fig. 1. A method of representing contours.

tank of water. When the water just covers the bottom of the tank, the contour is the line marking the outer circle in the plan view above Fig 1-*A*. As the water is raised by 1-foot increments the contours will be indicated by concentric circles. If the cone is inclined, the contours will be as in the plan view above Fig. 1-*B*. The evenly spaced contours at *X′* indicate a uniform slope; where farther apart, as at *Z′*, they indicate a more gentle slope;

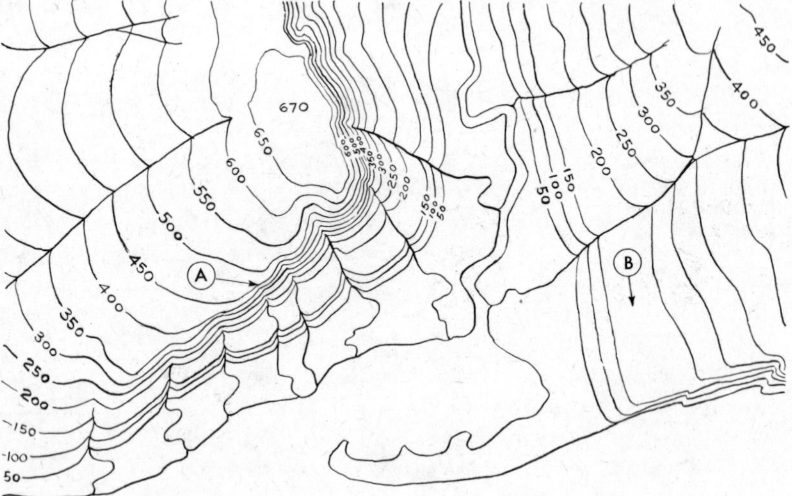

Fig. 2. Contour sketch.

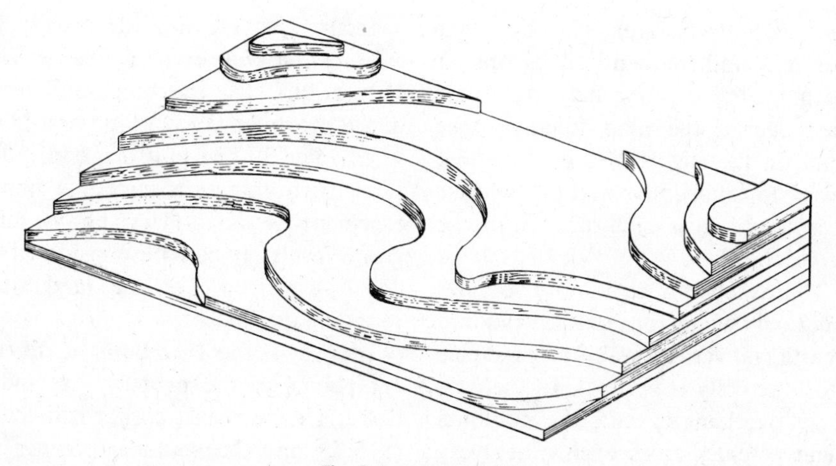

Fig. 3. A contour model.

and where closer together, as at *Y'*, they indicate a steep slope.

Fig. 2 shows a typical contour sketch in which the lower part describes typical contours and the upper part indicates the point of elevation. For example, the contours at *A* are close together and indicates the terrace shown at *A'*. The wider contour interval at *B* indicates the flat surface at *B'*.

Fig. 3 will further enable you to understand contours. If parts that are intended to show the various elevations (levels) of a given area are cut from cardboard and assembled successively, a contour model such as that shown in Fig. 3 is obtained. Each part is scaled to the level of the contour interval it represents. Plaster or clay in natural colors may be applied to the cardboard to give the model the finished appearance of a relief map.

Contour Interval. A certain uniform vertical increment, or contour interval, is always used in making a map, the interval varying with the scale of the map and the character of the country. For the closest of detail work on a larger scale, such as might be used for landscape gardening, a contour interval of 1 or 2 feet might be necessary. But for ordinary purposes 5 feet or more is used as the contour interval. The topography for preliminary railroad surveys usually is done with a 5-foot contour interval. The topographic maps made by the United States Geological Survey, which are plotted at a scale of approximately 1 or 2 miles per inch, are made with a contour interval varying from 20 to 100 feet, the 100-foot interval being necessary where steep mountain ranges are to be represented.

Principles of Contours. There are certain principles pertaining to contours which must be kept in mind by the estimator in order to avoid making errors in his work. Barring a few exceptional cases which are mentioned later, the following principles are observed: (1) a contour is always a continuous line enclosing an area; (2) since that area may and frequently does run off the map, the contour may run from one edge of the map to any other point on the edge; (3) a contour line never stops abruptly; and (4) contour lines do not cross each other or merge into each other.

The exceptions to the above rules occur only where the contours run into an artificial vertical wall or a precipice which actually is vertical. In the case of an overhanging cliff the contours might actually cross each other very slightly, but with rare exceptions (which are readily recognized where they occur) the above principles are used for most contour maps.

Contours are shown in the same manner on maps of large or small scale; the interval alone varies. The contour elevations are always given with reference to some datum plane, such as sea level, and the elevations are always a multiple of the contour interval. For example, if the contour interval is 20 feet and the datum plane is sea level, then each contour elevation will be a multiple of 20 even though the whole tract is far above sea level. The contours might be at elevations of 660, 680, 700, 740, etc. Then the 600-, 700-, or 800-foot contours would be made extra heavy and could be distinguished more readily, even in steep places where the contours would be very close together.

The contours should be numbered at frequent intervals so that the elevation of a contour at any point may be determined by following it for a very short distance.

Location Plan. Fig. 4 illustrates a typical *location plan* or *plot diagram* for HOUSE PLAN A, the single-unit residence shown on the working drawings in Chapter 3. The plot diagram indicates such things as property lines, easements, walks and driveways, and grade levels. It can be considered a "bird's-eye view" of the land surrounding the house.

In Fig. 4, the elevations at different points on the property are indicated. The northeast corner is highest (115.7′) and the southeast corner is lowest (110.7′), indicating a slight slope. The house itself is to be built at a level of 112.6′. The first floor level is raised almost two feet to 114.5′.

If Fig. 4 had been made in Chicago, using actual sea level elevations, the elevations shown would be increased by approximately 500 feet each.

With this type of map the estimator can easily compute the yardage of the lot (unless the surface is very irregular). Methods of calculation will be explained later in this chapter.

Determining Contours. One method of determining contours which is commonly used is called the *square method.* This method consists in establishing a grid of squares or rectangles over the desired lot or area and determining the elevation of each corner by means of a *level-transit,* a *wye-level,* or a *dumpy level.*

The points where the contours cross the side of the squares are next determined, usually with a hand level and a metal tape. Suppose one corner, Fig. 5, is at elevation 803.2 feet and

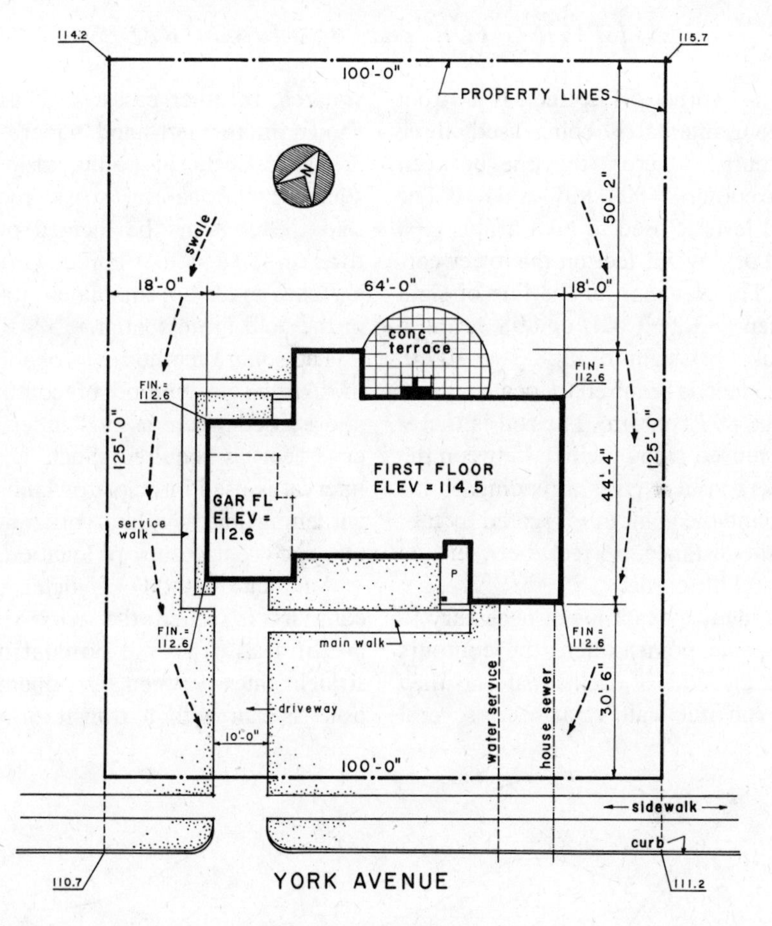

Fig. 4. Plot Plan of House Plan A.

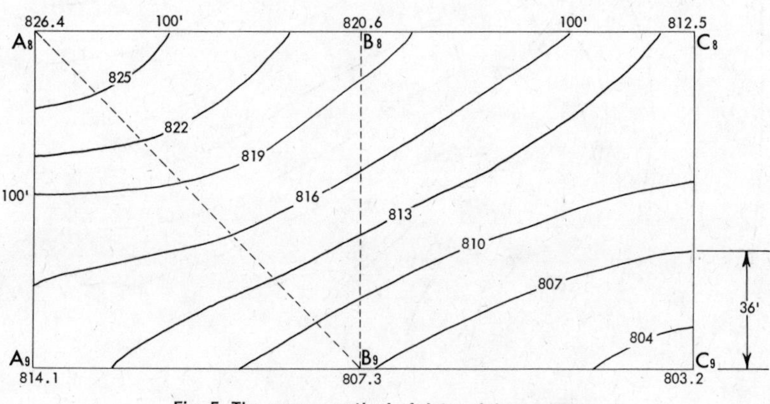

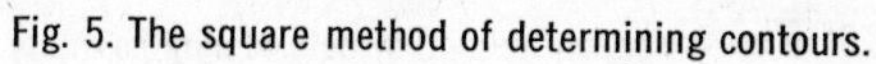
Fig. 5. The square method of determining contours.

the next corner, 812.5 feet. If a 3-foot contour interval is being used, three contours will cross the line between these corners—804, 807, and 810. The hand level is used to take a plus (+) sight of, say 5.1 feet on the lower corner. The elevation of the line of sight is then 803.2 + 5.1, or 808.3, and a minus (−) sight of 808.3 −807.0, or 1.3 feet, is required to locate a point on the 807 contour. The rod is therefore moved along the line between the corners until it gives a reading of 1.3 feet; and the contour is located by taking the distance, 36 feet, between the rod and the corner.

In many cases it is not necessary to locate the points where the contours cross the sides of each square, as they may run practically straight for several squares. In other cases, such as that shown in the left-hand square, Fig. 5, bends occur at points inside the squares, and careful work requires that the contours be located on the diagonal (dotted) lines. Careful sketches made approximately to scale in the field are essential.

The square method is probably the most accurate method of contouring and is used where a small interval, 2 or 3 feet, is required. Such a small interval is used in maps for landscape gardening and similar work and requires close accuracy of location.

Another method of determining contours is called the *cross-section* method. A traverse consisting of straight lines between instrument stations is run with a transit or other

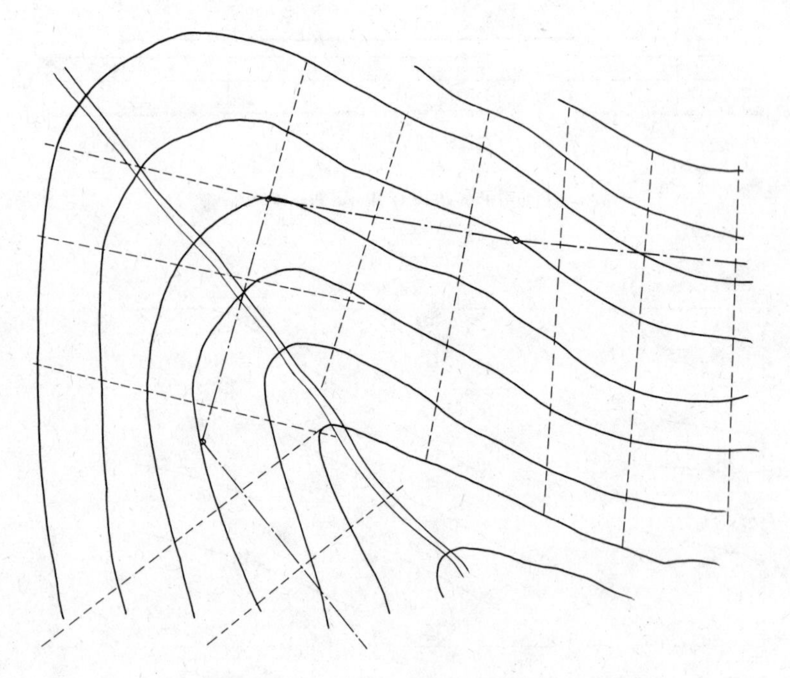

Fig. 6. The cross-section method of determining contours.

angle measurer and a tape. At certain intervals, stakes are placed and the elevations found with a level-transit. Cross lines are then run at right angles to the traverse lines as shown in Fig. 6. The elevation of the stations being known from the traverse lines, the contour points on the right-angle lines are determined in exactly the same way as in the square method.

Excavation

Briefly, excavation is the digging of a hole in the ground to provide room for engineering improvements. These improvements include footings, foundations, basements, sub-stories, ramps, exterior stairways, areaways, pipe trenches, pits for septic tanks, and traps.

The estimator, while considering his drawings and specifications, prior to his actual quantity take-off, should carefully study all the General Conditions. Once he has these clearly in mind he can start to do the quantity take-off. The quantity take-off for excavation is usually calculated in cubic yards. Should the excavation require the removal of a few additional inches of soil it would be calculated in square feet.

The estimator must not only do a very careful quantity take-off, but should study the job for such money-saving items as *topsoil*. This valuable substance is found on many building sites. He can then sell any unneeded soil to other contractors or landscapers. This is only one example of how a careful estimator can save or earn money for his contractor. Careful study of the site, topographic maps, plans, and specifications will reveal other opportunities for saving money.

Further savings can be realized with respect to *fill*. If the ground is lower in certain spots than the finished grade shown on the drawings, the contractor may have to buy fill to bring the level to the specified grade. This is not a usual occurrence but represents a possible cost to the contractor. The required fill will often take many more yards of earth than has been excavated. The estimator must consider this possibility in his calculations; failure to do so will lower the contractor's profit.

Other items which the estimator must take into acount are the costs involved in *backfilling, utility line excavation,* and *cradling*. (These are explained in the following section, "Special Terms.") This means that the estimator must carefully study *all* parts of the specifications and drawings and still realize that there might be other *unspecified* factors which must be taken into consideration.

Special Terms. The following terms should be understood by the estimator before he attempts to do a quantity take-off for excavation:

General Excavation. All excavation, other than rock and water removal, which can be done by available mechanical equipment or any general piece of equipment such as bulldozer, clam shell cranes, back hoes, scrapers, power shovels, and loaders, is considered as general excavation. The use of trucks to remove the excavated material is also included under this heading.

Special Excavation. All excavation done by hand, by special machine, by blasting, or by the use of special methods is considered as special excavation. One type of special excavation is a pipe trench for a water line. This would be done by a trencher, commonly called a "ferris wheel." Another type of special excavation is required for the installation of telephone, electrical, and utility poles. This type of excavation might require a vertical boring rig mounted on a bulldozer.

Sheet Piling. Vertical retaining walls of wood or steel are used when excavating adjacent to an existing structure or when excavating near a heavily traveled roadway where there is vibration from passing vehicles. These vertical retaining walls are called sheet piling; they retain any horizontal thrust of earth which might cause the excavation walls to slide, Fig. 7.

Wood planking placed closely together and steel sheets (often of an interlocking construction) are the most common forms of sheet piling. In some cases, where the sheet piling is to become a permanent part of the structure, precast concrete tongue-and-groove planking is used.

There are several methods of installing sheet piling:

1. If the earth is soft enough to permit driving, a *pile driver* with a drop hammer rig or single stroke hammer is used to drive the piles to a "point of refusal" which will prevent the sheet piles from being pushed over by any horizontal thrust of earth.
2. Another method of placing sheet piling is by *pneumatic ham-*

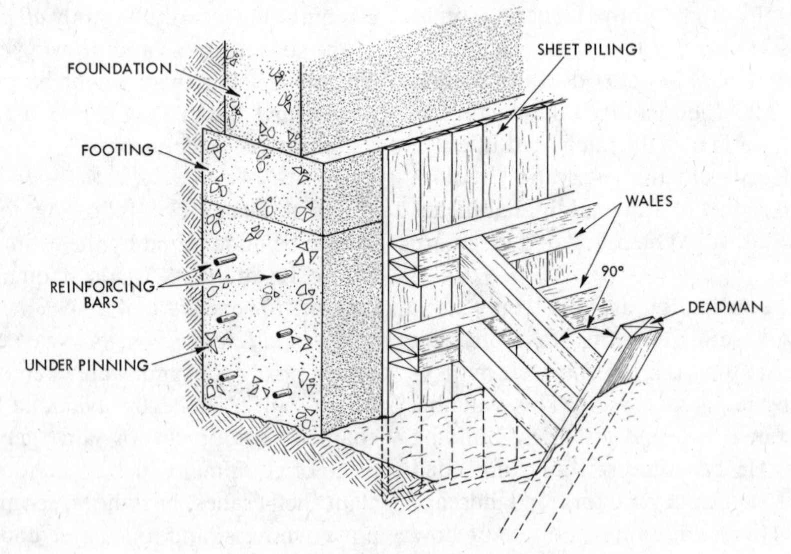

Fig. 7. Sheet piling and underpinning.

mer. This is done by hand on short-driven piles.

3. Another method is to place the sheet pile into a hole which was hand dug. The piles are then *braced, shored,* or *backfilled.*

Underpinning. The laws of some states require underpinning when a building is constructed adjacent to an existing structure and when the foundation of the proposed building goes below the foundation line of the existing structure. It is necessary to bring the foundation footing of the existing structure down to the level of the proposed building. (Underpinning is illustrated in Fig. 7.) This means that when excavating for a building 10 feet lower than an existing structure, you must pour a concrete foundation footing, 10 feet deep, the entire length of the existing structure which faces the proposed building. This is a costly and time-consuming job.

The estimator should also know the local laws and the soil conditions at the site. *Remember that the contractor is held responsible for any damage to existing structures.*

Soil and Rock. For estimating purposes, soil and rock may be classified as follows:

Light soil is a granular substance which can be readily shoveled by hand without the aid of machines. This includes gravel, sand, coarse sand, and fine sand.

Medium soil includes all soils which can be loosened by picks, shovels, and drag line scraper. This includes the cohesive soils such as clay and adobe.

When clay is encountered, the excavation is usually made about 4 inches deeper than for sand, sandy soil, or gravel. This allows for fill material such as crushed gravel.

Heavy soil includes all soils which can be loosened by picks, but are hard to loosen with shovels. Machinery such as back hoes and dippers must be used. Compacted gravel, small stones, and boulders are in this classification.

Hard pan includes boulders, clay, cemented mixtures of sand and gravel, and other substances which are difficult to loosen with a pick. It is advisable to use light blasting to loosen any type of cemented soil, thereby making it easy for a power shovel to pick it up and load it on a truck.

Rock includes soft rock (*shale*), medium hard rock (*slate*), and hard sound rock (*schist*). Rocks require heavy blasting which is expensive. Sub-surface drawings should be studied to determine if rock is present under a site; its presence may mean additional excavation costs.

Angle of Repose. The slope which a given material will maintain without sliding is called the angle of repose. Soils vary greatly in their ability to hold in place at the edge of an excavation. A clay soil, unless subjected to the action of running water, will stand practically vertical at the edge of an excavation; a soil with a high sand content slides so that the bank of an excavation becomes a long slope. Table I lists angles of repose for some common soils.

TABLE 1. ANGLES OF REPOSE

SOIL	DEGREES FROM HORIZONTAL		
	DRY	MOIST	WET
SAND	20–35	30–45	20–40
EARTH	20–45	25–45	25–30
GRAVEL	30–48	.....	.....
GRAVEL, SAND, AND CLAY	20–37	.....	.....

Cradling. Cradling is the temporary supporting of any existing utility lines in or around the excavated area. On many construction jobs it is not possible to remove, cut, or demolish utility lines; the lines must be properly protected so that the utilities of the surrounding area are not interrupted. Cradling is accomplished by placing timbers under the utility lines (usually pipes) and then jacking up the timbers as the excavation goes deeper; more timbers are placed under the pipe until there is a tower of timbers from the bottom of the excavation to the underside of the utility line. The utility lines are thus kept at their original level.

Well Points. This system is used to *dewater* an excavation by means of horizontal header pipes which are connected by a swing joint to vertical pipes which are driven into the lower ground water table. These vertical or riser pipes will then drain the ground water table by means of the well point

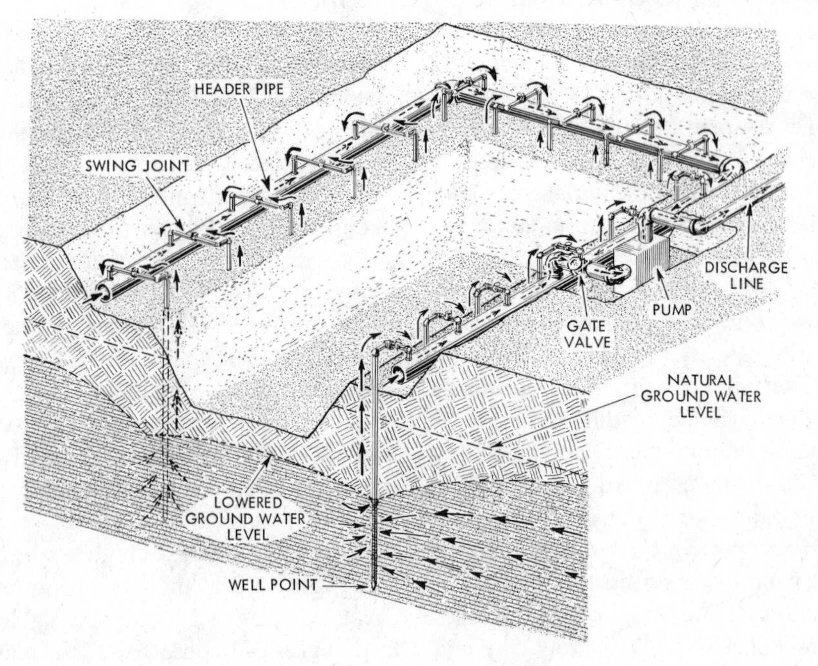

Fig. 8. The well point system of dewatering.

pump. Fig. 8 shows how such a system works.

Ground water is a great problem and unless the water is properly controlled it can cause sand excavations to be much more costly than rock excavations. Improper handling of ground water can slow the construction work and even bring it to a halt. Engineers realize that removing ground water is costly and when preparing an estimate, they utilize the best available advice on the practical and economical methods of dewatering. It is advisable, should the contractor discover water beneath his proposed excavation, to contact reputable dewatering contractors to bid on and dewater the building site.

Payline. The payline establishes the exact amount of excavation for which the contractor is responsible and for which he will be paid. The way to accurately compute the amount of excavation is outlined in some specifications as follows: "The payline shall be 2 feet from the top edge of the footing out, and then vertically up." This, of course, would not apply to soft soil. In soft soil the payline statement would read "two feet from the top edge of the footing and then up on a slope." This is done because soft soil cannot stand vertically; it might slide into the excavation.

Soil Swell. Soil, in its natural state, is generally closely compacted. When this soil is disturbed by excavating, and loaded into dump trucks or piled, it increases its volume due to *swell*. Such swelling can occur to the extent of 50 per cent. Table II shows the average increase in volume for various soils.

Neat Cut. A vertical-walled excavation is called a neat cut. It is made if, in the opinion of the contractor, the soil is well compacted and the excavation shallow enough to eliminate the possibility of slides or cave-ins.

Slope Cut. The term slope cut is applied to an inclined or sloped wall excavation. The slope cut is used for deep excavations in soft, shifting soil. The angle of a slope cut is determined by the angle of repose for soil in the area.

Backfill. Backfilling is the process by which the excavator replaces the earth around the sub-structure or foundation. In backfilling, it is necessary for the estimator to make sure the ground doesn't contain large boulders or extremely large masses of earth or debris. These would increase the cost of backfilling since it would require more cautious methods of pushing back the earth.

Compaction. Compaction is attempting to place the disturbed earth back into its original compact condition. This means trying to remove the

TABLE II. SOIL SWELL

SOIL	BEFORE EXCAVATION	AFTER EXCAVATION
EARTH CLAY	1	1 1/4
SAND AND GRAVEL	1	1 1/2
BROKEN STONES	1	1 1/2
FREE STONE	1	1 1/4
ROCK	1	1 1/2

air spaces so that the particles are as closely together as possible. This is usually achieved by wetting the earth while it is being backfilled and then continually going over it with a large piece of equipment, such as a bulldozer, power shovel, or power crane.

Rough Grading. The excavation specifications will probably call for rough grading of the site. Rough grading is done over that portion of the site which is not occupied by the building, roads, or walks. This is done before spreading of top soil and *finish grading.*

Excavating, Filling, and Rough Grading. In regard to this section of the specifications the contractor for general construction (and in some cases the excavator) must closely follow the written specifications.

Topsoil. Excavating for a building requires many types of work. For example, before actual excavating can begin, it is profitable to remove the existing brush and small trees. If there is any topsoil on the site, it is financially desirable to scrape as much of it as possible and stockpile it for future use; if there is an excess of topsoil, it can be sold to landscape gardeners or other builders. Topsoil is usually spread 6 to 12 inches deep.

If the building site is lower than the proposed grade or the proposed finished grade of the building, it is necessary for the estimator to calculate the amount of fill which is required to bring the land around the building up to the proposed grade. If the development, building, or structure is to be near existing structures, it might be necessary to shore, brace, and protect the adjacent structures with either temporary shoring or permanent supports, in which case they would require underpinning. To avoid injury to men and equipment, the excavating contractor must shore all areas where excessive vibrations (due to traffic and movement of equipment) might affect the side banks of the excavation.

Some of the different types of excavation for common structures are as follows:

1. Excavating and/or filling for all foundations of the building to the bottom of the piers, footings, wall beams, grade beams, and buttresses.
2. Excavating of the cellar, interior beams, and walls.
3. Excavating for all pits except housetrap pits which are usually estimated by the plumbing contractor.
4. Excavation for ramps, ramp walls, exterior stairs and walls, areaways, or exterior and interior retaining walls as detailed in the plans is another distinct part of the excavating contract which the excavator must study carefully since (in most cases) the special excavation will be in confined areas.
5. Excavation for the boiler stack and incinerator foundations are two special excavations. The boiler stack and incinerator foundations are usually the deepest part of the structure and might have to sustain extreme loads. The excavator must be careful not to excavate too deeply because filling in with loose soil will reduce the load-bearing capacity of the soil.
6. Backfilling around all wall

beams, slabs on the ground, foundation and interior walls, and wherever specified in the drawings or written specifications.

7. Excavation for utility supports for heating and sanitation lines, storm sewers, house sewers, as well as for water, fire, and gas lines is sometimes considered as a special excavation under a separate part of the contract—the heating and plumbing section. Regardless of responsibility, the general contractor and estimator should read and understand what is required by the excavation specifications for this particular portion of the work.

There are many other things which might be specified under the excavation section:

8. If there are large trees they are usually mentioned in the specifications. The method of removal and disposal of the trees can vary significantly. For example, in some areas you may cut the trees into small logs and bank them on the site. In other areas, you may cut the trees into small logs and use them as a portion of your backfill on the general site. In yet other areas, you might be required to cut the trees into small logs and transport them to a public or private rubbish dump. Each of these removal methods is expensive; the estimator and/or excavating contractor must therefore visit the site, understand the local requirements, and price the job in accordance with the methods required for the removal of the trees.

If the trees are to remain as a permanent part of the site, it is usually required that the excavating contractor protect these trees. Protection might be in the form of tying the roots, tying the trees with guy wires, building fences around the trees, marking them, blocking them out, or even pruning the trees. Regardless of what is necessary to protect the trees, the estimator must be sure to calculate its cost.

9. As previously discussed, the excavating of a building is divided into general and special excavation. General excavation can be done with equipment which is readily available and accessible to the construction site. Special equipment and hand excavation is considered as special excavation. In quantity take-offs for excavation, it is necessary to estimate and check-off each item separately since general excavation will usually cost less than special excavation.

10. Excavation estimates for a site are frequently approximate estimates because it is seldom possible to determine the exact geological conditions at the site. Regardless of the number of test borings, the spacing between bores is usually so great that what lays between the borings is seldom known for sure. In such cases there are certain items which the excavation estimator must submit as *unit prices.* (Unit prices are stated as price per yard, per hour, per ton, or per truckload.) These items are considered as special excavation. For example, the removal of an old concrete foundation, where the plans for the demolished building no longer exist, must be taken as special excavation and the contractor must be given a unit price. Without a unit price, the

estimate is a gamble to all concerned parties.

11. The removal of rock is special excavation because it is not always possible to determine how large or deep the boulders are at the construction site. When removing a boulder, it is not always possible to break it—leaving a portion below the grade line. The complete removal of rocks can be very expensive. The estimator should either set a unit price for rock excavation or the owner or contractor should give the company a unit price. The unit price is based upon the amount of rock by tonnage, boulder, truckload, or pieces, depending on the area and accessibility of the site.

12. Removal of underground water is one of the most costly and important items. You cannot work under or around water without special equipment. Dewatering is a special excavation item and an allowance is made to the contractor for pumping or installing a wellpoint system.

After the hole is dug and the concrete workers, the steel workers, and the other tradesmen who work below grade level have finished their work, it is necessary to backfill. For a large building, you must not only backfill from the exterior walls to the sidewalk but also to the pier frames of the structure. The cellar slabs must be backfilled and leveled. The estimator can easily see that the cost of backfilling the outside of a structure is usually less per yard than backfilling the interior or the interior foundation walls, where the working radius of the equipment is shortened considerably because of the pillars, pilasters, columns, and other parts of the sub-structure.

Now you are almost ready to do an excavating estimate. You are aware of what is involved in an excavating estimate and should see the need for carefully reading the excavation portion of the specifications to acquaint yourself with the various jobs and items. After acquainting yourself with these jobs and items, you should list them in the proper sequence, on an estimating sheet, and price each one. Then estimate for the various pieces of equipment.

First you will learn how the estimate may be done in several ways. Then the use of excavating equipment will be explained in more detail to give you a better understanding of haulage, the cost of loading material, and the cost of the machinery itself. This does not mean a useable dollar-and-cents value for the job because costs vary greatly among areas. However, the time that each piece of equipment takes to do a given job varies little, and these figures are therefore accurate and can be applied to most jobs.

Quantity take-offs

Profiles. Prior to the actual excavation, the area is staked out or *profiled* by one of several methods, depending on the size of the structure. On a very small excavation, a 6-8-10 triangle is used at each corner, and on larger jobs a transit is used to establish the lines. After the lines are established they are transformed to *batter boards,* Fig. 9, which are 1″ × 6″

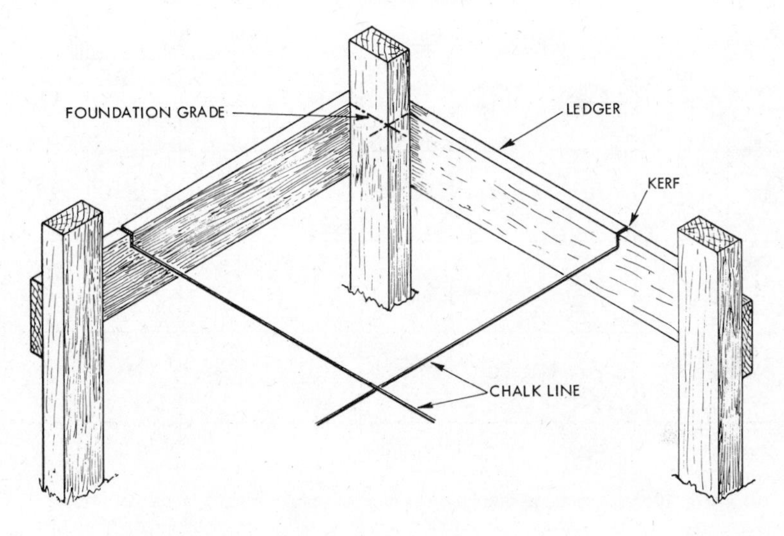

Fig. 9. Batter boards are used to mark the boundaries of the construction site.

boards nailed at right angles in a horizontal position to three 2″ × 6″ boards. The batter boards thus nailed form a corner in the shape of an L. Batter boards can also be used at the sides of the cut without the L leg to establish a line for special interior foundations and walls. The main purpose of the batter boards is to maintain excavation lines, foundation lines, and building lines.

Ordinarily so little lumber or other material is required that the contractor supplies it without consideration in the quantity take-off or estimate. On exceptionally large jobs where a great many 2 × 4's and planking are required, the amount of lumber is estimated by actual count of the pieces. Since batter boards are reuseable, the small material cost is not considered.

Gridding. Suppose that an airport is to be constructed on land which varies in elevation. The large portion of land which is needed might have hills, valleys, and gullies. The terrain must be cut (excavated) or filled until the final grade is uniform.

To estimate this job you must use the topographical site map and divide it into small squares, or *grids*. This operation is called *gridding* or *grilling*.

The grids are usually drawn on an overlay sheet of tracing paper. The size of the grids is determined by the nature of the terrain. If the terrain is gradually sloped, the grids might represent squares of 100 feet on each side. When the terrain is very irregular the grid size is decreased to 25 feet square; this gives a more accurate picture of the cut and fill needed for the particular site.

In Fig. 10, assume that the length and width of each grid is 100 feet; thus each grid is 100′ × 100′, or

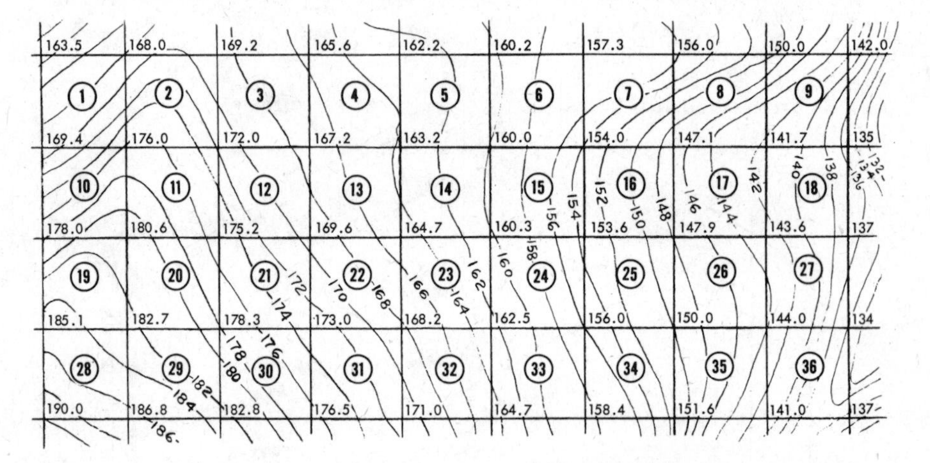

Fig. 10. An example of gridding. The contour plan is divided into equal areas.

10,000 square feet. The approximate elevation at each corner is established by using the nearest contour line. (These elevations are given on the drawing.)

Look at Grid 32 and note which contour lines pass close to the corners of the grid. The corners are approximately at elevations of 168.2′, 162.5′, 171.0′, and 164.7′. Adding together these four elevations and then dividing by four gives the mean or average elevation of the grid:

$$\frac{666.4'}{4} = 166.6'$$

The final grade is to be 162.0′; therefore, the existing grade for this grid is at an average of 4.6′ above the final grade. The excavators must cut 4.6 × 10,000, or 46,000 cubic feet from this particular grid.

Fig. 11 shows the cut (or fill) required for each grid. The differences of the totals for the cut and the fill indicates how much fill (if any) must be brought onto the site. In this case, you would have to haul 26,400 cubic feet, or 978 cubic yards of fill.

The excavated earth is stockpiled on the site and represents a further cost. The cost of moving this earth around the site is added to the cost of the additional fill. The unit costs and the total cost depend on the type, the size, and the efficiency of the equipment used for this job.

General Excavation. A quantity take-off for the excavation of a house foundation (in two different types of soil) will help you to understand better the method of computing general excavation yardage.

The house foundation is to be 40 feet long, 20 feet wide, and 5 feet deep for both examples. The footings extend 6 inches beyond the foundation.

Example 1. The foundation is in light soil and the payline is 2 feet out from the edge of the footing and then up

GENERAL EXCAVATION

QUANTITY TAKE-OFF AIRPORT SITE

No.	Average Elevation ft.	Avg. Cut ft/sq. ft.	Quantity cu. ft.	Avg. Fill ft/sq. ft.	Quantity cu. ft.	Total Cut cu. ft.	Total Fill cu. ft.	Difference cu. yds.
	0 FT. GRADE LEVEL=162.0'							
1	169.22	7.22	72,000					
2	171.30	9.30	93,000					
3	168.50	6.50	65,000					
4	164.55	2.55	25,000					
5	161.40			0.60	6,000			
6	157.87			4.13	41,300			
7	153.60			8.40	84,000			
8	148.70			13.30	133,000			
9	142.17			19.83	198,300			
10	176.00	14.00	140,000					
11	175.95	13.95	139,500					
12	171.00	9.00	90,000					
13	166.17	4.17	41,700					
14	162.05	0.05	500					
15	156.97			5.03	50,300			
16	150.65			11.35	113,500			
17	145.07			16.93	169,300			
18	139.33			22.67	226,700			
19	181.60	19.60	196,000					
20	179.20	17.20	172,000					
21	174.02	12.02	120,200					
22	168.88	6.88	68,800					
23	163.92	1.92	19,200					
24	158.10			3.90	39,000			
25	151.87			10.13	101,300			
26	146.37			15.63	156,300			
27	139.65			22.35	223,500			
28	186.15	24.15	241,500					
29	182.65	20.65	206,500					
30	177.65	15.65	156,600					
31	172.17	10.17	101,700					
32	166.60	4.60	46,000					
33	160.40			1.60	16,000			
34	154.00			8.00	80,000			
35	146.65			15.35	153,500			
36	139.00			23.00	230,000			
						1,995600	2,022000	978

Fig. 11. A quantity take-off for the general excavation of an airport site.

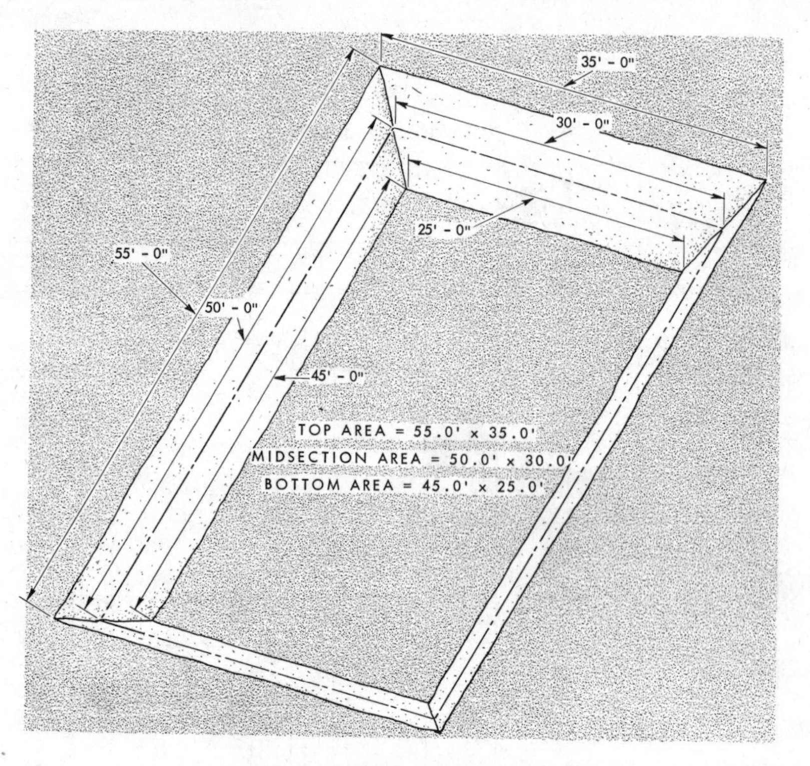

Fig. 12. A pictorial view of an excavation.

on a 1:1 slope. Figs. 12 to 15 illustrate this simple excavation. Since a sloped payline is required, it is necessary to use a formula for the volume of a frustum of a right pyramid:

$$V = h/3\ (A_b + A_t + \sqrt{A_bA_t})$$

$$or\ V = h/6\ (A_b + A_t + 4M)$$

where A_b = bottom area
A_t = top area
M = mid-section area

The second formula will be used for the volume calculations; thus the areas of the bottom, top, and the mid-section must be found. Since the footings extend 6 inches beyond the foundation, 1 foot must be added to the length and to the width. Figs. 14 and 15 are cross-sectional views of this excavation. The payline is 2 feet beyond the footings, so 4 feet must be added to both the length and width.

Bottom length =
$40.0' + 1.0' + 4.0' = 45.0'$
Bottom width =
$20.0' + 1.0' + 4.0' = 25.0'$
A_b =
$25.0' \times 45.0' = 1{,}125.0$ sq. ft.

Since the soil is on a slope, the length and width of the top area are the bottom dimensions increased by the horizontal dimensions of the slope. A 1:1 slope means that for every foot gained horizontally there is one foot of depth. The depth here is 5 feet so

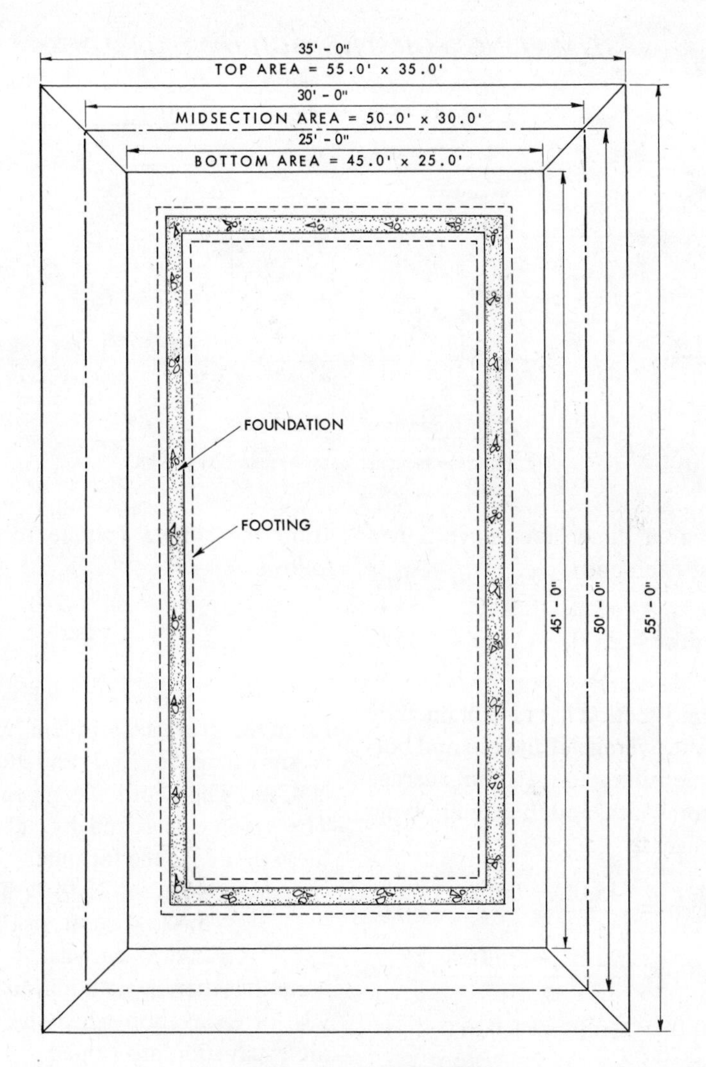

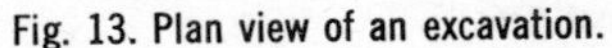
Fig. 13. Plan view of an excavation.

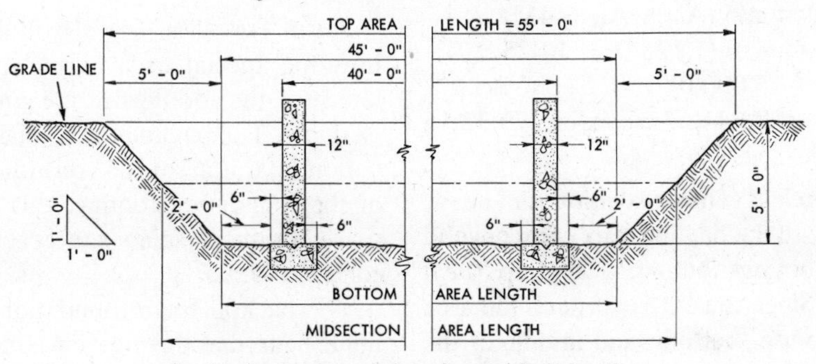

Fig. 14. Cross-sectional view of excavation's length.

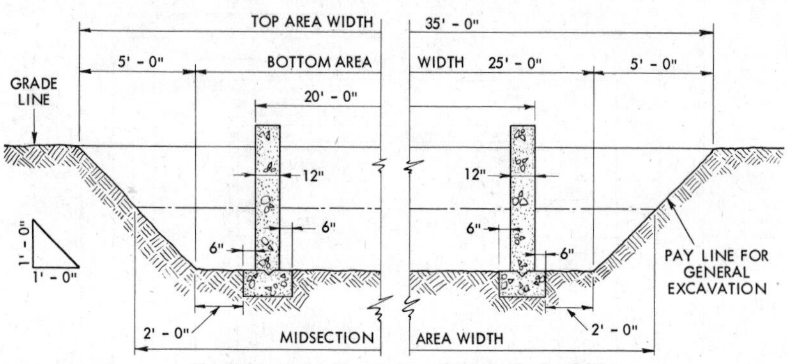

Fig. 15. Cross-sectional view of excavation's width.

the top area dimensions have 5 feet added at each end:

Top length = 45.0′ + 10.0′ = 55.0′
Top width = 25.0′ + 10.0′ = 35.0′
A_t = 35.0′ × 55.0′ = 1,925.0 sq. ft.

The mid-section area is obtained by finding the average of the top and bottom dimensions to get the average length and width, and then multiplying the averages:

$$\text{Av. length} = \frac{45.0' + 55.0'}{2} = 50.0'$$

$$\text{Av. width} = \frac{25.0' + 35.0'}{2} = 30.0'$$

M = 50.0′ × 30.0′ = 1,500.0 sq. ft.

Substituting in the formula:

$$V = h/6\,(A_b + A_t + 4M)$$
$$= 5/6\,(1{,}125.0 + 1{,}925 + 6{,}000.0)$$
$$= 7{,}541.67 \text{ cu. ft. or } 279.32 \text{ cu. yds.}$$

Example 2. The excavation is in heavy soil with the payline also 2 feet beyond the footings but vertically up *(neat cut)*. Since the cut is vertical, the area of the top, bottom, and middle of the cut are equal, and the volume is found from the simple volume formula as follows:

$$V = a \times b \times h$$

where a = length
b = width
h = depth

From the previous problem, the width is known to be 25.0′ and the length is 45.0′. The depth was given as 5.0′. The volume is found by substituting these figures in the formula:

$$V = 45.0' \times 25.0' \times 5.0'$$
$$= 5{,}625.0 \text{ cu. ft. or}$$
$$208.33 \text{ cu. yds.}$$

Note that there is an additional 71 cu. yds. of excavation when the walls of the excavation are sloped.

Special Excavation. Fig. 16 illustrates how the special excavation is calculated for the footings in the previous example. The payline for special excavation (in this case) is 6″ on either side of the foundation footing; this leaves space for the placing and bracing of concrete forms.

The backfill for a foundation footing is figured as the difference between the volume of concrete needed for the

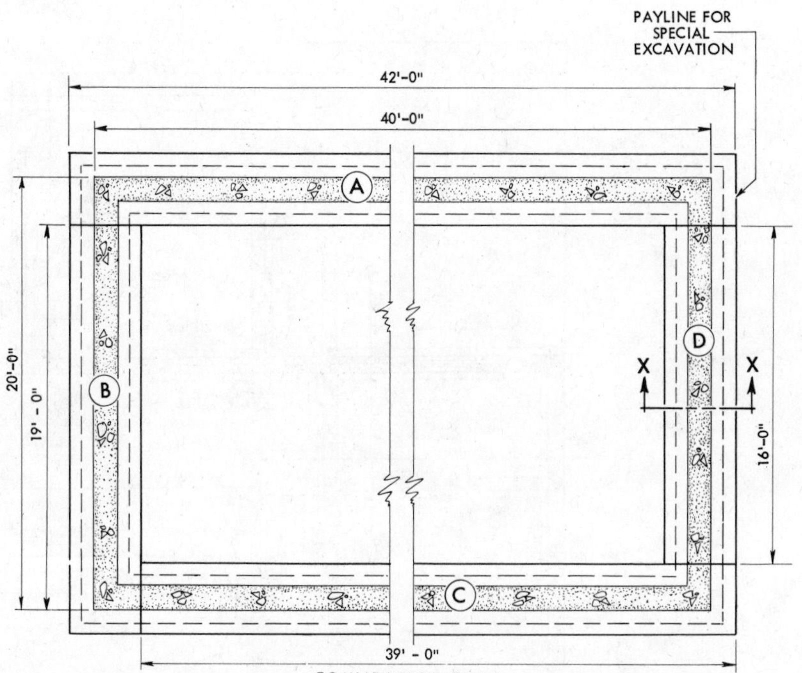

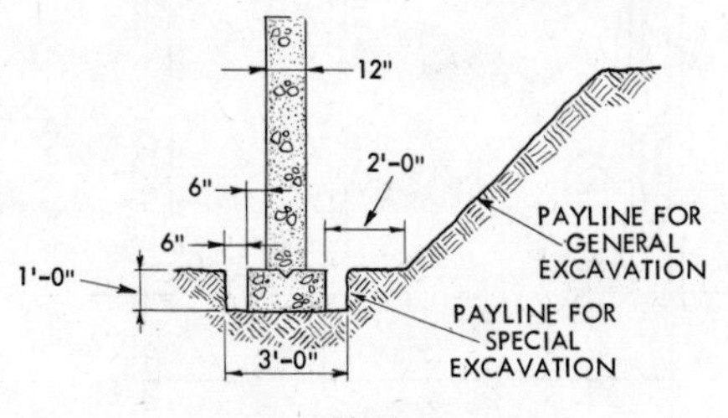

SPECIAL EXCAVATION - FOOTINGS

FOOTING	LENGTH	WIDTH	DEPTH	CU. FT.
A	42.0'	3.0'	1.0'	126.0
B	19.0'	3.0'	1.0'	57.0
C	39.0'	3.0'	1.0'	117.0
D	16.0'	3.0'	1.0'	48.0
			TOTAL	348.0 = 12.9 cu. yds.

Fig. 16. The method of calculating special excavation of footings.

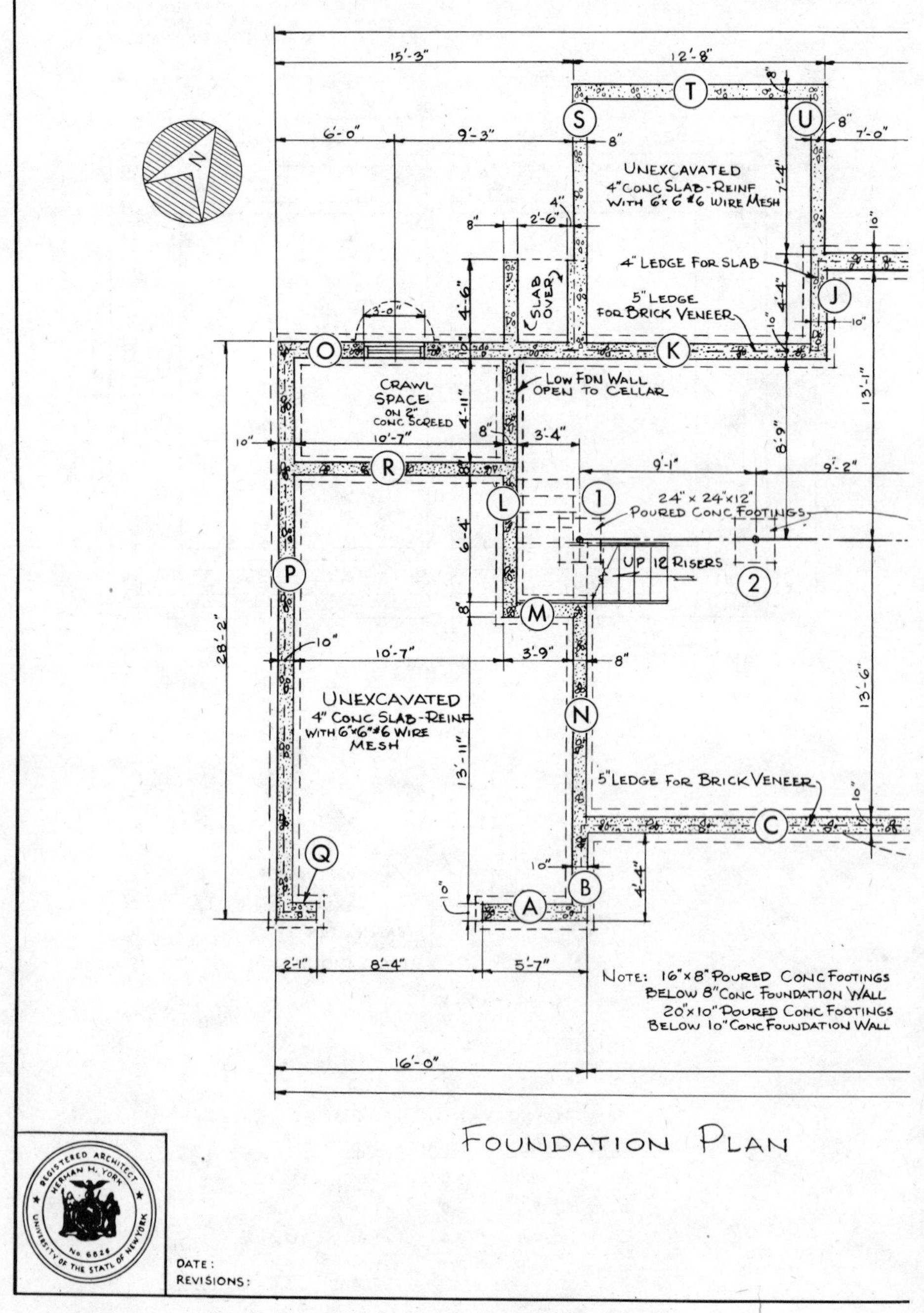

Fig. 17. House Plan A coded for the excavation take-off.

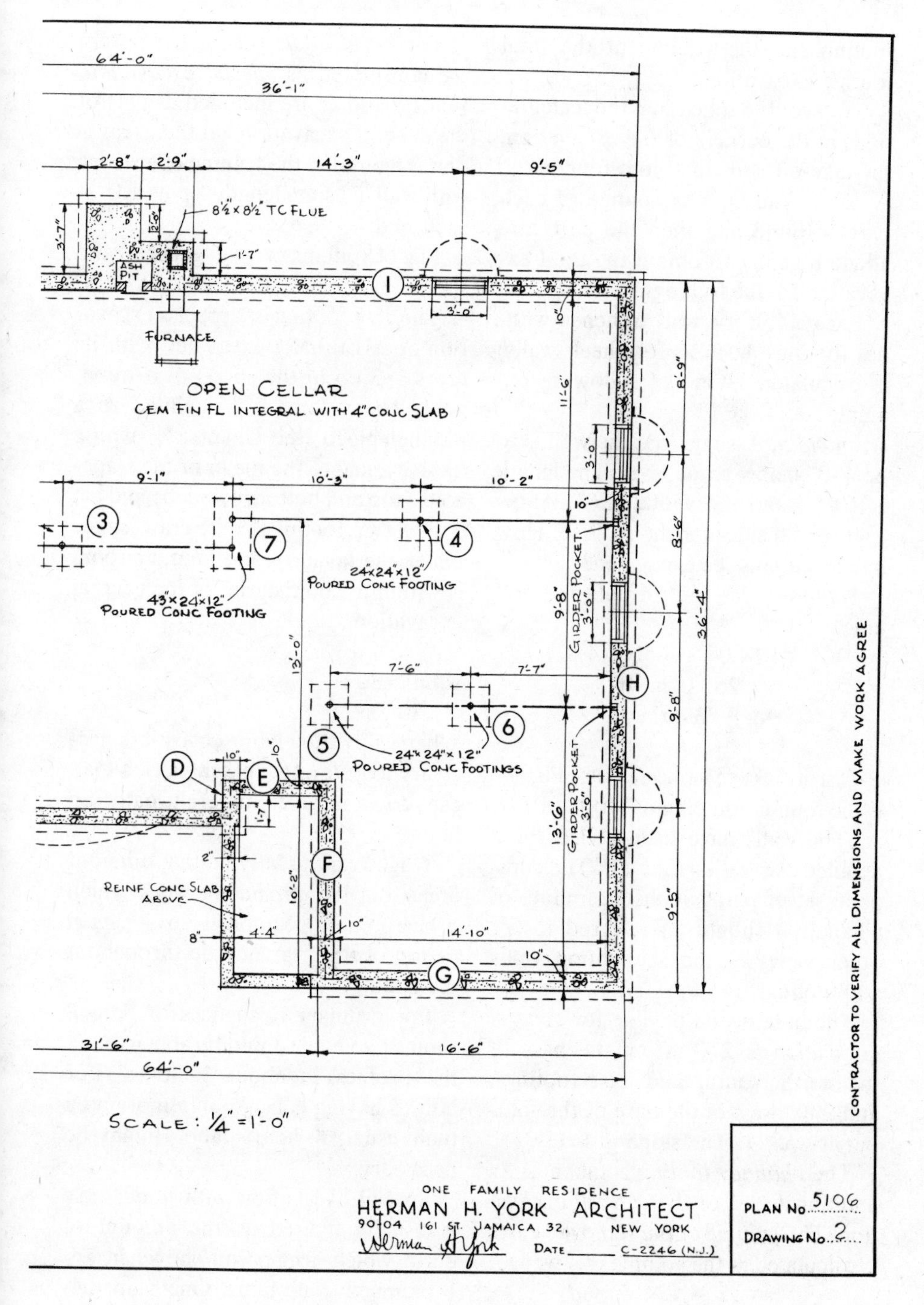
64'-0"
36'-1"
2'-8"
2'-9"
14'-3"
9'-5"
8½" x 8½" TC FLUE
3'-7"
1'-7"
ASH PIT
I
3'-0"
FURNACE
OPEN CELLAR
CEM FIN FL INTEGRAL WITH 4" CONC SLAB
10"
11'-6"
8'-9"
3'-0"
9'-1"
10'-3"
10'-2"
10"
3
7
4
8'-6"
24"x24"x12"
POURED CONC FOOTING
43"x24"x12"
POURED CONC FOOTING
GIRDER POCKET
9'-8"
3'-0"
36'-4"
13'-0"
7'-6"
7'-7"
H
9'-8"
5
6
24"x 24"x 12"
POURED CONC FOOTINGS
D
E
10"
1'-7"
GIRDER POCKET
3'-0"
13'-6"
2"
F
REINF CONC SLAB ABOVE
10'-2"
9'-5"
8"
4'-4"
10"
14'-10"
10"
G
31'-6"
16'-6"
64'-0"
SCALE: ¼" = 1'-0"
CONTRACTOR TO VERIFY ALL DIMENSIONS AND MAKE WORK AGREE
ONE FAMILY RESIDENCE
HERMAN H. YORK ARCHITECT
90-04 161 ST. JAMAICA 32, NEW YORK
DATE C-2246 (N.J.)
PLAN No. 5106
DRAWING No. 2

footing and the volume of the total footing excavation.

To avoid duplicating the calculations at the corners of the excavation, the take-off is divided into four parts, A, B, C, and D. The volume of each part is found and then the parts are added together to obtain the total excavation for the footings.

Excavation for wall trenches without footings (such as for cheek walls) is calculated as in the following example.

Example 3. Assume that the wall is to be 4′-0″ below grade level. The length is 10′-0″ and the width is 6″. Allow 2′-0″ on all sides for the payline. How much soil must be excavated?

Solution: Use the formula for volume.

$$V = 14.0' \times 4.5' \times 4.0'$$
$$= 252.0 \text{ cu. ft.}$$
$$= 9.33 \text{ cu. yds.}$$

Quantity Take-off for House Plans. Fig. 17 is the foundation plan of HOUSE PLAN A. The walls have been coded for a detailed excavation take-off. The complete set of plans at the beginning of Chapter 3 should be referred to for other views of the foundation walls and footings.

The base of the payline for *general excavation* is 2′-0″ from the base of the north, south, and east footings, and 2′-0″ west of the base of the footing at wall *A*. The slope is 1:1.

The *chimney footing* is taken as an additional part of the general excavation. The volume of the removed earth is calculated as the volume of a wedge:

$$V = \tfrac{1}{2} \times l \times w \times d$$

The *cheek walls* (under the porch) are considered as *special excavation.* Walls *S* and *U* are included as part of the general excavation, so the excavation is figured as the volume of a wedge with wall *T* as the length—plus 1.0′ at each end.

Fig. 18 itemizes the general and special excavation for HOUSE PLAN A.

The step footing is a special excavation and is usually excavated with its cross-section in the shape of a trapezoid. The volume is calculated as a parallelepiped (see Chapter 4) whose base is equal to the mean of the trapezoid's top and bottom bases, b_1 and b_2.

The step footing for the crawl space under the lavatory and laundry room is within the boundaries of the general excavation.

Labor costs

When estimating labor costs, it is necessary to know how much work a man can do in a given time—usually an hour.

Wages vary greatly among different areas, but the *amount* of work which a man can accomplish in a given period of time varies little throughout the country.

For ordinary residences of 5 to 8 rooms two men should be able to erect the profiles in about 3 hours. For houses having considerably more area than usual, 4 hours labor might be necessary.

Special excavation with hand tools is usually figured as the amount of earth which a man can shovel in approximately one hour. One man dig-

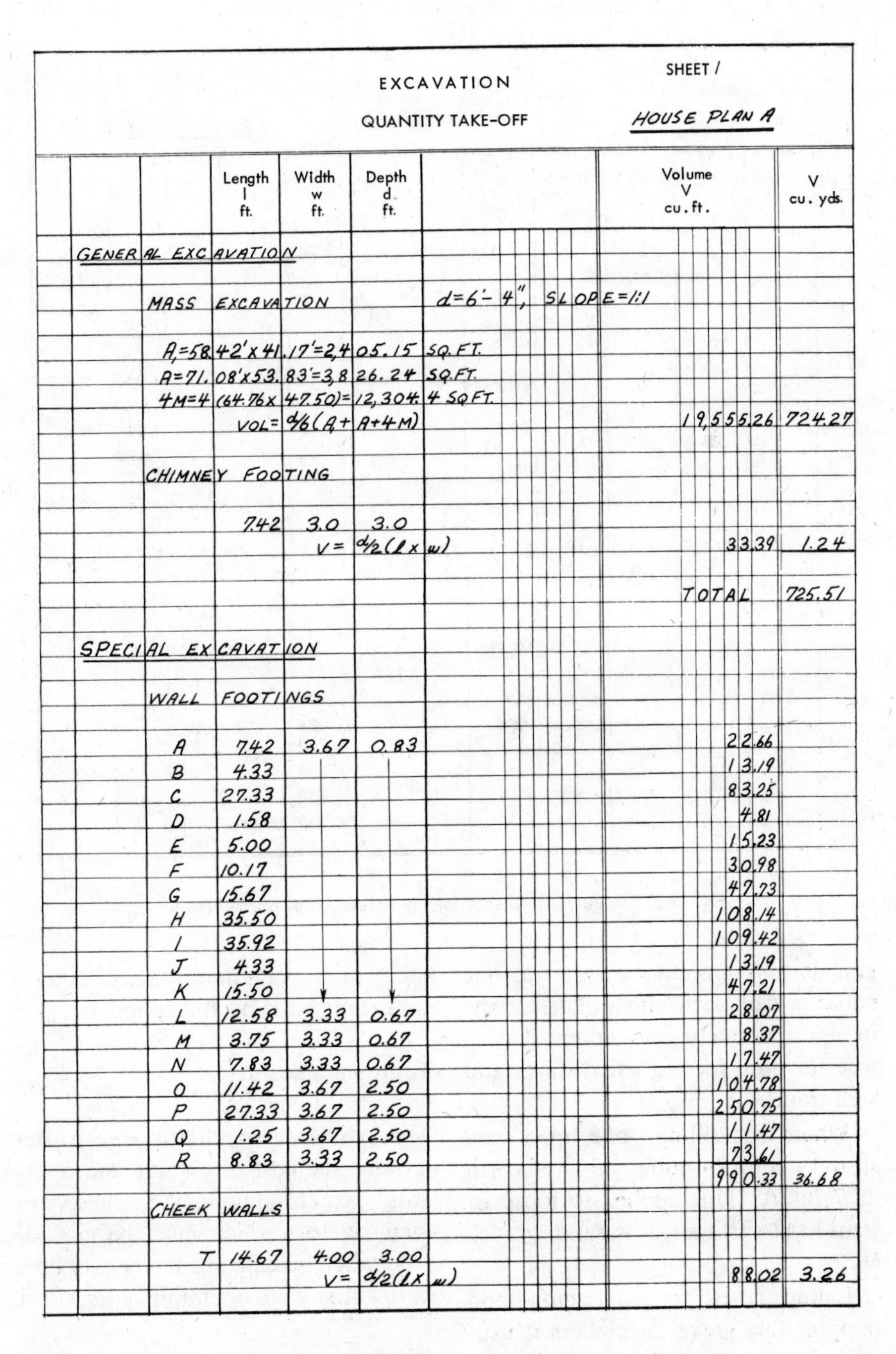

EXCAVATION

QUANTITY TAKE-OFF

SHEET /

HOUSE PLAN A

Item	Length l ft.	Width w ft.	Depth d ft.		Volume V cu. ft.	V cu. yds.
GENERAL EXCAVATION						
MASS EXCAVATION				d=6'-4", SLOPE=1:1		
A,=58.42'x41.17'=2,405.15 SQ.FT.						
A=71.08'x53.83'=3,826.24 SQ.FT.						
4M=4(64.76x47.50)=12,304.4 SQ.FT.						
VOL= d/6(A,+A+4M)					19,555.26	724.27
CHIMNEY FOOTING						
	7.42	3.0	3.0			
		V=	d/2(l x w)		33.39	1.24
					TOTAL	725.51
SPECIAL EXCAVATION						
WALL FOOTINGS						
A	7.42	3.67	0.83		22.66	
B	4.33	↓	↓		13.19	
C	27.33				83.25	
D	1.58				4.81	
E	5.00				15.23	
F	10.17				30.98	
G	15.67				47.73	
H	35.50				108.14	
I	35.92				109.42	
J	4.33				13.19	
K	15.50	↓	↓		47.21	
L	12.58	3.33	0.67		28.07	
M	3.75	3.33	0.67		8.37	
N	7.83	3.33	0.67		17.47	
O	11.42	3.67	2.50		104.78	
P	27.33	3.67	2.50		250.75	
Q	1.25	3.67	2.50		11.47	
R	8.83	3.33	2.50		73.61	
					990.33	36.68
CHEEK WALLS						
T	14.67	4.00	3.00			
		V=	d/2(l x w)		88.02	3.26

Fig. 18-A. The quantity take-off for the excavation of House Plan A.

EXCAVATION

QUANTITY TAKE-OFF

SHEET 2

HOUSE PLAN A

	Length l ft.	Width w ft.	Depth d ft.		Volume V cu. ft.	V cu. yds.
PIER FOOTINGS						
1	4.00	4.00	1.00		16.00	
2					16.00	
3					16.00	
4					16.00	
5					16.00	
6					16.00	
7	5.58				22.32	
					118.32	4.38
STEP FOOTING—PART OF FOUNDATION WALL FOOTINGS						
					TOTAL	44.32
	SUMMARY					
	MASS EXCAVATION			724.27	CU. YDS.	
	CHIMNEY FOOTINGS			1.24		
	FOUNDATION WALL FOOTINGS			36.68		
	CHEEK WALLS			3.26		
	PIER FOOTINGS			4.38		
				769.83	CU. YDS.	

Fig. 18-B. The quantity take-off for the excavation of House Plan A.

ging in soft ground can remove one cubic yard of earth with a wheelbarrow in approximately one hour. This is true for both footing excavations and wall trenches.

When backfilling, one man can shovel about 24 cubic yards of earth in 7 hours. This applies to loose or semi-hard earth such as sand, gravel, etc.

Labor costs for spreading and tamping fine gravel or cinders is usually given in a combined material and labor price per square inch or square foot *(unit cost)* by the supplier.

Equipment costs

The estimator should be able to determine the cost of owning and operating power equipment.* There are many factors which must be considered when calculating the hourly cost of owning and operating equipment

*Material in this section courtesy of Caterpillar Tractor Co.

and the cost per yard—the true measure of equipment performance.

Owning costs are the "fixed costs" which are always present regardless of whether or not the equipment is in operation. These include depreciation, interest, insurance, and taxes.

Operating costs are "variable costs." These include fuel and lubricants, repairs, and tires.

Operator's wages must be added to the owning and operating costs to obtain a true figure for hourly costs of operation. When renting equipment, the operator's salary is often included in the hourly rate. An example of this is the cost of renting a truck. Trucks are usually rented by the day. A 10 cubic yard water level truck, which can carry 12.5 yards of heaped earth, can be rented as a unit complete with driver for a fixed cost per day.

Overhead should not be considered with the owning and operating costs. This item includes indirect costs such as supervision, transportation, storage, etc., and should be placed under *General Operations*.

Power Shovels. When trucks are power shovel loaded, it is necessary to estimate the number of *hauling units* (vehicles) required to keep the shovel working efficiently. For estimating purposes, it is first assumed that the shovel operates at an ideal rate. The number of required hauling units is determined by the following formula:

$$\text{Hauling Units} = \frac{\text{shovel production at } 100\% \text{ efficiency}}{\text{hauling unit production}}$$

Note: *Bank-cubic yards* refers to the earth's volume before excavation. After being dug, the earth will usually swell and occupy a greater volume.

TABLE III. ESTIMATED HOURLY PRODUCTION OF POWER SHOVELS AT 100% EFFICIENCY.

	SHOVEL DIPPER SIZES IN CUBIC YARDS																	
MATERIAL CLASS	$\frac{3}{8}$	$\frac{1}{2}$	$\frac{3}{4}$	1	$1\frac{1}{4}$	$1\frac{1}{2}$	$1\frac{3}{4}$	2	$2\frac{1}{2}$	$2\frac{3}{4}$	3	$3\frac{1}{2}$	4	$4\frac{1}{2}$	5	$5\frac{1}{2}$	6	$6\frac{1}{2}$
Moist loam or Sandy clay	85	115	165	205	250	285	320	355	405	435	454	525	580	635	685	740	795	840
Sand and gravel	80	110	155	200	230	270	300	330	390	420	450	505	555	600	645	695	740	785
Common earth	70	95	135	175	210	240	270	300	350	380	405	455	510	560	605	645	685	725
Clay, tough, hard	50	75	110	145	180	210	235	265	310	335	360	405	450	490	530	570	605	640
Rock, well-blasted	40	60	95	125	155	180	205	230	275	300	320	365	410	455	500	540	575	610
Common with rock	30	50	80	105	130	155	180	200	245	270	290	335	380	420	460	500	540	575
Clay, wet & sticky	25	40	70	95	120	145	165	185	230	250	270	310	345	385	420	455	490	520
Rock, poorly blasted	15	25	50	75	95	115	140	160	195	215	235	270	305	340	375	410	440	470

NOTE: The above table is based on cubic yards (bank measure) per hour, 90-degree swing, best digging depth, material loaded into hauling units, and no delays.

Table III gives the estimated hourly production of power shovels operating at 100% efficiency.

The following conditions affect shovel production:

1. bucket capacity
2. type of material being excavated
3. angle through which the boom must move in order to reach the hauling unit
4. the deepest cut which will fill the bucket without spilling its contents

Example 4. A 2½ yard power shovel is working in common earth. The shovel loads into hauling units, each of which can haul 70 bank-yards per hour under the present hauling conditions.

(a) How many units can be served by the power shovel?

(b) If shovel efficiency is estimated as 85%, what is the estimated hourly production of the spread of equipment.

Solutions: For part (a) use the formula for Hauling Units:

$$\text{Hauling Units} = \frac{350 \text{ bank-cu. yds./hour}}{70 \text{ bank-cu. yds./hour}} = 5 \text{ units}$$

(b) Estimated hourly production is 85% of ideal efficiency:

$$350 \times 0.85 = 298 \text{ bank-cu. yds./hr.}$$

TABLE IV. OPTIMUM DEPTH OF CUT FOR POWER SHOVELS (IN FEET).

	POWER SHOVELS		
Size dipper, c.y.	Light, free flowing materials like loam, sand, gravel	Medium materials such as common earth	Harder materials such as rough and hard, or wet and sticky clay.
3/8	3.8	4.5	6.0
1/2	4.6	4.7	7.0
3/4	5.3	6.8	8.0
1	6.0	7.8	9.0
1 1/4	6.5	8.5	9.8
1 1/2	7.0	9.2	10.7
1 3/4	7.4	9.7	11.5
2	7.8	10.2	12.2
2 1/2	8.4	11.2	13.3

TABLE V. CORRECTION FACTORS FOR DEPTH OF CUT AND ANGLE OF SWING ON POWER SHOVEL OUTPUT.

Depth of cut in % of optimum	Angle of Swing, Degrees 45	60	75	90	120	150	180
40	.93	.89	.85	.80	.72	.65	.59
60	1.10	1.03	.96	.91	.81	.73	.66
80	1.22	1.12	1.04	.98	.86	.77	.69
100	1.26	1.16	1.07	1.00	.88	.79	.71
120	1.20	1.11	1.03	.97	.86	.77	.70
140	1.12	1.04	.97	.91	.81	.73	.66
160	1.03	.96	.90	.85	.75	.67	.62

TABLE VI. ESTIMATED HOURLY PRODUCTION OF DRAGLINES.

MATERIAL CLASS	DRAGLINE BUCKET SIZES IN CUBIC YARDS*													
	3/8	1/2	3/4	1	1 1/4	1 1/2	1 3/4	2	2 1/2	3	3 1/2	4	5	6
Moist loam or Sandy clay	70	95	130	160	195	220	245	265	305	350	390	465	540	610
Sand and gravel	65	90	125	155	185	210	235	255	295	340	380	455	530	600
Good, common earth	55	75	105	135	165	190	210	230	265	305	340	375	445	510
Clay, hard & tough	35	55	90	110	135	160	180	195	230	270	305	340	410	475
Clay, wet & sticky	20	30	55	75	95	110	130	145	175	210	240	270	330	385

*Dragline bucket size may be larger than machine rating. Thus, a 2 yd. dragline bucket might be used on a 1 1/2 yd. excavator.

The above table is based on bank yds., suitable depth of cut for maximum effect, no delays, 90 % swing, and all material loaded into hauling units.

Cost Estimating Sheet

MACHINE & MODEL DW 20-Series G-No.486 H.P. 320

ATTACHMENTS No. 27 Cable Control TIRES SIZES 29.5-29 (28 ply) Rock / 14.0-24 (16 ply) Rock

FOB FACTORY PRICE $52,905.00 FREIGHT ---- TOTAL $52,905.00

CONDITIONS Average

OWNING COST: 52,905 − 10,135

Depreciation @ 10,000 Hours. 42,770 $4.28

Interest, Insurance, Taxes @ $0.03 /$1000. 1.59

Total Owning Cost 5.87 5.87

OPERATING COST:

Fuels and Lubricants

Diesel Fuel 9.0 gph x $0.14 /gal 1.26

Gasoline 0.03

Lubricating Oil 0.19 gph x $1.00 /gal 0.19

Transmission Oil 0.04 gph x $0.70 /gal 0.03

Hydraulic Oil 0.04 gph x $1.00 /gal 0.04

Filters 0.04

Grease 0.15 lbs/hr x $0.20 /lb. 0.03

Repairs Incl. Labor 90% of Depreciation 4.28 x 0.9 3.85

Tires @ 45,000 Miles @ 10 M.P.H. 4,500 Hours

10,135 Replacement Cost 10,135/4500 2.26

Other —

Total Operating Cost 7.73 7.73

Total Hourly Owning and Operating Cost (Excl. Operator) 13.60

Operator's Wages 2.50

Total Hourly Owning and Operating Cost $16.10

Fig. 19. Sample cost estimating sheet for a tractor and scraper.

Table IV is based upon the ideal situation where the shovel is operating at its ideal depth of cut with a 90° swing from dig to dump. Since this ideal is seldom achieved in actual operation, the estimated production must be corrected for other than ideal depth of cut and a different degree of swing. When this is the case, Table V is used.

Dragline production is estimated in the same manner as shovel production, using Table VI.

Graders. Mechanical graders are used to maintain roads and to finish final grades. The estimator must be aware of the many conditions which affect the grader's production. The experience and skill of the operator and the type of soil have a greater influence on grader production than on that of most excavating machines. Graders can involve considerable costs.

To estimate the time required for a job, the number of *passes* must be known and an estimate made of the *average speed* and *efficiency* of the grader:

$$\text{Time (in hours)} = \frac{\text{number of passes} \times \text{distance (in miles)}}{\text{average speed (miles per hour)} \times \text{efficiency factor}}$$

The average speed is found by the following type of calculation:

$$\begin{array}{lr} \text{8 passes at 4.0 mph} = & 32 \\ \underline{\text{6}}\text{ passes at 5.0 mph} = & \underline{30} \\ 14 & 62 \end{array}$$

$$\frac{62}{14} = 4.43 \text{ mph average}$$

Example 5. To maintain a working road, a No. 12 Motor Grader must make 1 pass in second gear (3.6 mph) and 2 passes in third gear (5.5 mph). The efficiency of the grader is estimated as 75%. How long will it take to complete the job if the working road is 6.0 miles long?

Solution: Find the average speed and substitute into the formula to find the time.

$$\begin{array}{lr} \text{1 pass at 3.6 mph} = & 3.6 \\ \underline{\text{2}}\text{ passes at 5.5 mph} = & \underline{11.0} \\ 3 & 14.6 \end{array}$$

$$\frac{14.6}{3} = 4.87 \text{ mph average}$$

$$\text{Time} = \frac{3 \times 6.0}{4.87 \times 0.75} = 4.93 \text{ hours}$$

Equipment Cost Records. Fig. 19 is a sample estimate for a tractor with an attached scraper. Its purpose is to determine the hourly cost of operating a typical piece of excavation and grading equipment.

Questions and Problems

1. How is the amount of materials for profiles figured?
2. How is the amount of excavation for a basement figured when the surface is level?
3. How is the amount of backfill figured for a foundation footing?
4. How is the amount of excavation figured for wall trenches when the wall doesn't have a footing (such as a cheek wall)?
5. How much excavation is figured for a pier footing with the following dimensions: 3.0′ × 3.0′ × 1.0′? Allow a 1.0′ payline on all sides. (Answer: 25.0 cu. ft.)
6. Why must soil swell be considered by the estimator?
7. How is the total amount of backfill calculated for a job site?
8. A house occupies 1,500 sq. ft. of a 3,000 sq. ft. site. The garage and walks occupy 350 sq. ft. The remainder of the lot is to be covered with topsoil to a depth of 6″. How much topsoil is needed?
   (Answer: 43 cu. yds.)
9. Foundation wall footings for a house are to be backfilled by hand. How much earth can 2 men backfill in 2 days?
   (Answer: 96 cu. yds.)
10. How many trucks can a half-yard payloader fill in 7 hours if it is able to load one truck in 12 minutes? (Answer: 35)
11. It is necessary to remove 500 cubic yards of heaped earth from a building site. How long will it take 5 trucks to remove the earth if each truck takes 2 hours to complete a cycle and each truck has a capacity of 12½ yards of heaped earth?
   (Answer: 16 hours)
12. Why are excavation costs higher when hard earth is encountered?

6

CONCRETE

Almost all buildings have some form of concrete foundation and many have walls and floors of concrete. This chapter discusses common and recommended types of concrete footings, foundations, floors, and walls and explains how their costs are estimated. Some small buildings are built on posts or blocks, but these are not common and are not discussed in this chapter.

Concrete and mortar mixes

Concrete. In laboratory testing, the proportioning of sand and cement is done by weight. In theory, this is the way it should be done. The volume of a given weight of concrete will be different if it is packed loosely or thrown in a pile from the same weight packed tightly. The same is true of sand. A barrel of Portland cement will increase in volume from 10 to 30 per cent merely by being dumped from the barrel into a pile or from being shoveled into a measuring box. On the construction site, the cement should be measured in the original container as it comes from the manufacturer. This gives each container the same or nearly the same weight. Sand and stone, of course, must be measured loose as it is thrown in the measuring boxes. Sand will not vary quite as much in its weight-to-volume ratio as does Portland cement, but the dryness of the sand does have an effect on the volume of sand. Loose, dry sand occupies less space than does wet sand.

The *absolute* volume of concrete is easily explained by the following experiment. The materials needed are readily available and it is recommended that you do the experiment.

Take four quart-size jars and a one-gallon jar and fill the quart jars as follows:

1 with beads
1 with marbles
1 with flour
1 with water

First pour the marbles into the gallon jar. Then add the beads and shake well. You can see that there is less than two quarts volume in the gallon jar.

This is because the *voids* (spaces) between the marbles have been occupied by the beads—so you have a denser mixture, but the volume is less.

Now pour the flour into the gallon jar. Again shake the jar until the three ingredients are thoroughly mixed. By measuring, you can see that now you have less than the three quarts of ingredients which were added. Finally add the quart of water. This too will find its way into the spaces between the other ingredients.

Thus from the original four quarts of ingredients, you get a volume that is much less in volume than the expected gallon. This loss is caused by the displacement of the air spaces between the marbles, beads, and flour.

The same effect is encountered when mixing concrete. In the experiment above, the marbles represent (and act like) the coarse aggregate; the beads represent sand; the flour represents cement.

Concrete is measured in cubic yards or cubic feet. Although sidewalks are specified in square yards or square feet, their thickness is also given so that the volume of concrete can be calculated.

Ready-Mixed Concrete. Most concrete used in building construction is ready-mixed. The concrete is delivered to the construction site in special trucks which dump it directly into forms or into a hopper for distribution by wheelbarrows, mechanical concrete *buggies,* or crane-hoisted buckets.

Cinder Concrete. Cinder concrete is made in much the same manner as ordinary concrete except that cinders are used in place of stone as aggregate. Cinder concrete can be used for fireproofing and for short reinforced floor and roof slabs. It's principal advantage is its lightness in weight. The cinders should be from well burned hard coal, reasonably free of sulphides. Sulphides, occurring in considerable quantity, cause reinforcing steel bars to rust.

Reinforced Concrete. Concrete which has steel bars or mesh embedded in it is stronger than plain concrete. This concrete is obviously more expensive than plain concrete because of the additional cost of the steel bars or mesh and the tie wire. The labor cost per cubic yard of concrete will also be greater.

Prestressed Concrete. In recent years, prestressed concrete structural members have been used more frequently in building construction. The American Concrete Institute defines prestressed concrete as, "Concrete whose stresses resulting from external loadings are counterbalanced by prestressing reinforcement placed in the structure." Some of the concrete structural members made by this method are double and single tees, channel sections, beams, hollow slabs, and flat planks.

On a large construction project where many prestressed concrete members are of identical size, the cost of each will be lowered.

Weight of Concrete. Contrary to common belief, the weight of concrete varies considerably. The weight usually is between 105 and 160 pounds per cubic foot, depending on the materials used. The type of concrete generally used weighs about 145 pounds per cubic foot. This figure is an average which the estimator may use safely.

Cinder concrete, however, weighs between 75 and 115 pounds per cubic foot.

Cement may be specified at 4.00, 3.80, etc., cubic feet to the barrel. If it is specified at 4.00 cubic feet per barrel, then 1 barrel weighs 376 pounds and contains 4 bags. Each bag weighs 94 pounds and occupies a cubic foot.

Proportions. For ordinary work, the concrete ratio is stated in the specifications. The ratio is usually given in the form 1:2:4:8. This ratio means 1 part cement, 2 parts sand (or fine aggregate), 4 parts stone (or coarse aggregate), and 8 parts water. The ratio will vary according to the architect's purposes and local conditions.

Sometimes the ratio is given in a shorter form, such as 1:2:4. In this case the amount of water is not stated in the specifications and the contractor must use his own judgment. The amount of water varies with the type of structure, degree of exposure, and compressive strength of the concrete.

The voids in ordinary broken stone are somewhat less than half of the volume, and it is common practice to use one-half as much sand as the volume of the broken stone. The proportion of cement is varied according to the strength required in the structure, and according to the desire to economize. On this principle you have the familiar ratios 1:2:4, 1:2½:5, 1:3:6, and 1:4:8. Note that in each of these cases the ratio of the sand to the broken stone is a constant, and the ratio of the cement alone is variable.

Tables I and II give the ingredients in 1 cubic yard of concrete at various mixes, and the volume factors of various mixes.

Ideal Conditions in Preparing the Mix. The general principle to be adopted in mixing concrete is that the amount of water used should be only enough to crystallize the cement paste; the amount of paste should be no more than is sufficient to fill the voids between the particles of sand; and the mixture thus produced should be only enough to fill the voids between the broken stones. If this ideal could be realized, the total volume of mixed concrete would exactly equal the total volume of the stone involved.

No matter how thoroughly the ingredients are mixed—and no matter how careful the mixer—some of the particles of cement will get between the grains of sand and the volume of the mixture will be greater than the volume of the sand. The grains of sand will get between the smaller stones and the smaller stones will get between the larger stones and separate them.

Actual Conditions. Because perfect mixing is highly impractical, the amount of water used in mixing concrete is always a little excessive. The cement paste is also usually more than is necessary to fill the spaces between the sand particles. The amount of *mortar* (cement, sand, and water), therefore, is also considerably in excess of the amount required for filling the spaces between the stones. Even allowing some excess in all particulars, however, there is much variation in the percentage of voids in the sand as well as the stone. Because of this variation, an experimental determination of the voids must be made. Even in the best concrete work there is a periodic re-testing of the mixture. For less careful work, the proportions ordinarily adopted in practice are considered sufficiently accurate.

Concrete Problem. Compute the amount of materials needed to obtain 15 cubic yards of concrete. Use a 1:3:7:8 ratio and the *absolute volume* method of finding the amount of concrete.

Solution: The absolute volume method of finding the amount of concrete is very accurate. The 8 in the ratio means there is to be 8 gallons of water to every sack of cement.

Each of the materials used to make the concrete mix has an absolute volume constant. The product of the ratios and the constants are computed and then added together. The total gives the amount of concrete obtained from each bag.

Absolute Volume	Constants
cement	0.485
sand	0.567
aggregate	0.567
water	0.134

The sand and cement have the same constant because their absolute volumes are the same, even though they differ in size and shape.

$$\begin{aligned} 1 \times 0.485 &= 0.485 \\ 3 \times 0.567 &= 1.701 \\ 7 \times 0.567 &= 3.969 \\ 8 \times 0.134 &= \underline{1.072} \\ & 7.227 \text{ cu.ft. concrete per bag cement} \end{aligned}$$

Since the problem is given in terms of cubic yards, you must divide the number of cubic feet per bag into the number of cubic feet per yard in order to obtain the required number of bags per cubic yard.

$$7.227 \overline{)27.00} \quad = 3.7 \text{ bags per cu. yd.}$$

Multiplying each of the ratios by 3.7 gives the amount of each material needed for one cubic yard of concrete:

cement: $1 \times 3.7 = 3.7$
sand: $3 \times 3.7 = 11.1$ cu. ft.
aggr: $7 \times 3.7 = 25.9$ cu. ft.
water: $8 \times 3.7 = 29.6$ U.S. gallons

To find the amount of materials needed for 15 cu. yds. of concrete, multiply by 15:

cement: $15 \times 3.7 = 55.5$ bags
sand: $15 \times 11.1 = 166.5$ cu. ft.
aggr: $15 \times 25.9 = 388.5$ cu. ft.
water: $15 \times 29.6 = 444.0$ U.S. gals.

If each bag weighs 94 pounds and if water weighs 8.35 lb./gal., the weight of the materials is as follows: cement, 5,217 lbs.; sand, 15,651 lbs.; aggregates, 36,519 lbs.; and water, 3,707 lbs.

Mortar. Mortar is a mixture of sand, cement, and water. The ratio of cement to sand (by volume), doesn't exceed 3:1 for most applications. See Table I.

TABLE I. MATERIALS REQUIRED FOR 1 CUBIC

MIXTURE	3/4" GRAVEL		
	BARRELS CEMENT	CUBIC YARD SAND	CUBIC YARD STONE
MORTAR			
1:1 1/2	3.61	0.80	USING
1:2	3.02	0.90	"
1:2 1/2	2.60	0.96	"
1:3	2.28	1.01	"
CONCRETE			
1:1:2	2.30	0.35	0.74
1:1 1/2:3	1.71	0.39	0.78
1:1 3/4:2 3/4	1.75	0.43	0.75
1:2:3	1.54	0.47	0.73
1:2:3 1/2	1.44	0.44	0.77
1:2:4	1.34	0.41	0.81
1:2 1/2:4	1.24	0.47	0.75
1:2 1/2:4 1/2	1.16	0.44	0.80
1:2 1/2:5	1.10	0.42	0.83
1:3:4	1.15	0.52	0.72
1:3:5	1.03	0.47	0.78
1:3:6	0.92	0.42	0.84

TABLE II. VOLUME FACTORS OF

KIND OF CONCRETE WORK	VOLUME CEMENT BAGS	MIX BY JOB DAMP SAND CU. FT.	MATERIAL STONE GRAVEL CU. FT.
FOOTINGS, HEAVY FOUNDATIONS	1	3.75	5
WATERTIGHT CONCRETE FOR CELLAR WALLS AND WALLS ABOVE GROUND	1	2.5	3.5
DRIVEWAYS, FLOORS, WALKS — ONE COURSE	1	2.5	3
DRIVEWAYS, FLOORS, WALKS — TWO COURSE	1 1	TOP 2 BASE 2.5	0 4
PAVEMENTS	1	2.2	3.5
WATERTIGHT CONCRETE FOR TANKS, CISTERNS AND PRECAST UNITS (PILES, POSTS, THIN REINFORCED SLABS, ETC.)	1	2	3
HEAVY DUTY FLOORS	1	1.25	2

YARD OF MORTAR AND CONCRETE

1" STONE DUST OUT			2 1/2" STONE DUST OUT		
BARRELS CEMENT	CUBIC YARD SAND	CUBIC YARD STONE	BARRELS CEMENT	CUBIC YARD SAND	CUBIC YARD STONE
VERY	FINE	SAND	3.87	0.86	
"	"	"	3.21	0.95	
"	"	"	2.74	1.01	
"	"	"	2.39	1.06	
2.57	0.39	0.78	2.63	0.40	0.80
1.85	0.42	0.84	1.90	0.43	0.87
1.85	0.47	0.80	1.93	0.46	0.84
1.70	0.52	0.77	1.73	0.53	0.79
1.57	0.48	0.83	1.61	0.49	0.85
1.46	0.44	0.89	1.48	0.45	0.90
1.35	0.52	0.82	1.38	0.53	0.84
1.27	0.48	0.87	1.29	0.49	0.88
1.19	0.46	0.91	1.21	0.46	0.92
1.26	0.58	0.77	1.28	0.58	0.78
1.11	0.51	0.85	1.14	0.52	0.87
1.01	0.46	0.92	1.02	0.47	0.93

VARIOUS MIXES

WORK-ABILITY OR CON-SIST-ENCY	A ONE BAG BATCH MAKES THIS VOLUME OF CONCRETE CU. FT.	TOTAL WATER PER BAG GALLONS	MATERIALS FOR ONE CUBIC YARD OF CONCRETE		
			CEMENT BAGS	SAND CU. FT.	STONE GRAVEL CU. FT.
STIFF	6.2	8.00	4.3	16.3	21.7
MEDIUM	4.5	6.00	6.0	15.0	21.0
STIFF	4.1	5.50	6.5	16.3	19.5
STIFF	2.14	...	12.6	25.2	...
STIFF	2.8	6.00	5.7	14.2	22.8
STIFF	4.2	5.25	6.4	14.1	22.4
MEDIUM	3.8	5.00	7.1	14.2	21.3
WET	3.9	5.75	6.9	13.8	20.7
STIFF	2.8		9.8	12.3	19.6

TABLE III. BARRELS OF PORTLAND CEMENT PER CUBIC YARD OF MORTAR

(VOIDS IN SAND BEING 45 PERCENT, AND 1 BBL. CEMENT YIELDING 3.4 CUBIC FEET OF CEMENT PASTE.)

Proportions of Cement to Sand	1:1 Bbls.	1:1.5 Bbls.	1:2 Bbls.	1:2.5 Bbls.	1:3 Bbls.	1:4 Bbls.
Bbl. specified to be 3.5 cu. ft. ...	4.62	3.80	3.25	2.84	2.35	1.76
Bbl. specified to be 3.8 cu. ft. ...	4.32	3.61	3.10	2.72	2.16	1.62
Bbl. specified to be 4.0 cu. ft. ...	4.19	3.46	3.00	2.64	2.05	1.54
Bbl. specified to be 4.4 cu. ft. ...	3.94	3.34	2.90	2.57	1.86	1.40
Cu. yds. sand per cu.yd. mortar...	0.6	0.8	0.9	1.0	1.0	1.0

TABLE IV. BARRELS OF PORTLAND CEMENT PER CUBIC YARD OF MORTAR

(VOIDS IN SAND BEING 35 PERCENT AND 1 BBL. CEMENT YIELDING 3.65 CU. FT. OF CEMENT PASTE.)

Proportion of Cement to Sand	1:1 Bbls.	1:1.5 Bbls.	1:2 Bbls.	1:2.5 Bbls.	1:3 Bbls.	1:4 Bbls.
Bbl. specified to be 3.5 cu. ft. ...	4.22	3.49	2.97	2.57	2.28	1.76
Bbl. specified to be 3.8 cu. ft. ...	4.09	3.33	2.81	2.45	2.16	1.62
Bbl. specified to be 4.0 cu. ft. ...	4.00	3.24	2.73	2.36	2.08	1.54
Bbl. specified to be 4.4 cu. ft. ...	3.81	3.07	2.57	2.27	2.00	1.40
Cu. yds. sand per cu. yd. mortar..	0.6	0.7	0.8	0.9	1.0	1.0

Table III shows barrels of cement per cubic yard of mortar when the voids in the sand are 45% of its volume and 1 barrel of cement yields 3.4 cubic feet of cement paste. Table IV shows the same thing when the voids are 35% of its volume and the yield is 3.65 cubic feet of cement paste.

Such mortar is commonly used as topping for concrete floors.

Fireproofing. Experience and tests have shown that the fire-resisting qualities of concrete are greater than those of any other known type of building material. During a fire, the temperature may reach 1,900° F and injure the concrete to a depth of ¾-inch but the body of the concrete is not affected.

Two inches of concrete will safely protect an I beam. Reinforced concrete beams and girders should have a clear thickness of 1½ inches of concrete outside the steel on the sides and 2 inches on the bottom. Slabs should have at least 1 inch below the bars, and columns should have 2 inches.

Fig. 1 shows a typical steel beam cross section and the fireproofing concrete surrounding it. The concrete is sometimes reinforced by heavy wires wound around the column. This also helps to maintain the bond between the concrete and steel.

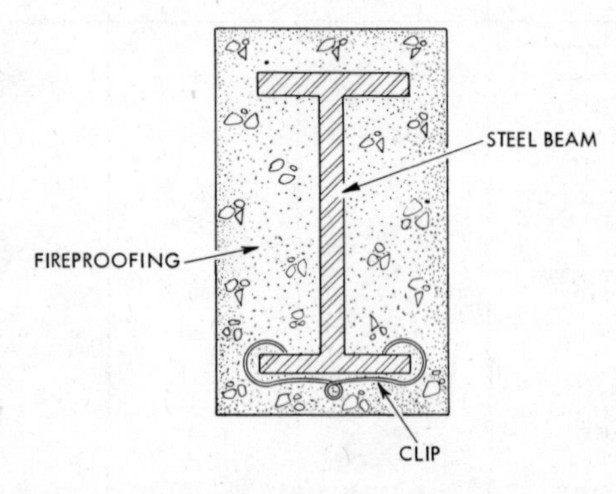

Fig. 1. Cross section of fireproof steel beam. The clip helps to keep the concrete bonded to the steel.

Theory. Portland cement concrete is made of sand, stone, cement paste, and water—none of these materials are combustible and therefore Portland cement concrete is used as fireproofing material. Furthermore, the finished concrete is porous and retains 12 to 18 per cent of its water, thus it will not readily transfer heat. This water is chemically combined and is not given off at the boiling point. On heating, a part of the water is given off at 500°F, but dehydration does not take place until 900°F is reached. The mass is kept for a long time at comparatively low temperature by the vaporization of water which absorbs heat. A steel beam imbedded in concrete is thus cooled by the vaporization of water in the surrounding concrete.

Resistance to the passage of heat is offered by the porosity of concrete. Air is a poor conductor, and an air space is an efficient protection against conduction. The outside of the concrete may reach a high temperature; but the heat only slowly and imperfectly penetrates the mass, and reaches the steel so gradually that it is carried off by the metal as fast as it is supplied.

Porosity of Cinder Concrete. Porous substances, such as asbestos, are always used as heat-insulating material. For this reason, cinder concrete, being highly porous, is a much better insulator than a dense concrete made of sand and gravel or stone.

Footings, foundations and floors

Footings. Footings distribute the load transmitted to them, by foundations, over a large ground area. The reason for this is that a narrow or pointed object will pierce the ground more readily than a blunt or broad object. A stick of wood, for example, which has a point one inch square will pierce the ground more readily than a stick of wood which has a two inch square point. This same principle applies to foundations.

Look at the foundation shown in Fig. 2. This foundation is 12 inches thick and 20 feet long. In other words,

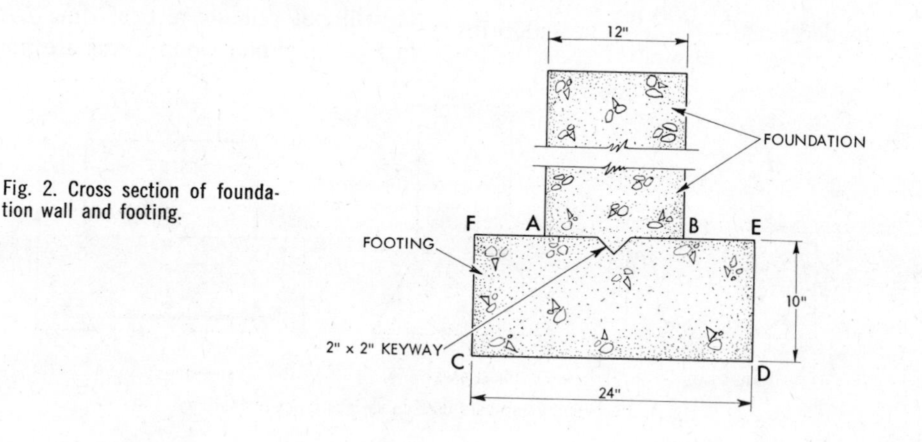

Fig. 2. Cross section of foundation wall and footing.

the foundation is 20 square feet (1 ft. × 20 ft. = 20 sq. ft.). Now suppose that the foundation will support 2,000 pounds per square foot. The total supportable load for this foundation then will be 2,000 × 20 = 40,000 lbs.

Remember that the various kinds of soil have varying load-bearing capacities. One kind might easily support 1,500 pounds per square foot while another might support 2,000 pounds per square foot. If the soil can support a maximum of 1,500 pounds per square foot, then that kind of soil has a capacity of 1,500 pounds per sq. ft.

If the foundation in Fig. 2 carries 2,000 pounds per square foot and it rests on soil which can only carry 1,500 pounds per square foot, the soil will be overloaded and the foundation will settle. Thus, a footing is needed to distribute the load over a wider area. The footing shown in Fig. 2 is 24 inches wide and 20 feet long, or 40 square feet (2 ft. × 20 ft. = 40 sq. ft.). Thus, the 40,000 pound foundation is now distributed over 40 square feet. Now the soil is not overloaded since $\frac{40{,}000 \text{ lbs.}}{40 \text{ sq. ft.}} = 1{,}000$ lbs. per sq. ft. This is below the limit of the soil's load-bearing capacity.

Shape of the Footing. A footing should have a perfectly flat edge in contact with the soil (*CD* in Fig. 2). This edge should also be horizontal to prevent sliding.

Composition of Footings. Although in rare instances, stone, blocks, and other materials are used, footings are usually made of concrete. Concrete is the most desirable material for two reasons: (1) when dry, the concrete becomes one solid piece— this helps to avoid the possibility of settling; (2) concrete requires less labor than other kinds of footings—thus lowering the labor cost.

Steel reinforcing bars or rods, Fig. 3, add to the strength of the footings and are often used. The written specifications and the working drawings give the size, spacing, and other necessary data about the reinforcing steel.

Footing Forms. A footing should have a constant area in its rectangular cross section. To assure this, wood forms are sometimes necessary — especially where the soil is loose or sandy or will not remain vertical (line *DE* in Fig. 2). When wood forms are not

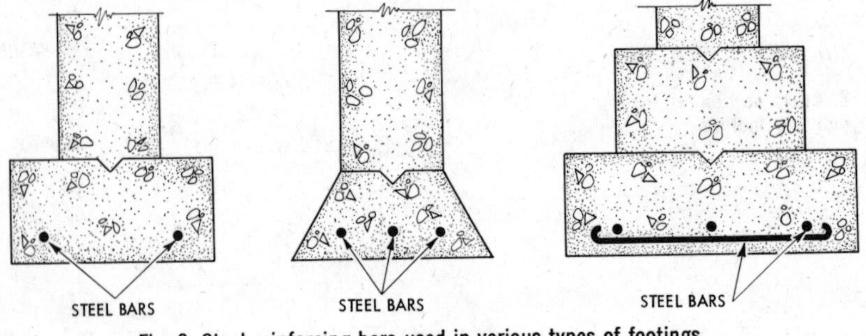

Fig. 3. Steel reinforcing bars used in various types of footings.

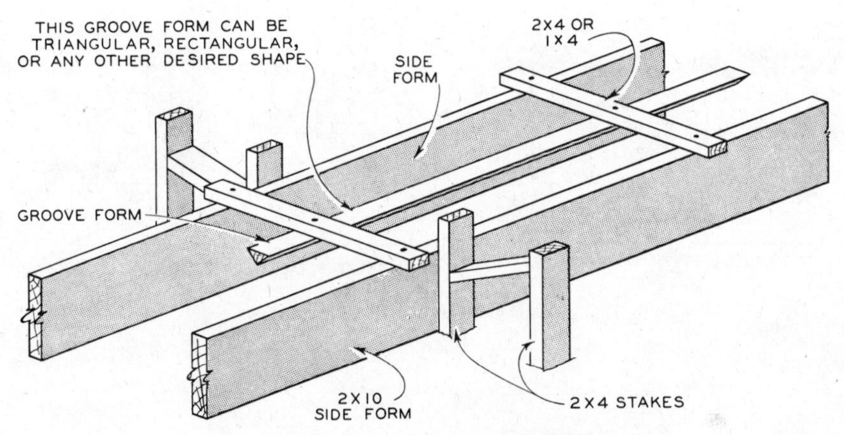

Fig. 4. Typical formwork for a concrete footing.

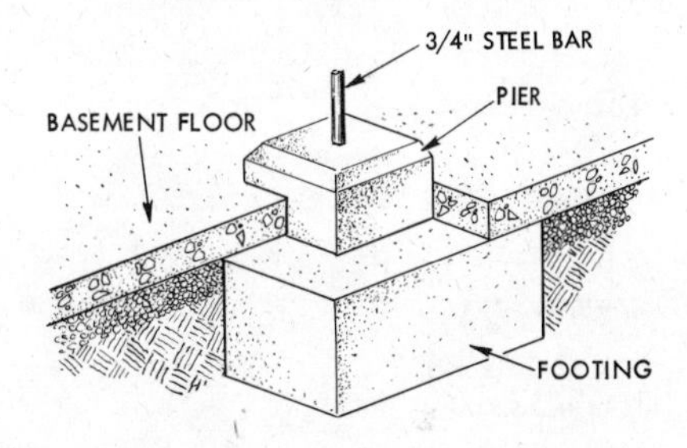

Fig. 5. Column footing with concrete pier. The pier keeps a wood post above the floor level to protect it from moisture. The steel bar holds the post in place.

used, the soil must be wetted so that it doesn't absorb water from the wet concrete. Fig. 4 shows typical form work for concrete footings.

Column Footings. Inside areas of a building are usually supported by interior partitions. The partitions are in turn supported by I beams which run along the basement ceilings. At one or more places these I beams are supported by columns. These columns will each require a footing, Fig. 5. The size of the footing depends upon the total load transmitted by the column and the load capacity of the soil.

Foundations. There are a great many types of foundations. The most common types are shown in Fig. 6. Concrete is by far the most popular of these. Concrete blocks are next in popularity and rubble is third. Rubble is not often used except where stone is abundant; it is then economical despite the additional labor involved. Brick and concrete is sometimes used where a basement is to be used as a recreation room. In such a case, the bricks are glazed on the edges facing the room and may be colored.

Of course the common types of foundations shown in Fig. 6 can be varied. Variations include foundations built of concrete part way up and block, brick, or some other material for the remainder. An example of this

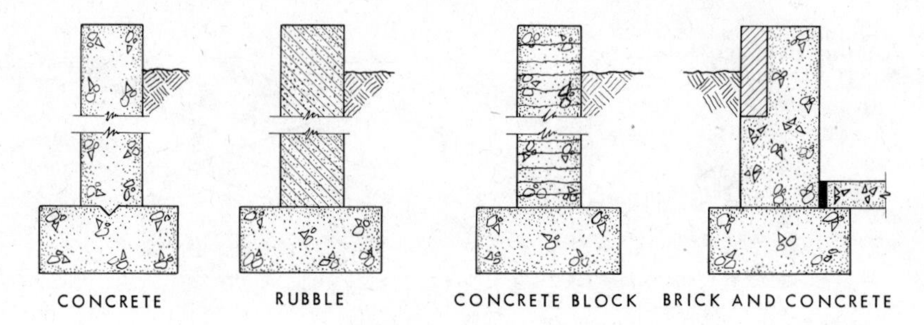

Fig. 6. Common types of concrete foundations.

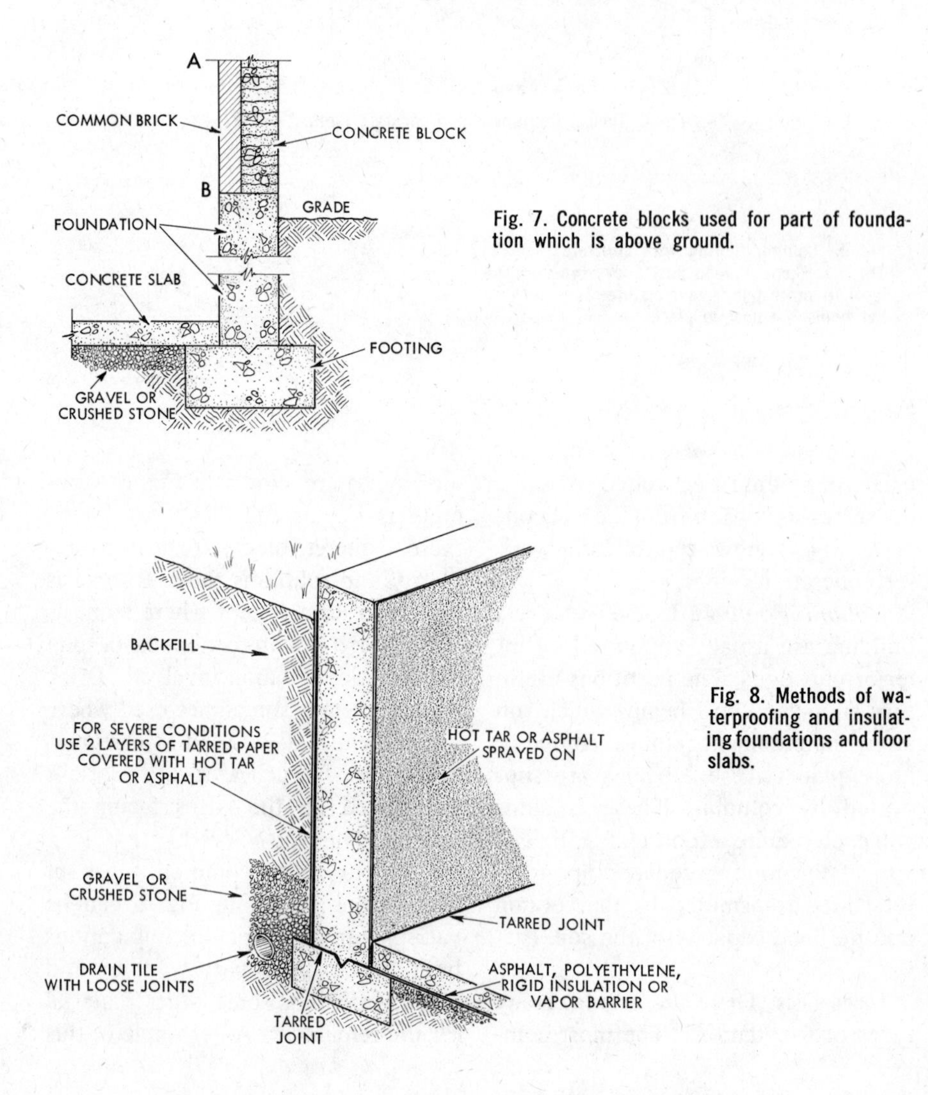

Fig. 7. Concrete blocks used for part of foundation which is above ground.

Fig. 8. Methods of waterproofing and insulating foundations and floor slabs.

method is shown in Fig. 7. The basement shown has a shallow excavation and the portion *AB* is above grade. Concrete blocks are used above grade to give the outside a better appearance; the interior is face brick. Many variations of this are possible. For example, the wall above grade *(AB)*, could be entirely brick with face brick on the outside and common brick on the inside. For a stucco house, as another example, wall *AB* could be of hollow tile.

Anchor bolts are embedded in the top of the foundation walls before the concrete has set. These bolts hold wood *sills* securely in place on top of the wall. (Wood sills are discussed in the next chapter.)

Waterproofing. In areas where the soil does not provide adequate natural drainage, it is advisable to waterproof foundations. As shown in Fig. 8, asphalt or some other waterproofing agent, such as polyethylene film, can be used. The agent to be used will be on the drawing or in the specifications. In addition to the waterproofing agent, it is helpful to place loosely connected drainage tiles as shown in Fig. 8. This is an especially good practice here, since much of the backfill is gravel and small rocks; seepage water can find its way through this type of material to the drain tile. The tile is connected to a sewer or to some other natural means of disposal. Exceptionally wet conditions may require a sump pump.

There are a number of ways to waterproof a foundation. The estimator should therefore check carefully the drawings and specifications for each building.

Cheek Walls. These are foundation walls which support light structures and structures which impose relatively little load, such as porches, areaways, and entrance steps. Such structures rarely need footings. The term derives from the smoothing of the face of the wall after the concrete has set.

Floors. Concrete floors generally vary from 3 to 8 inches in thickness. The specifications and drawings will indicate how thick the floor is to be. As

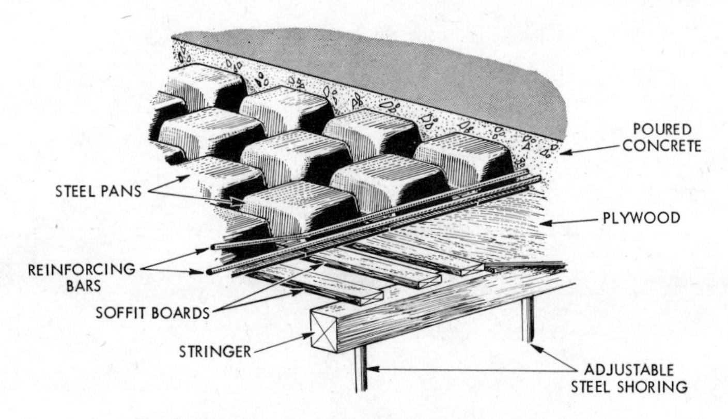

Fig. 9. Metal pans used for concrete floor construction.

indicated in Fig. 8, the concrete foundation floor is generally laid over a well-tamped layer of gravel, crushed stone, or cinders. Sometimes a cement topping (¼ to 1-inch thick) is specified for the concrete base.

Floors are sometimes constructed with metal or plastic pans, Fig. 9. These create a flat slab on the top surface and a ceiling below which looks like a waffle — hence they are often called "waffle-slab" systems.

Another type of floor system uses a layer of wire mesh over a metal decking which rests upon the structural steel. See Fig. 10. A thin concrete slab (2 to 4 inches thick) is formed by pouring concrete over the mesh and metal decking.

Still another system uses the *reinforced* and *prestressed concrete plank method* to construct the floors.

Beams, columns and stairs

Beams. Beams are used for horizontal support of floors and roofs. They are usually rectangular or T-shaped in cross section. Concrete beams are usually an integral part of the floor which

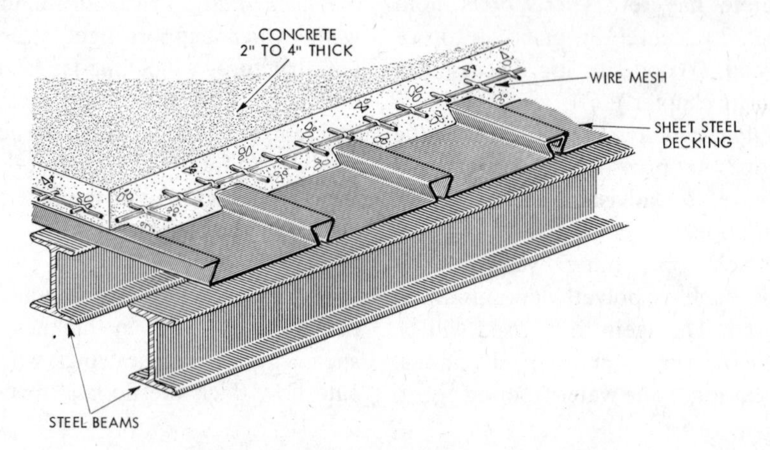

Fig. 10. Reinforced concrete floor slab with metal decking.

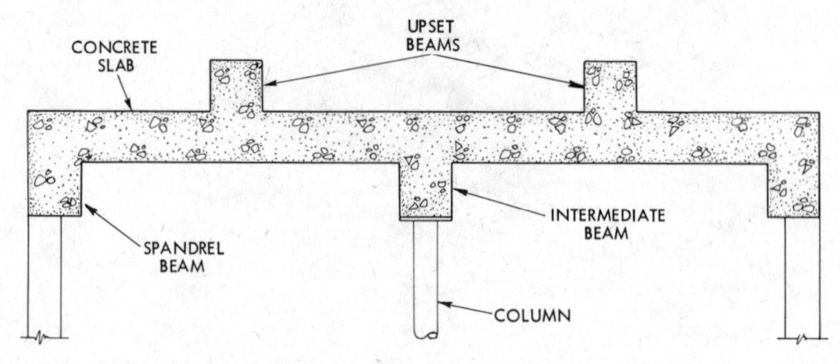

Fig. 11. Concrete beams as integral parts of a concrete slab.

they support. Fig. 11 shows cross sections of some concrete beams.

Spandrel beams run around the inside edge of the outer walls of a building; they are integral parts of the floor slab and must be able to support masonry. The depth of the outside face of the beam is noted on the drawings or in the specifications.

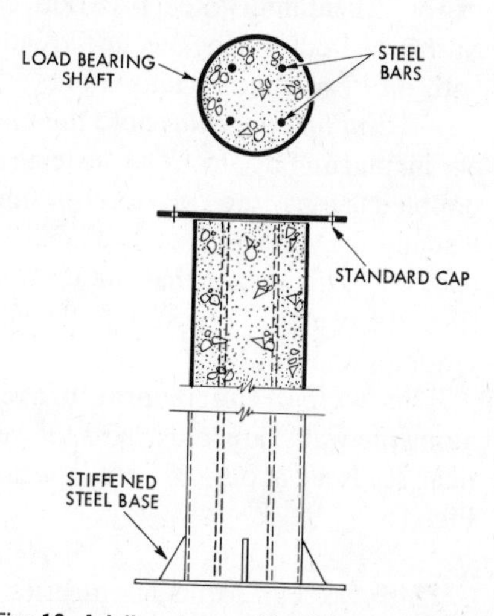

Fig. 12. A lally column with reinforcing steel bars.

Spandrel inserts are placed within the forms for the spandrel beams and project from the spandrel beam face. These inserts take the bolts which hold up the lintels which support the brickwork on the building face.

Intermediate beams also form and support a part of the floor slab but are not on the outside edge of the building.

Free-standing beams span an open space and frame into slabs only at the beam ends. An example of this is a beam across an elevator shaft.

Upset beams are found on the roofs of high-rise buildings. They form the underside supports for water towers and air conditioning units.

Columns. Columns are vertical supports which support beams which in turn support floor slabs. Sometimes a column directly supports a floor slab.

Lally columns are iron or steel pipes filled with concrete and set on plates on top of footings, Fig. 12. They are used to support steel beams (I or wide flange) and are bolted to the beams.

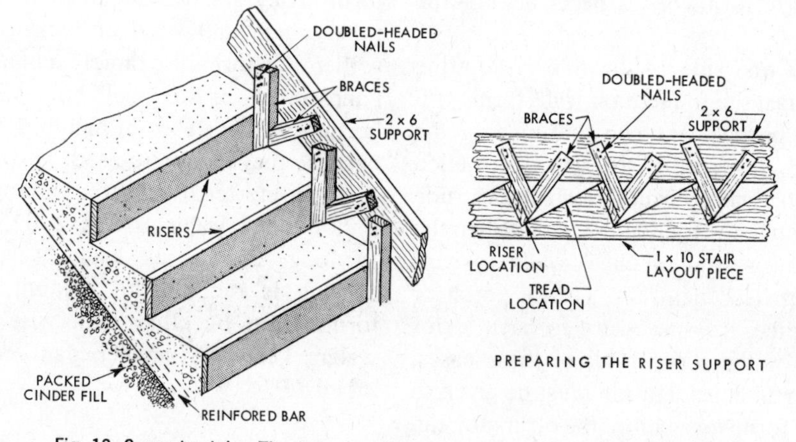

Fig. 13. Concrete stairs. The stairs are formed by risers held in place with braces.

Dovetail slots are metal strips which are placed inside the columns to form slots on which hang the brick tie inserts.

Stairs. Fig. 13 shows a typical concrete stairway. The concrete is poured into forms which are removed after the concrete has hardened. Steel reinforcing bars are embedded in the concrete to increase its load-bearing capacity.

Formwork

Concrete formwork is usually figured separately for each section of the concrete work; "foundation walls," "interior column footings," "cheek walls," etc., are some of the separate listings for formwork.

When a cellar slab is poured, the wall acts as the form, thereby eliminating any need for framing members around the edges of the slab. However, when an *exterior slab* is poured, (such as a stair platform, a terrace, or a patio), formwork is necessary for the sides.

If one side of the slab is abutting against the foundation wall, forms will not be necessary on that side of the slab. The face of the slab is dowelled to the wall by iron ties which protrude through the foundation wall. When the slab is resting on cheek walls or spanning steel members, as in a porch or terrace, it is necessary to have forms under the slab. In all of these cases, careful consideration must be given to the formwork before the estimator can begin his calculations.

Panel forms. Panel forms are the most common forms in use today and consist of 4′ × 8′ plywood panels which are cut to form 2′ × 8′ panels or 4′ square panels, Fig. 14. These smaller sizes allow the panels to be quickly moved about the site by the carpenters.

The panels are made of either ⅝″ or ¾″ sheathing grade plywood cut to the desired dimensions and treated with oil or a commercial sealant.

Vertical bracing studs hold the panels in place, Fig. 15. The bracing is doubled along the plywood seams. Usually the studs are set 2′-0″ *on centers,* o.c. This means that the distance from the center of one stud to the center of another is 2′-0″.

The *walers* (horizontal bracers along the walls) are laid across the vertical studs at about 16″ o.c. spacing, Fig. 15.

Pan Forms. Pan forms are a series of interlocking metal panels which are set up to form the desired wall for the form. They are usually 2′ × 2′ sections which are wired or locked together. Pan forms are rapidly replacing panel forms in the small house field because of their ease in handling and because they can be re-used.

The preferred bracing for pan forms is horizontal walers set at 2′-0″ o.c. In long, continuous walls, vertical bracing is required to prevent the forms from buckling; otherwise the walers keep the smaller pan forms intact.

Angular Braces. Angular braces help

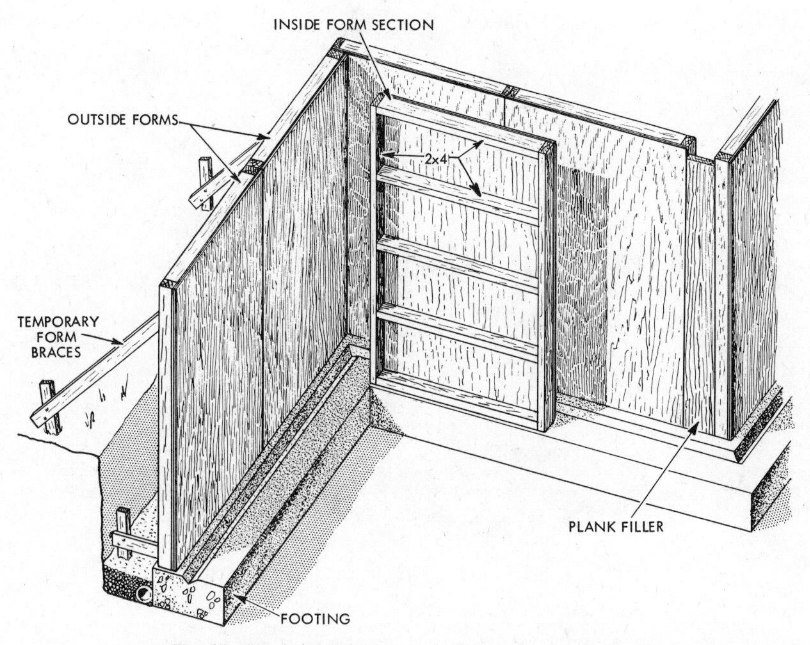

Fig. 14. Wood panel forms can be re-used many times.

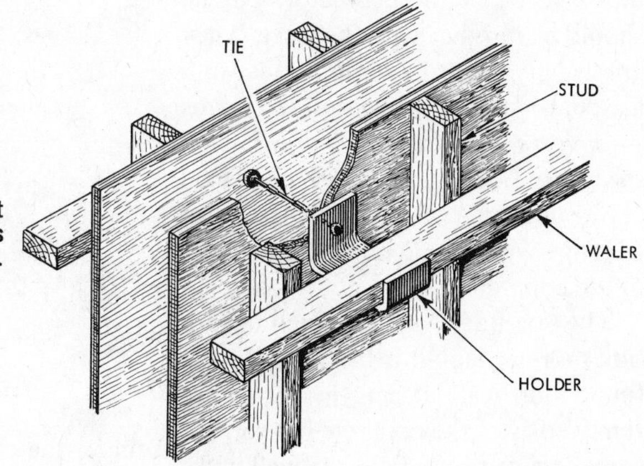

Fig. 15. Walers are held against the studs by patented form ties and holders. Allenform Corporation.

to keep the foundation wall form vertical. They run at a 45° angle to the walls from the earth in front of the forms, Fig. 16. *Stakes* are driven into the ground to secure them.

Kickers. Kickers are placed along the ground perpendicular to the wall and secured to the brace at the stake, Fig. 16. They aid in bracing the bottom of the forms and prevent buckling or kicking out.

Spacing Bars and Spreaders. Spacing bars

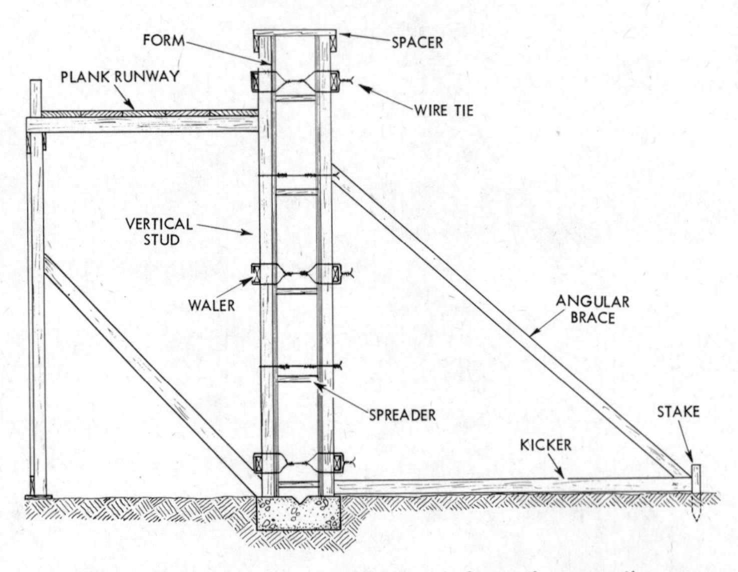

Fig. 16. Angular braces and kickers shown in formwork cross section.

or rods and spreaders are not usually the concern of the estimator, but he should be familiar with their use. They are usually made of scrap lumber and are cut to fit over the tops of the forms — *spacers,* or inside the forms — *spreaders*. See Fig. 16. They are placed at varying intervals so as to be most effective without affecting the strength of the concrete.

The spreaders are one-inch square blocks; their length is the width of the foundation wall. When the forms are removed after the concrete has set, the spreaders can remain in the wall without damaging the wall.

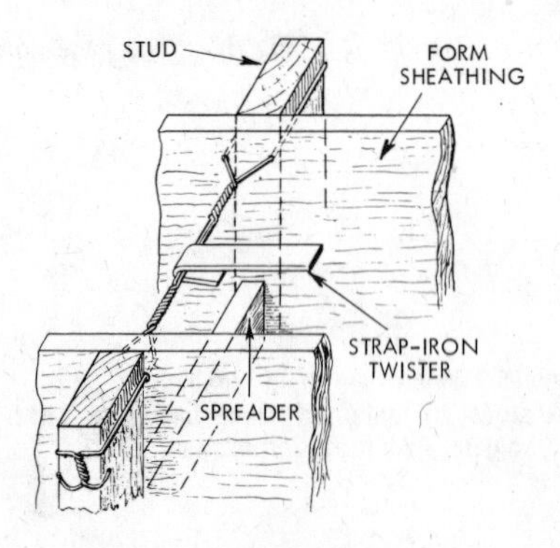

Fig. 17. Wire ties.

Ties. Ties are metal or plastic wires which are placed through the width of the foundation walls and secured to the outside ends of the forms before the concrete is poured. They prevent the buckling out of the forms under the pressure of the poured concrete. These ties are placed throughout the forms at specified intervals and clamped to the outside braces, Fig. 17.

Snap ties are special ties which are twisted and snapped off when the forms are removed, leaving the ties embedded about ½-inch inside the foundation wall, Fig. 18.

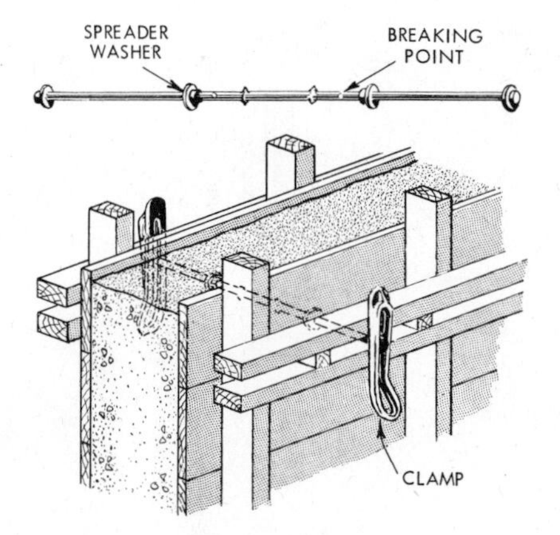

Fig. 18. Snap ties.

Finishing concrete

After the concrete has set and the forms are removed, it may be necessary to do several things to the concrete.

Troweling. This is necessary to cover over any rough spots or holes in the face of the concrete. Such items as noticeable honeycombs or the impressions of the snap ties must be patched.

Rubbing. Usually, any surface impressions of the formwork that appear on the concrete after the forms are stripped must be removed. Rubbing is the method used to smooth these rough surfaces.

Quantity take-offs

Steel. Steel beams, columns, and *lintels* (framing over windows) are being used in many small homes. Sometimes they are used alone as in HOUSE PLAN A, but often they are used with reinforced concrete.

Reinforcing steel is estimated by the pound or ton (2,000 lbs.). To find the amount of reinforcing steel bars needed for the structure, first find the number of pieces and lengths of each size bar. Then multiply by the weight per linear foot. A 5% waste factor is usually added; this is sufficient to take care of waste and overlap.

Steel bars must be wired in place (horizontally and vertically) in retaining walls, cantilever girder frames, foundation walls, etc. *Tie wire* must therefore be taken into consideration in the quantity take-off.

Footings. The amount of concrete needed for footings is estimated by adding the volumes of each section of footing, each section being one of the foundation walls. Where corners exist, you must be careful not to overlap the lengths and thereby increase the quantity of concrete.

Example 1. Find the amount of concrete needed for the wall footings of the house foundation in Fig. 19. (Note that the cheek walls have no footings.)

Solution: Set up a table as follows:

Width = 2.0 Height = 1.0'

Footing	Length	Volume
A	41.0'	82.0 cu. ft.
B	19.0'	38.0
C	39.0'	78.0
D	17.0'	34.0
	116.0'	232.0 cu. ft. or 8.6 cu. yds.

Example 2. Find the amount of concrete needed for the chimney footings in Fig. 19.

Solution: The chimney footing is

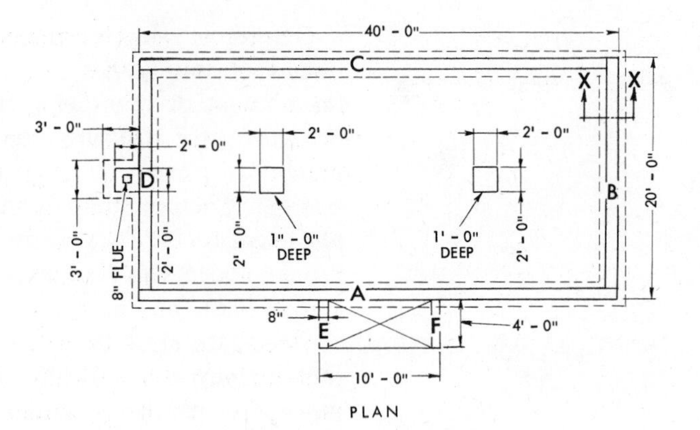

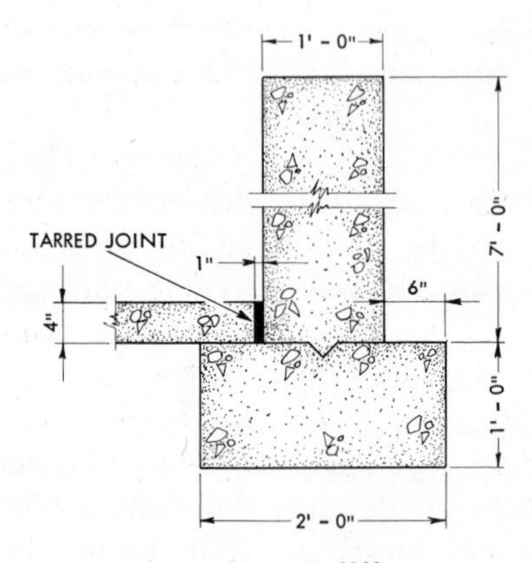

Fig. 19. Plan and section view of a house foundation.

3.0′ by 2.5′; the depth is 1.0′. This is a simple volume problem.

$V = 3.0' \times 2.5' \times 1.0' = 7.5$ cu. ft.

Example 3. Find the volume of concrete needed for the column footings in Fig. 19.

Solution: There are two column footings and they are equal in size.

$V = 2' \times 2' \times 1' = 4$ cu. ft. for each footing

$2 \times 4 = 8$ (cu. ft.)

Foundation Walls. The amount of concrete needed for foundation walls is estimated in the same manner as the footings.

Example 4. Find the volume of concrete needed for the foundation walls of the house in Fig. 19.

Solution: Use a table as in Example 1.

Width = 1.0' Height = 7.0'

Wall	Length	Volume
A	40.0'	280.0 cu. ft.
B	19.0'	133.0
C	39.0'	273.0
D	18.0'	126.0
	116.0'	812.0 cu. ft. or 30.1 cu. yds.

Note that the total length of the foundation wall is equal to the total length of the wall footings (116.0′). Since the center lines of the footings and walls are the same, their lengths should be the same. This serves as a useful check on your calculations.

Example 5. Find the volume of concrete needed for the chimney foundation in Fig. 19. The chimney foundation is 7.0′ deep.

Solution: Find the total volume of the chimney foundation and subtract the volume of the flue.

Chimney foundation volume =
$2.0' \times 2.0' \times 7.0' = 28.0$ cu. ft.

Flue volume =
$0.67' \times 0.67' \times 7.0' = 3.14$ cu. ft.

$V = 28.0 - 3.14 = 24.86$ cu. ft.

Example 6. Find the volume of concrete needed for the porch cheek walls. The cheek walls are 3.0′ deep.

Solution: The cheek walls are figured in the same manner as the foundation walls.

Length = 4.0′ Width = 0.67′ Height = 3.0′

Wall	Volume
E	8.0 cu. ft.
F	8.0
	16.0 cu. ft.

Floor Slabs. The concrete for all floor slabs is figured as the volume of the particular slab; the volumes of large openings such as stairwells are deducted from the floor volume.

Example 7. Find the volume of concrete needed for the basement floor slab in Fig. 19.

Solution: The volume is the area of the basement (minus the area of the column footings) multiplied by the slab thickness.

Area of basement =
$38.0' \times 18.0' = 684.0$ sq. ft.

Area of columns =
$2 \times 2.0' \times 2.0' = 8.0$ sq. ft.

Slab area =
$684.0 - 8.0 = 676.0$ sq. ft.

Slab volume =
$676.0 \times 0.33 = 225$ cu. ft.,
or 8.33 cu. yds.

Example 8. Find the volume of concrete needed for the porch slab in Fig. 19.

Solution: This problem is figured as a simple volume problem.

$10.0' \times 4.0' \times 0.33' = 13.2$ cu. ft.

Note that the floor slab in Fig. 19 has a 1″ space around it to allow for expansion. This *expansion joint* is usually filled with tar or other expansion material. The amount of material needed is found by finding the area which comes in contact with the slab.

Example 9. Find the amount of expansion material needed for the foundation in Fig. 19.

Solution: Find the inside perimeter of the foundation wall.

$38.0' + 38.0' + 18.0' + 18.0' =$
$112.0'$

The joint is to be only 1″ wide and 4″ deep, so the volume of needed material is found by converting these dimensions to feet:

$0.08' \times 0.33' \times 112.0' = 2.96$ cu. ft.

Beams. The concrete for a beam is computed as the volume of the beam, or that portion apart from the floor slab. The lengths of beams with identical cross-sectional area are added

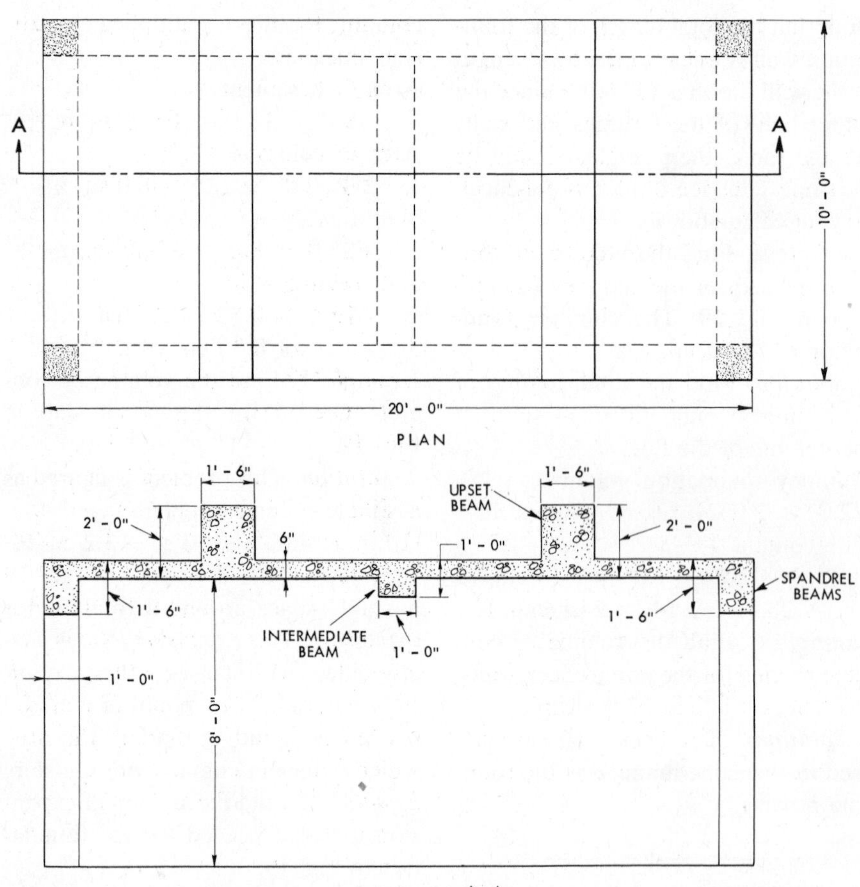

Fig. 20. Concrete beams, slabs, and columns.

together before finding the total beam volume.

Example 10. Find the amount of concrete needed for the spandrel beams in Fig. 20.

Solution: There are 2 beams which are 8.0′ long and 2 beams which are 18.0′ long, each framing into columns at the corners of the structure. The beams have identical cross-sectional areas.

$2 \times 8.0' = 16.0' \quad 2 \times 18.0' = 36.0'$

$$16.0' + 36.0' = 52.0'$$

The beams are 1.0′ wide and 1.0′ deep (after 6″ for the floor slab), or 1 sq. ft. in cross-sectional area.

$$1.0 \times 52.0 = 52.0 \text{ (cu. ft.)},$$

or 1.93 cu. yds.

Example 11. Find the amount of concrete needed for the intermediate beam in Fig. 20.

Solution: The beam is 8.0′ long and 1.0′ wide. The depth is 1.0′ (1′-6″ —0′-6″).

8.0′ × 1.0′ × 1.0′ = 8.0 cu. ft., or 0.3 cu. yds.

Example 12. Find the amount of concrete needed for the upset beams in Fig. 20.

Solution: There are 2 upset beams of equal size.

2 × 10.0′ × 1.5′ × 1.5′ = 45.0 cu. ft. or 1.67 cu. yds.

The volume of the slab in Fig. 20 is found by multiplying its surface area by its thickness:

10.0′ × 20.0′ × 0.5′ = 100.0 cu. ft., or 3.7 cu. yds.

Columns. The volume of concrete needed for columns is usually computed by multiplying the cross-sectional area by the height which is taken as the total distance from floor to floor. Thus in Fig. 20 the heights of the columns are each 8′-6″. The 6″ thickness of the floor slab is neglected since the concrete will probably *shrink* as it hardens.

Example 13. Find the amount of concrete needed for the columns in Fig. 20.

Solution: There are 4 columns of equal height and cross-sectional area.

4 × 8.5′ × 1.0′ × 1.0′ = 34.0 cu. ft. or 1.2 cu. yds.

Concrete Constant. When finding the amount of concrete needed for a column, it is convenient to find first the average cross-sectional area. The number thus obtained is called the *concrete constant*. When multiplied by the height of the column, the concrete constant will give the volume of the column.

Example 14. Find the concrete constant and volume of a column which has a square base 2′-0″ on each side, and whose height is 9′-0″.

Solution: First find the area of the base.

Area = 2.0′ × 2.0′ = 4.0 sq. ft.

The concrete constant is 4.

Volume = 4 × 9.0 = 36.0 (cu. ft.), or 1⅓ cu. yds.

Stairs. The volume of stairs is figured by multiplying the number of treads by the width of the stairway. You are generally safe to assume 1 cu. ft. of concrete per linear foot of tread.

Precast treads are figured by the piece or by the linear foot.

Example 15. Find the amount of concrete needed for a concrete stairway with 10 treads each 3′-6″ wide.

Solution: Multiply the number of treads by the width of the stairway.

10 × 3.5′ = 35.0′

Assuming 1 cu. ft. per linear foot of tread, 35.0 cu. ft. of concrete is needed.

Concrete Fill. The estimator must consider the concrete fill necessary for concrete floors, stairs, and roofs.

Floor and roof fills are best estimated using cubic yards (instead of square feet). Since roofs usually slope to a drain, their thicknesses will vary. When using the square foot (of surface area) method to find the concrete, the *average* thickness of the roof is used in the calculations.

Fills for stairs and platforms are figured by the square foot of tread area in relation to the thickness, or by the linear foot (of treads).

Estimating formwork

Each class of formwork is figured separately. In general, formwork is figured in *square feet of contact area, (SFCA).*

The following sections first discuss the amount of formwork needed for each class of construction—footings, foundations, columns, etc. Then the bracing and ties are discussed separately.

Footings. The two main types of footings which the estimator figures are for foundations and for columns.

Foundation footings. These footings must have formwork on both sides. The lengths of the inside and outside forms are multiplied by the depth of the footing to obtain the *SFCA*.

Example 15. Find the *SFCA* of the foundation footings in Fig. 21. The footings are 1.0′ deep.

Solution: Multiply the length of each footing by its depth and add together the areas.

Depth = 1.0'

Outside Forms

Wall	Length	SFCA
A	41.0'	41.0
B	21.0'	21.0
C	41.0'	41.0
D	21.0'	21.0
		124.0

Inside Forms

Wall	Length	SFCA
A	37.0'	37.0
B	17.0'	17.0
C	37.0'	37.0
D	17.0'	17.0
		108.0

Footing Forms:

124.0 + 108.0 = 232.0 (sq. ft.)

Column footings. Formwork for the interior column footings is figured in the same manner as the foundation footings. Since column footings are usually solid, only an outside form will be required.

The notation *F4S* means that forms are necessary on 4 sides.

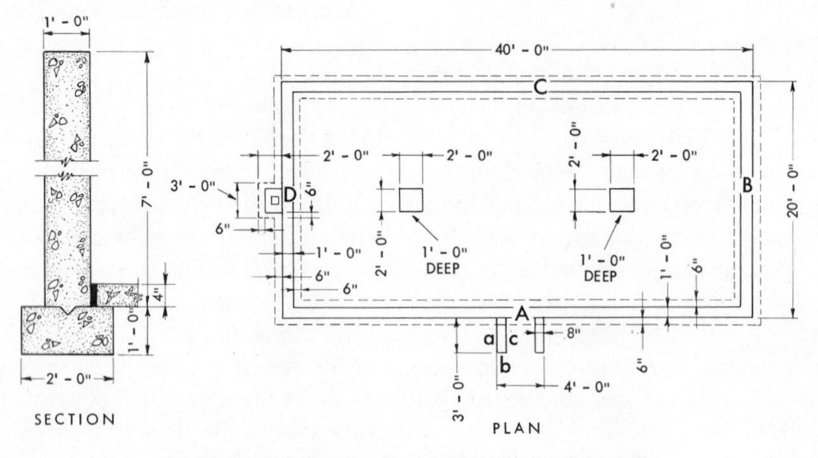

Fig. 21. House foundation plan used for formwork take-off.

Example 16. Find the amount of formwork required for the two column footings in Fig. 21.

Solution: The two forms are identical so the area of one is found and doubled. The area is the perimeter of the footing times the depth of the footing.

$$\text{Perimeter} = 2' + 2' + 2' + 2' = 8'$$
$$\text{Area} = 8' \times 1' = 8 \text{ sq. ft.}$$
$$\text{Total area} = 2 \times 8 = 16 \text{ (sq. ft.)}$$

Walls. Concrete foundation walls are sometimes grouped by the estimator, for convenience, into foundation walls with footings and those without footings (cheek walls).

Foundation walls are figured in the same manner as the foundation footings.

Example 17. Find the *SFCA* of the foundation wall in Fig. 21.

Solution: Set up tables for the outside and inside faces.

Depth = 7.0'

Outside Forms

Wall	Length	SFCA
A	40.0'	280.0
B	20.0'	140.0
C	40.0'	280.0
D	20.0'	140.0
		840.0

Inside Forms

Wall	Length	SFCA
A	38.0'	266.0
B	18.0'	126.0
C	38.0	266.0
D	18.0	126.0
		784.0

Total area =
$$840.0 + 784.0 = 1624.0 \text{ (sq. ft.)}$$

Note that there is no deduction for window openings in the amount of wall formwork. This allowance is made in the framing of the forms as an interior form fitting between the walls, Fig. 14.

The estimator doesn't allow for the *lap joints* at each corner of the walls. This extra inch at the butt-ends of the formwork is negligible and is only figured in the carpenter's layout.

Cheek walls. The forms for cheek walls are figured in the same manner as the foundation walls.

Example 18. Find the *SFCA* of the cheek walls in Fig. 21. The walls are 3.0' deep.

Solution: The two walls are identical so you need only find the area of one wall and then double the area.

Side	Length	Depth	SFCA
a	3.0'	3.0'	9.0
b	0.67'	3.0'	2.0
c	3.0'	3.0'	9.0
			20.0

Total area =
$$2 \times 20.0 = 40.0 \text{ (sq. ft.)}$$

Floor Slabs. Formwork for slabs depends on the type of slab to be poured and whether or not the sides of the slabs are to be open or framed.

If the slab is to be a *monolithic* pour, i.e., poured as part of the walls, then part of the slab formwork will be an extension of the exterior wall formwork and figured as such.

The formwork for slabs is usually figured as the *SFCA* of the underside of the slab.

Example 19. Find the amount of formwork required for the slab which goes over the cheek walls in Fig. 21. (The slab is not monolithic, i.e., it is poured separately from the cheek walls.) The depth of the slab is 3″. (Remember that 3″ = 0.25′.)

Solution: The formwork is figured in a manner similar to that of the column footings. The perimeter of the slab times its depth will give the *SFCA* of the sides of the slab.

Perimeter =
3.0′ + 4.0′ + 3.0′ = 10.0 ft.

Area of sides =
10.0′ × 0.25′ = 2.5 sq. ft.

Underside area =
3.0′ × 4.0′ = 12.0 sq. ft.

Total area =
2.5 + 12.0 = 14.5 (sq. ft.)

Beams. Formwork for beams is figured as the *SFCA* of open surface, beam sides, and beam bottom. Thus a spandrel beam has formwork on three sides—the outside beam side includes slab edge form. An upset beam has formwork only on its two sides, and an intermediate beam has formwork on three sides (beam sides and beam bottom).

Example 20. Find the amount of formwork needed for the spandrel beams in Fig. 20.

Solution: Each beam has three open sides. For each beam, add together the widths of these three open sides; multiply by the total length of the beams of that size. (Since there are two sets of two beams each—all with identical widths—it is only necessary to multiply the lengths of each size by two.) Add together the lengths.

Width of beams =
1.5′ + 1.0′ + 1.0′ = 3.5′
Lengths: 2 × 8.0′ = 16.0′
2 × 18.0′ = 36.0′
52.0′

Total area =
3.5′ × 52.0′ = 182.0 sq. ft.

Example 21. Find the amount of formwork needed for the upset beams in Fig. 20.

Solution: The beams are identical so the area of one need only be multiplied by 2. Note that these beams do not need formwork on the top or bottom.

Area = 2 × 10.0′ × (1.5′ + 1.5′)
= 60.0 sq. ft.

Columns. Each side of a column has formwork. Thus the *lateral* or *face area* must be calculated. If the average perimeter (or diameter) of the column is known, then it is easy to find the lateral area; it is the perimeter (or circumference) times the height of the column.

The perimeter is also called the *form constant.* Thus the form constant for columns with a perimeter of 2.0′ is 2. The form constant times the column width will give the lateral area of a column.

Example 22. Find the amount of formwork needed for the columns in Fig. 20.

Solution: The four columns are of equal size. Find the form constant and multiply it by four times the length of one column.

Perimeter = 1′ + 1′ + 1′ + 1′ = 4′
Form constant is 4
Area = 4 × 4 × 8 = 128 (sq. ft.)

Stairs. The *SFCA* of stairs is figured as the areas (in square feet) of the underside, ends, and risers of the stairs. (See Fig. 13.)

Angular Braces and Kickers. An angular brace forms the hypotenuse of a right triangle with the form as one leg and a kicker as the other leg. Since the angular brace is usually at a 45° angle, the legs of the triangle are equal. If the length of one of the three elements of a 45° right triangle is known, then the other elements can be found.

Example 23. Find the length of an angular brace which is at a 45° angle to the top of a 7.0′ form.

Solution: Since the form is one leg of a 45° right triangle, the kicker is also 7.0′ and the Right Triangle Law can be used to find the length of the angular brace (the hypotenuse).

$$c^2 = a^2 + b^2$$
$$= 7^2 + 7^2$$
$$= 98 \text{ sq. ft.}$$
$$c = 9.9', \text{ length of the angular brace.}$$

Snap Ties. Snap ties are figured as a specified number per square foot of wall area.

Example 24. Find the number of snap ties needed for the foundation walls in Fig. 21. One tie is to be used for each 9 square feet of wall area.

Solution: Find the outside area of each wall and divide by 9.

Wall	Outside Area	No. of Ties
A	40.0' x 7.0'	31
B	20.0' x 7.0'	16
C	40.0' x 7.0'	31
D	20.0' x 7.0'	16
		94

Vertical Bracing. Vertical bracing studs for the walls and column footings are figured as the length of the wall divided by the on-center distances of the studs. If there are two studs at each position, then the amount of studs is doubled.

Example 25. Find the number of pieces needed for the vertical braces on the outside walls of Fig. 21. There are two braces at each on-center distance. Each of the studs is 8.0′ long.

Solution: If the centers are two feet apart, then divide each wall length by two feet and double your answer to get the required number of braces.

Wall	Length ÷ OC Distance	No. of Pieces
A	40 ÷ 2	40
B	20 ÷ 2	20
C	40 ÷ 2	40
D	20 ÷ 2	20
		120

Total brace lengths =
$$120 \times 8.0' = 960.0'$$

Walers. The walers or horizontal braces which run the length of the wall are usually spaced 16″ apart, o.c. distance. The first few rows of walers may be doubled because of the greater pressure at the bottom of the forms. The greater the height of the form, the greater will be the number of double rows of walers.

In Fig. 21 the number of walers needed is computed as follows:

No. of walers

$$= \frac{\text{Height of wall}}{\text{On-center distance}} + 1$$

$$= \frac{7.0'}{1.33'} + 1$$

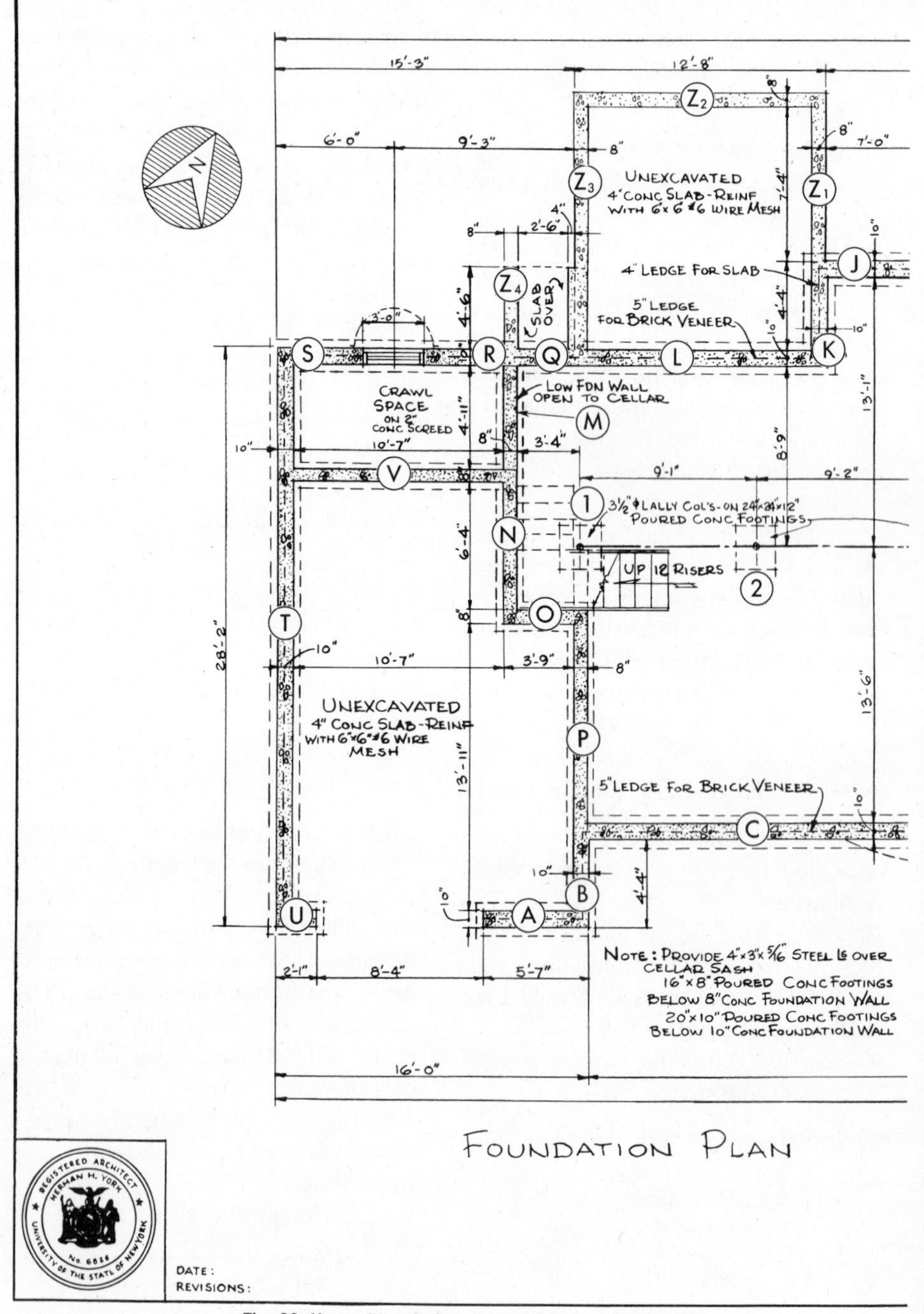

Fig. 22. House Plan A. Concrete and formwork take-off.

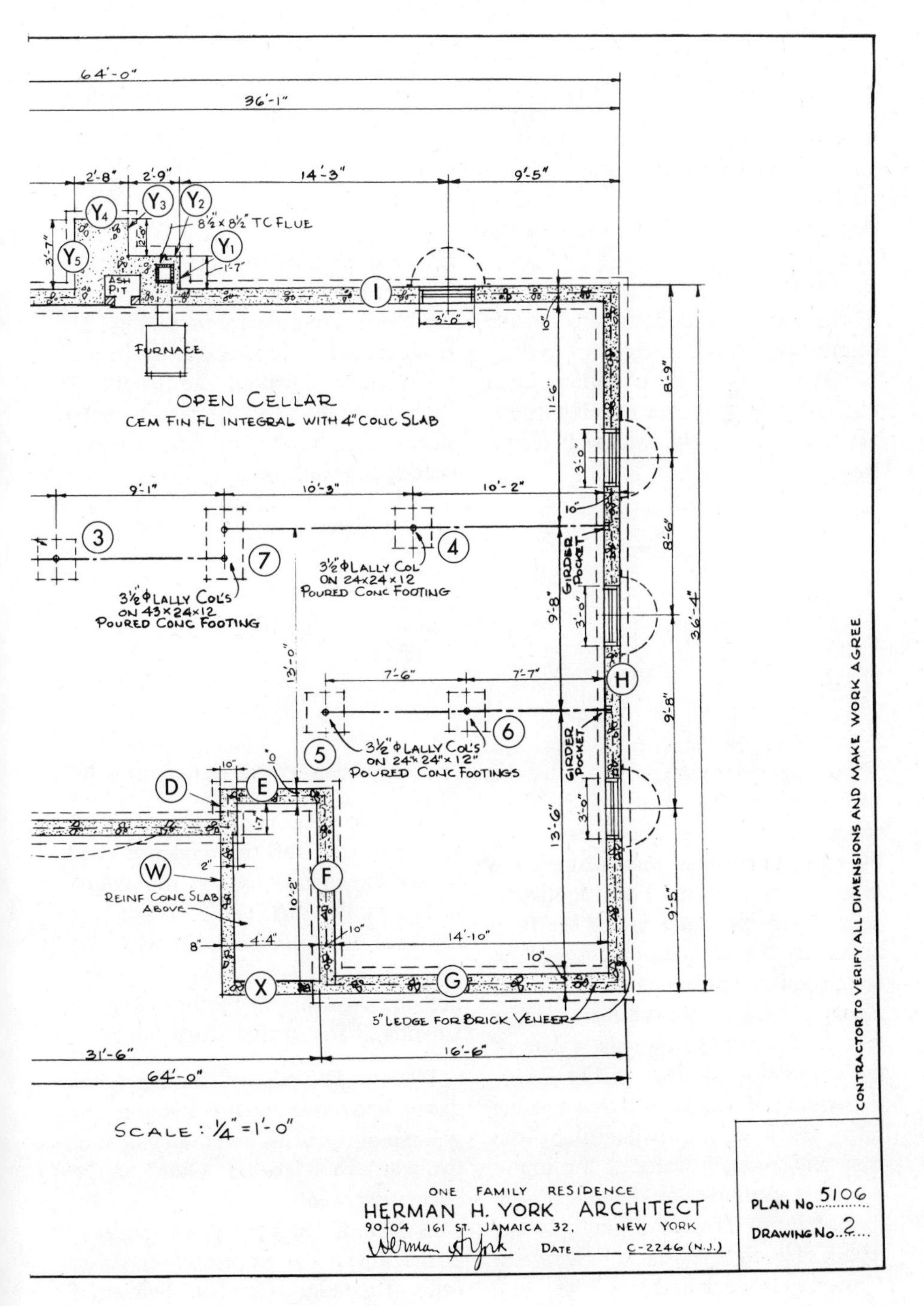
64'-0"
36'-1"
2'-8"
2'-9"
14'-3"
9'-5"
8½" x 8½" TC FLUE
ASH PIT
FURNACE
OPEN CELLAR
CEM FIN FL INTEGRAL WITH 4" CONC SLAB
9'-1"
10'-3"
10'-2"
3½" ϕ LALLY COL'S ON 43 x 24 x 12 POURED CONC FOOTING
3½" ϕ LALLY COL ON 24 x 24 x 12 POURED CONC FOOTING
GIRDER POCKET
7'-6"
7'-7"
3½" ϕ LALLY COL'S ON 24" x 24" x 12" POURED CONC FOOTINGS
REINF CONC SLAB ABOVE
4'-4"
14'-10"
5" LEDGE FOR BRICK VENEER
31'-6"
16'-6"
64'-0"
36'-4"
CONTRACTOR TO VERIFY ALL DIMENSIONS AND MAKE WORK AGREE
SCALE: ¼" = 1'-0"
ONE FAMILY RESIDENCE
HERMAN H. YORK ARCHITECT
90-04 161 ST. JAMAICA 32, NEW YORK
DATE
C-2246 (N.J.)
PLAN No. 5106
DRAWING No. 2

= 6.26 or approximately
6 walers

This means that each wall has 6 walers running the length of the wall. In addition to these 6 walers, each wall will have 3 extra walers for additional bracing near the bottom of the form.

Example 26. Find the linear footage of the walers needed for the outside foundation walls in Fig. 21.

Solution: As shown earlier, each wall will have 9 walers running its entire length; multiply each wall length by 9.

Outside Wall	Linear Ft. (+ 1' overhang)	Total Feet
A	41'	369
B	21'	189
C	41'	369
D	21'	189
		1116 ft.

Foundation footings also must be braced so that the form will hold its shape after pouring. This is best done by using horizontal walers along the length of the footings. For a footing 1′ deep, a double waler along the base and a single waler along the top of the form should be adequate.

Example 27. Find the number of linear feet needed for the outside walers of the foundation footings in Fig. 21.

Solution: Each foundation footing form has a double waler along the base and a single waler at the top—3 walers running the length of the footing form. Therefore, multiply the length of each form by 3. (There is a 1′ overlap at each end.)

Outside Wall	Waler Length	Linear Feet
A	42'	126
B	22'	66
C	42'	126
D	221	66
		384 ft.

Example 28. Find the number of linear feet needed for the vertical stakes for the outside footing forms in Fig. 21. Each stake is 4′ long and 2′ apart, o.c.

Solution: Dividing the length of each wall by the distance between stakes will give the number of stakes needed for each wall.

Wall	Length ÷ 2 ft.	Linear Ft.
A	21	84
B	11	44
C	21	84
D	11	44
		256 ft.

Quantity take-off for House Plan A

Figs. 23 to 30 contain detailed quantity take-offs for the concrete and formwork needed for the foundation of HOUSE PLAN A. Fig. 22 is used for this take-off. In general, the walls are labeled counter-clockwise. (Use the detailed drawings at the beginning of Chapter 3 for further clarification).

Concrete. The take-off for the foundation and cheek walls is taken as the volume of the walls less the volume of five windows—labeled "Outs" on the estimating sheet.

A 5″ wide ledge for brick veneer is at the top of the exterior foundation walls. The ledge extends 6″ below the

CONCRETE
QUANTITY TAKE-OFF

SHEET 1

HOUSE PLAN A

Wall Code	Length l ft.	Width w ft.	Height h ft.	Area A sq.ft.	Volume V cu.ft.	V cu. yds.
	EXTERIOR FOUNDATION WALLS					
A	5.58	0.83	7.00		31.25	
B	4.33	↓	↓		24.25	
C	27.33				153.05	
D	1.58				8.85	
E	4.33				24.25	
F	11.00				61.39	
G	14.83				83.05	
H	36.33				203.45	
I	22.83				127.64	
J	7.83				43.85	
K	4.33				24.25	
L,Q	15.50		7.00		86.80	
R	5.83		5.00		24.19	
S	6.00		4.00		19.92	
T	27.33		4.00		90.74	
U	1.25	0.83	4.00		4.15	
	196.22				1011.08	
OUTS (WINDOWS)	3.00	0.83	1.67		-20.83	
				TOTAL	990.25	37—
	INTERIOR FOUNDATION WALLS					
M,N,O,P	26.00	0.67	7.00		121.94	
V	10.58	0.67	7.00		28.35	
	36.58				150.29	6—
	EXTERIOR CHEEK WALLS					
Z4	4.50	0.67	7.00		21.11	
W	8.58	0.67	7.00		40.24	
X	4.33	0.67	7.00		20.31	
Z1	7.33	0.67	4.00		19.64	
Z2	12.67	0.67	4.00		33.96	
Z3	7.17	0.67	4.00		19.22	
Z3	4.50	1.00	4.00		18.00	
					172.48	7—
	CHIMNEY FOOTING					
	6.25	5.25	1.00		32.81	
	2.75	2.00	1.00		-5.50	
					27.31	1—

Fig. 23-A. House Plan A. Concrete take-off.

CONCRETE

QUANTITY TAKE-OFF

SHEET 2

HOUSE PLAN A

Wall Code	Length l ft.	Width w ft.	Height h ft.	Area A sq.ft.	Volume V cu.ft.	V cu. yds.
	CHIMNEY STACK					
	2.67	4.42	7.00		82.60	
	2.75	2.42	7.00		46.62	
FLUE	0.71	0.71	3.00		1.51	
ASH PIT	1.50	1.83	1.00		2.75	
					124.96	5—
	EXTERIOR WALL FOOTINGS					
A	6.42	1.67	0.83			
B	4.33					
C	27.33					
D	→					
E	6.67					
F	10.17					
G	15.67					
H	35.50					
I	22.00					
J	7.67					
K	4.33					
L,Q	15.08					
R	5.75					
S	6.42					
T	27.33	↓	↓			
U	1.25	1.67	0.83			
	195.92				272.33	11—
	INTERIOR WALL FOOTINGS					
M,N,O,P	25.50	1.33	0.67		22.70	
V	9.83	1.33	0.67		8.76	
	35.33				31.46	2—
	PIER FOOTINGS					
1 TO 6	2.00	2.00	1.00		24.00	
7	3.58	2.00	1.00		7.17	
					31.17	2—
	CELLAR SLAB					
	14.83	34.67	0.33	514.16		
	5.83	24.58		142.84		
	26.50	26.58		704.37		
	3.75	11.92	↓	44.70		
	4.33	12.16	0.33	−52.65		
				1,353.42	446.63	17—

Fig. 23-B. House Plan A. Concrete take-off.

					SHEET 3
CONCRETE QUANTITY TAKE-OFF					HOUSE PLAN A

Wall Code	Length l ft.	Width w ft.	Height h ft.	Area A sq.ft.	Volume V cu.ft.	V cu. yds.
	MISC.	CONCRETE		SLABS		
FRONT. ENT.	10.17	5.00	0.33	50.85	17–	
REAR P.	12.67	12.33		156.22	52–	
REAR ENT.	4.50	3.50		15.75	5–	
GARAGE	13.92	14.33	↓	199.47	66–	
"	7.00	10.58	0.33	74.06	25–	
CRAWL SPACE	10.58	4.92	0.17	52.06	9–	
				548.41	174–	7–
					TOTAL CONCRETE	95–

Fig. 23-C. House Plan A. Concrete take-off.

top of the wall. This volume is subtracted from the total volume of each exterior foundation wall.

The additional concrete for walls *R, S,* and *V* is the volume of concrete of the stepped wall below the top of the normal footing.

The additional concrete for the cheek walls is the volume of a wedge of concrete. This wedge starts at the bottom of wall Z_2 and slopes to the top of the foundation walls.

The concrete for the chimney is taken as the volume of the chimney less the volumes of the flue stack and ash pit.

The cellar slab is divided into 4 parts for the quantity take-off. Working from right to left, the area of each part is calculated and added together. Any area outside the boundary of the cellar slab is then subtracted.

The volumes are converted to whole cubic yards and in almost all cases the amount of concrete is greater than the volume of the wall, column, or slab. This "extra" concrete can be counted as part of the wastage.

Formwork. Only the quantity of formwork, bracing, and ties is computed here. Any estimate made from these *minimum* figures would, of course, have to consider waste, transportation, and labor costs.

FORMWORK — QUANTITY TAKE-OFF — SHEET 1 — HOUSE PLAN A

Wall Code	Length l ft.-in.	Length l ft.	Height h ft.-in.	Height h ft.	Area A sq.ft.
	WALLS—OUTSIDE FACE				
A	5'-7"	5.58	7'-4"	7.33	40.90
B	4'-4"	4.33	↓	↓	31.74
C	26'-6"	26.50			194.25
D	1'-7"	1.58			11.58
E	4'-2"	4.17			30.57
F	9'-6"	9.50			69.64
G	16'-6"	16.50			120.95
H	36'-4"	36.33			266.30
I	23'-8"	23.67			173.50
Y_1	1'-7"	1.58			11.58
Y_2	2'-9"	2.75			20.16
Y_3	2'-0"	2.00			14.66
Y_4	2'-8"	2.67			19.57
Y_5	3'-7"	3.58			26.24
J	7'-0"	7.00	↓	↓	51.31
Z_1	8'-0"	8.00	7'-4"	7.33	58.64
Z_2	12'-8"	12.67	4'-0"	4.00	50.68
Z_3	12'-4"	12.33	7'-4"	7.33	90.38
Q	2'-6"	2.50	7'-4"	7.33	18.33
Z_4	9'-8"	9.67	7'-4"	7.33	70.88
R	5'-9"	5.75	4'-8"	4.67	26.85
S	6'-0"	6.00	2'-8"	2.67	16.02
T	28'-2"	28.17	2'-8"	2.67	75.21
U	2'-1"	2.08	2'-8"	2.67	5.55
W	8'-7"	8.58	7'-4"	7.33	62.89
X	5'-0"	5.00	7'-4"	7.33	36.65
					1,595.03
	WALLS—INSIDE FACE				
C	26'-8"	26.50	7'-4"	7.33	194.25
D	1'-7"	1.58	↓	↓	11.58
E	5'-10"	5.83			42.74
F	10'-2"	10.17			74.55
G	14'-10"	14.83			108.70
H	34'-8"	34.67			254.13
I,J	35'-1"	35.08			257.14
K	4'-4"	4.33	↓	↓	31.74
L,Q	15'-8"	15.67	7'-4"	7.33	114.86
M	4'-11"	4.92	4'-8"	4.67	22.98
N	7'-0"	7.00	7'-4"	7.33	51.31
O	3'-9"	3.75	7'-4"	7.33	27.49
P	9'-10"	9.83	7'-4"	7.33	72.05
					1,263.52

Fig. 24-A. House Plan A. Formwork take-off.

FORMWORK — SHEET 2

QUANTITY TAKE-OFF — HOUSE PLAN A

Wall Code	Length l ft.-in.	Length l ft.	Height h ft.-in.	Height h ft.	Area A sq.ft.
	WALLS—UNEXCAVATED INSIDE				
A	4'-9"	4.75	7'-4"	7.33	34.82
P, B	13'-11"	13.92	7'-4"	7.33	102.03
O	3'-9"	3.75	7'-4"	7.33	27.49
N	7'-0"	7.00	7'-4"	7.33	51.31
V	10'-7"	10.58	2'-8"	2.67	28.25
T	20'-11"	20.92	2'-8"	2.67	55.86
U	1'-3"	1.25	2'-8"	2.67	3.34
BUTTENDS	1'-8"	1.67	2'-8"	2.67	4.46
M	4'-11"	4.92	2'-8"	2.67	13.14
R	4'-11"	4.92	4'-8"	4.67	22.97
T	4'-11"	4.92	2'-8"	2.67	13.14
L	11'-4"	11.33	7'-4"	7.33	83.05
Z1	11'-8"	11.67	7'-4"	7.33	85.54
Z2	11'-4"	11.33	4'-0"	4.00	45.32
Z3	11'-8"	11.67	7'-4"	7.33	85.54
W	7'-11"	7.92	7'-4"	7.33	58.05
X	4'-4"	4.33	7'-4"	7.33	31.74
F	10'-2"	10.17	7'-4"	7.33	74.55
S	6'-8"	6.67	2'-8"	2.67	17.81
					838.41
	FOOTINGS—OUTSIDE FACE				
A	6'-5"	6.42	10"	0.83	5.33
B	4'-4"	4.33	↓	↓	3.59
C	27'-4"	27.33			22.68
D	1'-7"	1.58			1.31
E	3'-6"	3.50			2.91
F	10'-2"	10.17			8.44
G	17'-4"	17.33			14.38
H	37'-2"	37.17			30.85
I	23'-8"	23.67			19.65
Y1	1'-7"	1.58			1.31
Y2	2'-9"	2.75			2.28
Y3	2'-0"	2.00			1.66
Y4	3'-6"	3.50			2.91
Y5	3'-7"	3.58			2.97
J	7'-8"	7.67			6.37
K	4'-4"	4.33			3.59
L, Q	15'-1"	15.08			12.52
R	5'-9"	5.75			4.77
S	6'-5"	6.42	↓	↓	5.33
T	29'-0"	29.00	10"	0.83	24.07
					176.92

Fig. 24-B. House Plan A. Formwork take-off.

SHEET 3

FORMWORK

QUANTITY TAKE-OFF

HOUSE PLAN A

Wall Code	Length l ft.-in.	Length l ft.	Height h ft.-in.	Height h ft.	Area A sq.ft.
	FOOTINGS-INSIDE FACE				
A	4'-9"	4.75	10"	0.83	3.94
B,P	13'-2"	13.17	10"	0.83	10.93
O	3'-9"	3.75	8"	0.67	2.51
N	6'-8"	6.67	8"	0.67	4.47
V	9'-10"	9.83	8"	0.67	6.59
T	20'-2"	20.17	10"	0.83	16.74
U	1'-3"	1.25	↓	↓	1.04
BUTTENDS	3'-4"	3.33	↓	↓	2.76
C	25'-9"	25.75	↓	↓	21.37
D	1'-8"	1.67	↓	↓	1.37
E	6'-8"	6.67	↓	↓	5.54
F	10'-2"	10.17	↓	↓	8.44
G	14'-0"	14.00	↓	↓	11.62
H	33'-10"	33.83	↓	↓	28.08
I,J	34'-3"	34.25	↓	↓	28.43
K	4'-4"	4.33	↓	↓	3.59
L,Q	15'-11"	15.92	10"	0.83	13.21
M	4'-6"	4.50	8"	0.67	3.00
N	6'-8"	6.67	↓	↓	4.47
O	3'-9"	3.75	↓	↓	2.50
R	5'-5"	5.42	10"	0.83	4.50
S	4'-9"	4.75	10"	0.83	3.94
T	4'-2"	4.17	10"	0.83	3.46
					192.50
	INTERIOR COLUMN FOOTINGS				
8 COL.	8'-0"	8.0	1'-0"	1.0	64 —
	(FORM CONSTANT = 8)				
	TOTAL FORMWORK				4,130.38

Fig. 24-C. House Plan A. Formwork take-off.

				SHEET 4
2' - 0" OC— 6 walers per 7' - 4" wall wall length + 6" overhang at free ends		HORIZONTAL WALERS QUANTITY TAKE-OFF		HOUSE PLAN A

Wall Code	Waler Length ft.-in.	Waler Length ft.	No. of Walers		Total Length ft.
	FOUNDATION—EXT. FACE				
A	6'-7"	6.58	6—		39.48
B	4'-10"	4.83	↓		28.98
C	26'-6"	26.50			159—
D	2'-1"	2.08			12.48
E	4'-2"	4.17			25.02
F	9'-6"	9.50			57.00
G	17'-6"	17.50			105.00
H	37'-4"	37.33			223.99
I	24'-2"	24.17			145.02
Y_1	2'-1"	2.08			12.48
Y_2	3'-3"	3.25			19.50
Y_3	2'-6"	2.50			15.—
Y_4	3'-8"	3.67			22.02
Y_5	4'-1"	4.08			24.48
J	7'-0"	7.00	↓		42.—
Z_1	8'-6"	8.50	6—		51.—
Z_2	13'-8"	13.67	4—		54.68
Z_3	12'-10"	12.83	6—		76.98
Q	2'-6"	2.50	6—		15.—
Z_4	11'-8"	11.67	6—		70—
R	5'-9"	5.75	5—		28.75
S	6'-5"	6.42	4—		25.68
T	29'-2"	29.17	3—		87.51
U	3'-1"	3.08	3—		9.24
W	9'-1"	9.08	6—		54.48
X	6'-0"	6.00	6—		36—
					1,440.76
	INSIDE FACES—EXC. WALLS				
C	26'-8"	26.67	6—		160.02
D	2'-1"	2.08	↓		12.48
E	6'-10"	6.83			40.98
F	10'-8"	10.67			64.02
G	14'-10"	14.83			88.98
H	34'-8"	34.67			208.02
I, J	35'-1"	35.08			210.48
K	4'-10"	4.83			28.98
L, Q	15'-11"	15.92			95.52
M	4'-11"	4.92			29.52
N	7'-0"	7.00			42—
O	4'-3"	4.25	↓		25.50
P	10'-9"	10.75	6—		64.50
					1071—

Fig. 25-A. House Plan A. Horizontal walers.

SHEET 5

HORIZONTAL WALERS

QUANTITY TAKE-OFF

HOUSE PLAN A

2' - 0" OC— 6 walers per 7' - 4" wall
wall length + 6" overhang at free ends

Wall Code	Waler Length ft.-in.	Waler Length ft.	No. of Walers	Total Length ft.
UNEXCAVATED INSIDE				
A	5'-3"	5.25	6	31.50
P,B	13'-11"	13.92	6	83.52
O	4'-3"	4.25	6	25.50
N	7'-6"	7.50	6	45.—
V	10'-7"	10.58	3	31.74
T	20'-11"	20.92	3	62.76
U	1'-9"	1.75	3	5.25
V	10'-7"	10.58	3	31.74
M	4'-11"	4.92	3	14.76
R	6'-5"	6.42	4	25.68
T	4'-11"	4.92	3	14.76
L	11'-4"	11.33	6	67.98
Z1	11'-8"	11.67	6	70.02
Z2	11'-4"	11.33	3	33.99
Z3	11'-8"	11.67	6	70.02
W	9'-6"	9.50	6	57.—
X	4'-4"	4.33	6	25.98
S	5'-2"	5.17	3	15.51
BUTTEND	1'-10"	1.83	3	5.49
"	1'-10"	1.83	6	10.98
				729.18
			TOTAL	3,240.94

Fig. 25-B. House Plan A. Horizontal walers.

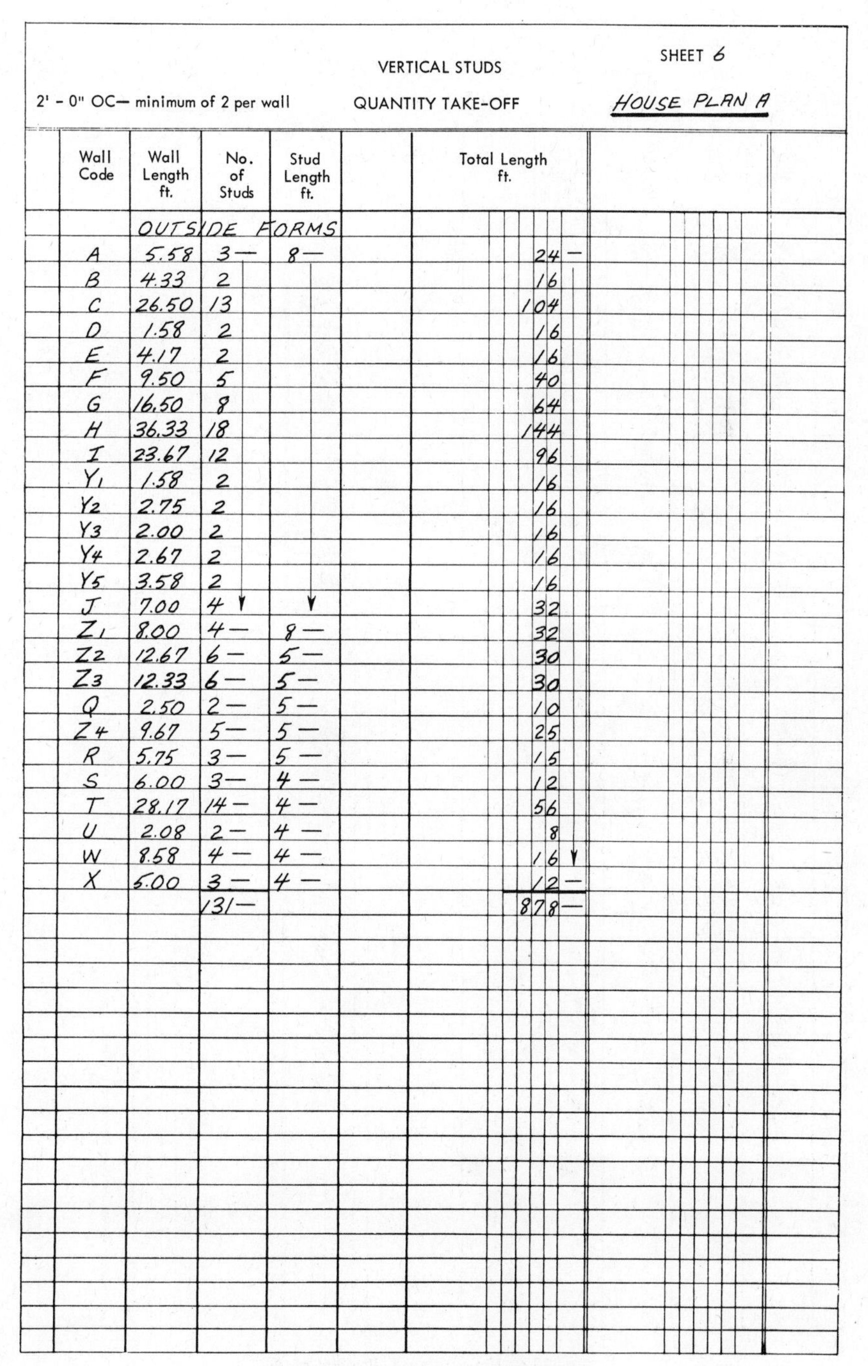

SHEET 6

VERTICAL STUDS

2' - 0" OC— minimum of 2 per wall QUANTITY TAKE-OFF HOUSE PLAN A

Wall Code	Wall Length ft.	No. of Studs	Stud Length ft.		Total Length ft.	
	OUTSIDE FORMS					
A	5.58	3—	8—		24—	
B	4.33	2			16	
C	26.50	13			104	
D	1.58	2			16	
E	4.17	2			16	
F	9.50	5			40	
G	16.50	8			64	
H	36.33	18			144	
I	23.67	12			96	
Y_1	1.58	2			16	
Y_2	2.75	2			16	
Y_3	2.00	2			16	
Y_4	2.67	2			16	
Y_5	3.58	2			16	
J	7.00	4			32	
Z_1	8.00	4—	8—		32	
Z_2	12.67	6—	5—		30	
Z_3	12.33	6—	5—		30	
Q	2.50	2—	5—		10	
Z_4	9.67	5—	5—		25	
R	5.75	3—	5—		15	
S	6.00	3—	4—		12	
T	28.17	14—	4—		56	
U	2.08	2—	4—		8	
W	8.58	4—	4—		16	
X	5.00	3—	4—		12—	
		131—			878—	

Fig. 26-A. House Plan A. Vertical studs.

VERTICAL STUDS

SHEET 7

2' - 0" OC— minimum of 2 per wall — QUANTITY TAKE-OFF — HOUSE PLAN A

Wall Code	Wall Length ft.	No. of Studs	Stud Length ft.	Total Length ft.
	INSIDE FACE			
C	26.50	13	8—	104—
D	1.58	2		16
E	5.83	3		24
F	10.17	5		40
G	14.83	8		64
H	34.67	17		136
I,J	35.08	17		136
K	4.33	2		16
L,Q	15.67	8		64
M	4.92	2		16
N	7.00	4		32
O	3.75	2		16
P	9.58	5	8—	40—
		88		704—
	UNEXC INSIDE			
A	4.75	3	8—	24—
P,B	13.92	7	8	56
O	3.75	2	8	16
N	7.00	4	8	32
V	10.58	5	3	15
T	20.92	10	3	30
U	1.25	2	3	6
BUTTEND	0.83	2	3	6
"	0.83	2	8	16
V	10.58	5	3	15
M	4.92	3	3	9
R	5.75	3	5	15
T	4.92	3	3	9
L	11.33	6	8	48
Z1	11.67	6	8	48
Z2	11.33	6	5	30
Z3	11.67	6	8	48
W	9.50	5	8	40
X	4.33	2	8	16
S	5.17	3	3—	9—
		85		488—

Fig. 26-B. House Plan A. Vertical studs.

WALERS–FOOTINGS — SHEET 8

3 per length, + 6" overlap at free ends — QUANTITY TAKE-OFF — HOUSE PLAN A

Wall Code	Waler Length ft.-in.	Waler Length ft.	No. of Walers	Total Length ft.
FOOTINGS–OUTSIDE FACE				
A	7'-5"	7.42	3—	22.26
B	4'-10"	4.83		14.49
C	27'-4"	27.33		81.99
D	2'-1"	2.08		6.24
E	3'-9"	3.75		11.25
F	10'-8"	10.67		32.01
G	18'-4"	18.33		54.99
H	38'-2"	38.17		114.50
I	24'-2	24.17		72.50
Y_1	2'-1"	2.08		6.25
Y_2	3'-3"	3.25		9.75
Y_3	2'-6"	2.50		7.50
Y_4	4'-6"	4.50		13.50
Y_5	4'-1"	4.08		12.25
J	8'-2"	8.17		24.50
K	4'-10"	4.83		14.49
L, Q	15'-7"	15.58		46.77
R	6'-3"	6.25		18.75
S	7'-1"	7.08		21.24
T	31'-0"	31.00	3—	93.00
				678.23
INTERIOR COLUMNS				
8 COLS.	8'-0"	8.0	3	192—

Fig. 27-A. House Plan A. Footing walers.

SHEET 9

WALERS-FOOTINGS

3 per length, + 6" overlap at free ends — QUANTITY TAKE-OFF — HOUSE PLAN A

Wall Code	Waler Length ft.-in.	Waler Length ft.	No. of Walers		Total Length ft.
	FOOTINGS—INSIDE FACE				
A	5'-3"	5.25	3—		15.75
P,B	13'-2"	13.17			39.51
O	4'-3"	4.25			12.75
N	7'-6"	7.50			22.50
V	9'-10"	9.83			29.49
T	20'-2"	20.17			60.50
U	2'-2"	2.17			6.50
BUTTENDS	5'-4"	5.33			16.00
C	25'-8"	25.67			77.00
D	2'-1"	2.08			6.24
E	7'-8"	7.67			23.00
F	10'-8"	10.67			32.00
G	14'-0"	14.00			42.00
H	33'-10"	33.83			101.49
I, J	34'-3"	34.25			102.75
K	4'-10"	4.83			14.50
L, Q	16'-11"	16.92			50.75
M	4'-6"	4.50			13.50
N	7'-6"	7.50			22.50
O	4'-3"	4.25			12.75
P	11'-1"	11.09			30.00
V	9'-10"	9.83			29.49
M	4'-2"	4.17			12.51
R	5'-4"	5.33			16.00
S	4'-9"	4.75	↓		14.25
T	4'-2"	4.17	3—		12.50
					816.23

Fig. 27-B. House Plan A. Footing walers.

SHEET 10

VERTICAL BRACING–FOOTINGS

4' – 0" OC; minimum of 2 per wall — QUANTITY TAKE-OFF — HOUSE PLAN A

Wall Code	Wall Length ft.	No. of Stakes	Stake Length ft.		Total Length ft.
	FOOTING STAKES—INSIDE FACE				
A	4.75	2	4'		8 —
P,B	13.08	3			12
O	3.75	2			8
N	7.00	2			8
V	9.75	2			8
T	20.75	5			20
U	1.67	2			8
BUTTENDS	3.33	2			8
C	25.17	6			24
D	.83	2			8
E	6.67	2			8
F	10.17	3			12
G	14.00	4			16
H	33.83	9			36
I,J	34.25	9			36
K	4.33	2			8
L,Q	15.17	4			16
M	4.50	2			8
N	6.67	2			8
O	3.75	2			8
P	9.50	2			8
V	9.88	2			8
M	4.17	2			8
R	5.75	2			8
S	6.50	2			8
T	4.17	2	4'		8 —
					316 —

Fig. 28-A. House Plan A. Vertical bracing for footings.

VERTICAL BRACING-FOOTINGS

SHEET 11

4' - 0" OC; minimum of 2 per wall

QUANTITY TAKE-OFF

HOUSE PLAN A

Wall Code	Wall Length ft.	No. of Stakes	Stake Length ft.		Total Length ft.
	FOOTING STAKES – OUTSIDE FACE				
A	6.42	2	4'		8 —
B	4.33	2			8
C	26.83	7			28
D	1.58	2			8
E	3.50	2			8
F	10.17	3			12
G	17.33	4			16
H	37.17	9			36
I	23.67	6			24
Y_1	1.58	2			8
Y_2	2.75	2			8
Y_3	2.00	2			8
Y_4	3.50	2			8
Y_5	3.58	2			8
J	7.67	2			8
K	4.33	2			8
L,Q	16.42	4			16
R	5.75	2			8
S	7.75	2			8
T	29.00	7	4'		28 —
					264 —
	FOOTING STAKES – COLUMNS				
8 COL.		8	4'		256 —
			TOTAL STAKES		836 —

Fig. 28-B. House Plan A. Vertical bracing for footings.

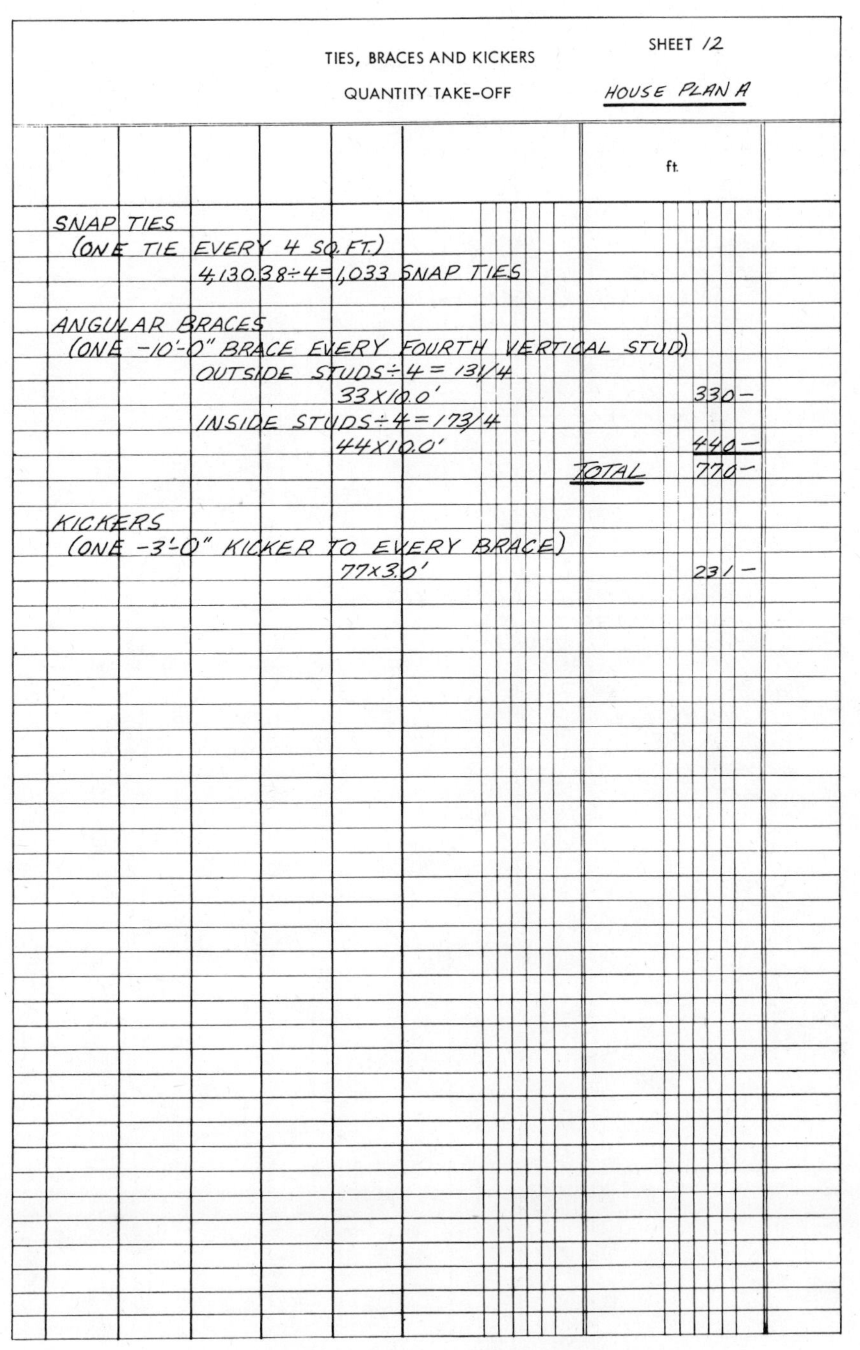

SHEET 12

TIES, BRACES AND KICKERS

QUANTITY TAKE-OFF

HOUSE PLAN A

	ft.
SNAP TIES	
(ONE TIE EVERY 4 SQ. FT.)	
4,130.38÷4=1,033 SNAP TIES	
ANGULAR BRACES	
(ONE -10'-0" BRACE EVERY FOURTH VERTICAL STUD)	
OUTSIDE STUDS÷4 = 131/4	
33X10.0'	330 —
INSIDE STUDS÷4 = 173/4	
44X10.0'	440 —
TOTAL	770 —
KICKERS	
(ONE -3'-0" KICKER TO EVERY BRACE)	
77X3.0'	231 —

Fig. 29. House Plan A. Ties, braces, and kickers.

CONCRETE AND FORMWORK SUMMARY

QUANTITY TAKE-OFF

Item	Length l ft.	Area A sq.ft.	Volume V cu. yds.
CONCRETE			
EXTERIOR FOUNDATION WALLS			37
INTERIOR FOUNDATION WALLS			6
EXTERIOR CHEEK WALLS			7
CHIMNEY FOOTING			1
CHIMNEY STACK			5
COLUMN FOOTINGS			2
WALL FOOTINGS			13
CELLAR			17
MISC. SLABS			7
			95
FORMS			
EXTERIOR FOUNDATION WALLS		2,858.55	
INTERIOR FOUNDATION WALLS		838.41	
WALL FOOTINGS		369.42	
COLUMN FOOTINGS		64.—	
		4,130.38	
SNAP TIES—1,033			
ANGULAR BRACES-77	770.—		
KICKERS-77	231.—		
WALERS	3,240.94		

Fig. 30. House Plan A. Concrete and formwork summary.

Estimating labor

The average time it takes a worker to do his part of the job will vary with the local conditions, but the variance should be small except for extreme conditions.

Forms and concrete equipment must be hauled onto the site before work is begun. This is an item of overhead expense. Some contractors add what they think the cost in time will total.

One carpenter can erect about 300 square feet of forms in 8 hours. Usually he has a helper whose time is figured as one-half of the carpenter's.

A man can mix and place about 110 cubic feet of concrete in about 7 hours. This includes charging the mixture and wheeling the concrete to the forms. For long hauls, some extra time should be allowed.

Concrete forms can be removed and piled (ready to be moved) by one man at the rate of 700 square feet in 8 hours.

The surface of the concrete must be rubbed when it is desired to remove all form marks. One man can rub about 100 square feet per hour.

One cement-finisher with a laborer-helper can *top dress* about 425 square feet of sidewalk in about 7 hours. For basement floors, 600 square feet can be done in the same time.

Questions and Problems

1. How many bags of cement are needed for the 95 cu. yds. of concrete foundation used in House Plan A? The concrete mixture is 1:2:4:8. Use the absolute volume method. (See the "Concrete Problem" in the section on Concrete and Mortar Mixes.)
   (Answer: 570 bags)
2. It is estimated that the following numbers and sizes of reinforcing bars will be needed for concrete reinforcement:
   20 #5 bars, each 20 ft. long at 1.043 lbs./lin. ft.
   20 #6 bars, each 15 ft. long at 1.502 lbs./lin. ft.
   Find the total amount of steel needed for this job.
   (Answer: 867.8 lbs.)
3. Find the concrete constant and volume of a concrete column with a circular base, 1′ in diameter and 10′ high. The cross-sectional area is constant.
   (Answer: 31.4 cu. ft.)
4. Find the volume of concrete needed for a concrete stairway with 12 treads which are each 4′ wide. (Answer: 1.8 cu. yds.)
5. Find the amount of formwork needed for the intermediate beam in Fig. 20.
   (Answer: 16.0 sq. ft.)
6. Find the form constant and

SFCA of the 2 columns in Fig. 21, each of the columns being 8.0′ high. (Ans.: 8,128 sq. ft.)

7. Find the length of board needed for 20 kickers if the angular braces are 7′ long. (Assume the braces are at a 45° angle.)
(Answer: 99 ft.)

8. Snap ties are to be used every 12 square feet in foundation walls whose outside areas total 1,500 square feet. How many snap ties are needed?
(Answer: 125)

9. Find the amount of lumber needed for the vertical bracing studs on the inside walls of Fig. 21. The on-center distances are 2′ and there are two 8′ braces at each on-center distance.
(Answer: 896 ft.)

10. Find the linear footage of the walers needed for the inside foundation walls of Fig. 21. Each wall has 9 walers running its entire length.
(Answer: 1,008 ft.)

7

CARPENTRY AND MILLWORK

This chapter is concerned with wood *framing and millwork. Wood framing is also known as* rough carpentry. *Some of the other materials used in house framing are discussed briefly.*

Millwork is wood work which is finished, machined, and partly assembled at a mill before delivery to the construction site.

Most residential buildings are constructed in a certain order: excavation will necessarily come before concrete work which must precede framing and millwork. Although the order of construction may vary, you should learn to think of the operations in a definite order. This is especially true of framing and millwork where it is easy to overlook small but costly items.

Lumber and board measure

Lumber is classified according to its use, size, and quality. It is usually furnished in standard sizes. See Table I.

Yard lumber is less than 5″ thick and is intended for general building purposes.

Dimension lumber is yard lumber which is not less than 2″ nor more than 5″ thick. It is also called *dimension stuff*. The narrower pieces are *scantlings* and wider ones are *planks*.

Yard lumber which is less than 2″ thick and from 4″ to 12″ wide is called *board* lumber.

Rough dimensions of yard lumber differ from the smoothed or *dressed* lumber which has one or more of its sides or edges *surfaced*. About ⅛″ is removed from each side of the thinner pieces and from ⅜″ to ½″ is removed from the sides of the thicker pieces. This smoothing process is often called *surfacing* and the letters *D* for dressed or *S* for surfaced are used to indicate the number of sides or edges of a piece to be dressed or surfaced. Thus *D1S* or *S1S* means dressed or surfaced on one side. *S4S* means surfaced on all 4 sides while *S1S1E* means surfaced on one side and one edge.

Shiplap and *dressed-and-matched* boards are listed in Table I. These

TABLE I. AMERICAN STANDARD SOFTWOOD YARD LUMBER SIZES.*

Product	Rough Green or Nominal Sizes (Board Measure)		Dressed Dimensions	
	Thickness	Width	Thickness — Standard Yard	Width (Face When Worked)
	Inches	Inches	Inches	Inches
Finish	—	3	5/16	2 5/8
	—	4	7/16	3 1/2 [a]
	—	5	9/16	4 1/2 [a]
	—	6	11/16	5 1/2 [a]
	1	7	25/32	6 1/2 [a]
	1 1/4	8	1 1/16	7 1/4 [a]
	1 1/2	9	1 5/16	8 1/4 [a]
	1 3/4	10	1 7/16	9 1/4 [a]
	2	11	1 5/8	10 1/4 [a]
	2 1/2	12	2 1/8	11 1/4 [a]
	3	—	2 5/8	—
Common boards and strips	1	3	25/32	2 5/8
	1 1/4	4	1 1/16	3 5/8
	1 1/2	5	1 5/16	4 5/8
	—	6	—	5 5/8
	—	7	—	6 5/8
	—	8	—	7 1/2
	—	9	—	8 1/2
	—	10	—	9 1/2
	—	11	—	10 1/2
	—	12	—	11 1/2
Dimension and heavy joist	2	2	1 5/8	1 5/8
	2 1/2	4	2 1/8	3 5/8
	3	6	2 5/8	5 5/8
	4	8	3 5/8	7 1/2
	—	10	—	9 1/2
	—	12	—	11 1/2
Bevel siding	—	4	7/16 by 3/16	3 1/2
	—	5	10/16 by 3/16	4 1/2
	—	6	—	5 1/2
Wide bevel siding	—	8	7/16 by 3/16	7 1/4
	—	10	9/16 by 3/16	9 1/4
	—	12	11/16 by 3/16	11 1/4
Rustic and drop siding (shiplapped)	—	4	9/16	3 1/8
	—	5	3/4	4 1/8
	—	6	—	5 1/16
	—	8	—	6 7/8
Rustic and drop siding (dressed and matched)	—	4	9/16	3 1/4
	—	5	3/4	4 1/4
	—	6	—	5 3/16
	—	8	—	7

*Recommended by the United States Department of Agriculture.
[a] Based on kiln-dried lumber.

TABLE I (continued)

Product	Rough Green or Nominal Sizes (Board Measure)		Dressed Dimensions	
	Thickness	Width	Thickness: Standard Yard	Width (Face When Worked)
	Inches	Inches	Inches	Inches
Flooring	—	2	5/16	1 1/2
	—	3	7/16	2 3/8
	—	4	9/16	3 1/4
	1	5	25/32	4 1/4
	1 1/4	6	1 1/16	5 3/16
Ceiling	—	3	5/16	2 3/8
	—	4	7/16	3 1/4
	—	5	9/16	4 1/4
	—	6	1 1/16	5 3/16
Partition	—	4	3/4	3 1/4
	—	5	—	4 1/4
	—	6	—	5 3/16
Shiplap	—	6	25/32	5 1/8
	—	8	—	7 1/8
	—	10	—	9 1/8
	—	12	—	11 1/8
Dressed and matched	1	4	25/32	3 1/4
	1 1/4	6	1 1/16	5 1/4
	1 1/2	8	1 5/16	7 1/4
	—	10	—	9 1/4
	—	12	—	11 1/4

*Recommended by the United States Department of Agriculture.

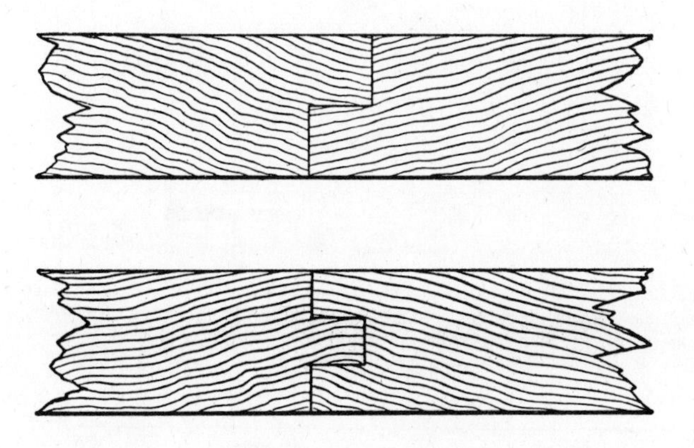

Fig. 1. Types of boards: shiplapped (top), tongued-and-grooved (bottom).

boards are illustrated in Fig. 1. Shiplap boards have their edges cut or *rabbeted* so that they fit together, while the other has its edges grooved at one end and tongued at the other end so that two boards can be fitted tightly together. These are often called *tongued-and-grooved* boards.

The purpose of having the edges of boards shiplapped or matched together is to prevent an open joint between any two boards after they shrink.

Framing and millwork

Outside Wall Framing. Plans do not often include framing drawings for outside walls. When they do, it simplifies the estimator's job. When they do not, the estimator must calculate the number of 2 × 4 studs needed for the wall and around the window and door openings.

Fig. 2 is a typical outside wall framing drawing. The vertical studs are represented by double lines. All window and door openings are shown and it is a simple matter to calculate the number of required 2 × 4 studs. The corner posts can be constructed like any of those shown in Fig. 3. The pictorial views of *western* and *balloon* framing, Figs. 4 and 5, will help you to visualize the outside wall framing details.

Steel buildings are composed of vertical wall members of steel which are comparable to 2 × 4 studs.

The following materials and methods of construction should be care-

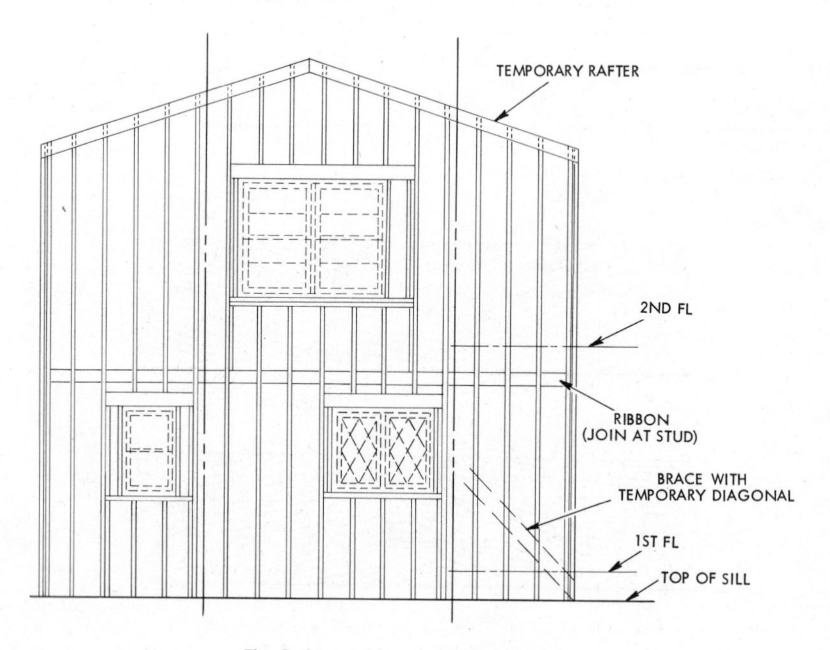

Fig. 2. An outside wall framing diagram.

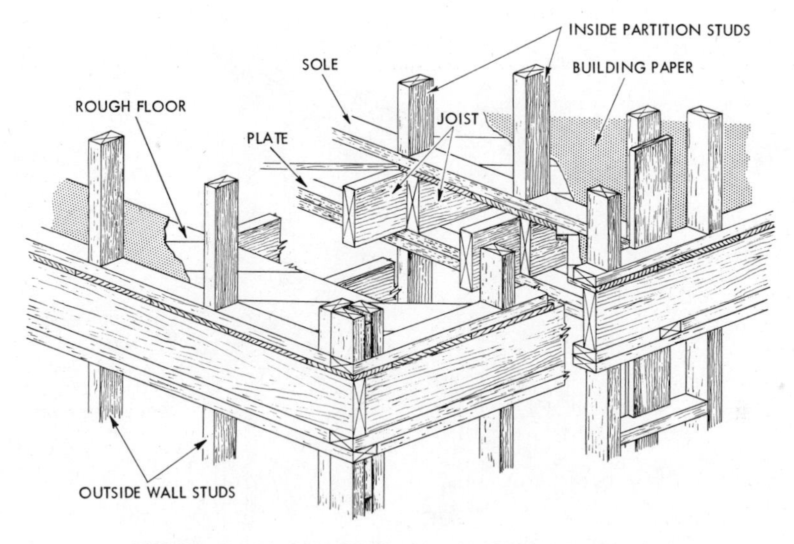

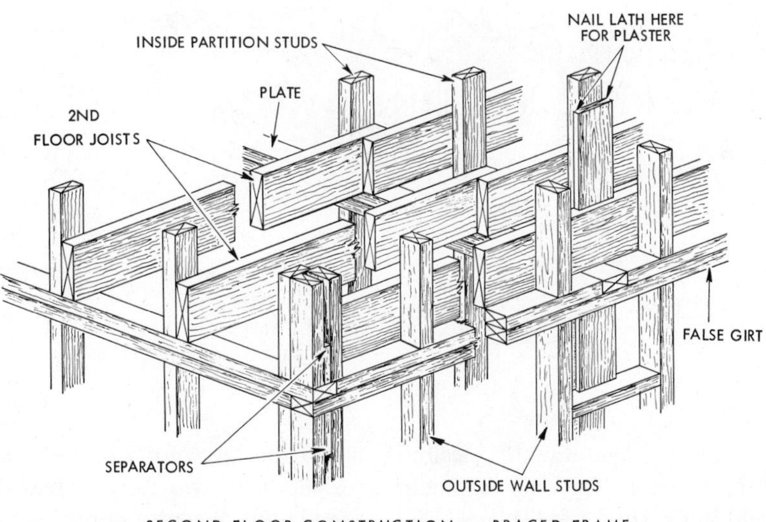

Fig. 3. Construction of corner posts for platform framing and braced frame construction.

fully studied and understood by the estimator. The framing illustrations are detailed so that terms such as *sill, sole,* and *stud* will not cause confusion when they appear in the specifications.

Western Framing. Fig. 4 is a pictorial view of western or *platform* framing. The letters *A* through *F* are near the load-bearing walls. For western framing, most of the studs will

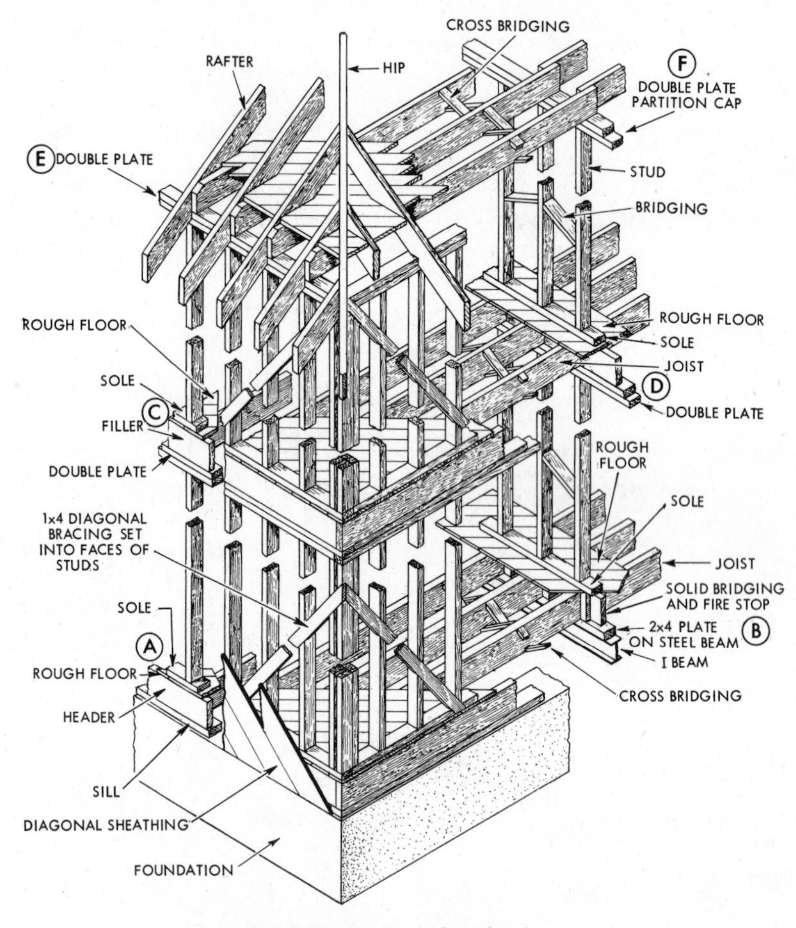

Fig. 4. Western or platform framing.

have been pre-cut to the desired lengths at the mill. The only cutting in the field is for sills, soles, plates, and headers.

Balloon Framing. Fig. 5 is a pictorial view of balloon framing. This method of construction is not as popular as the simpler western frame partly because it does not lend itself to either prefabrication or pre-cutting. There is also a shortage of dry lumber suitable for the long, two-story stud which is used in this type of framing.

Plank and Beam Framing. Fig. 6 is a pictorial view of plank and beam framing. This method of construction uses a few large members to replace the many small members used in typical wood framing. The floor and roof planks serve as structural (that

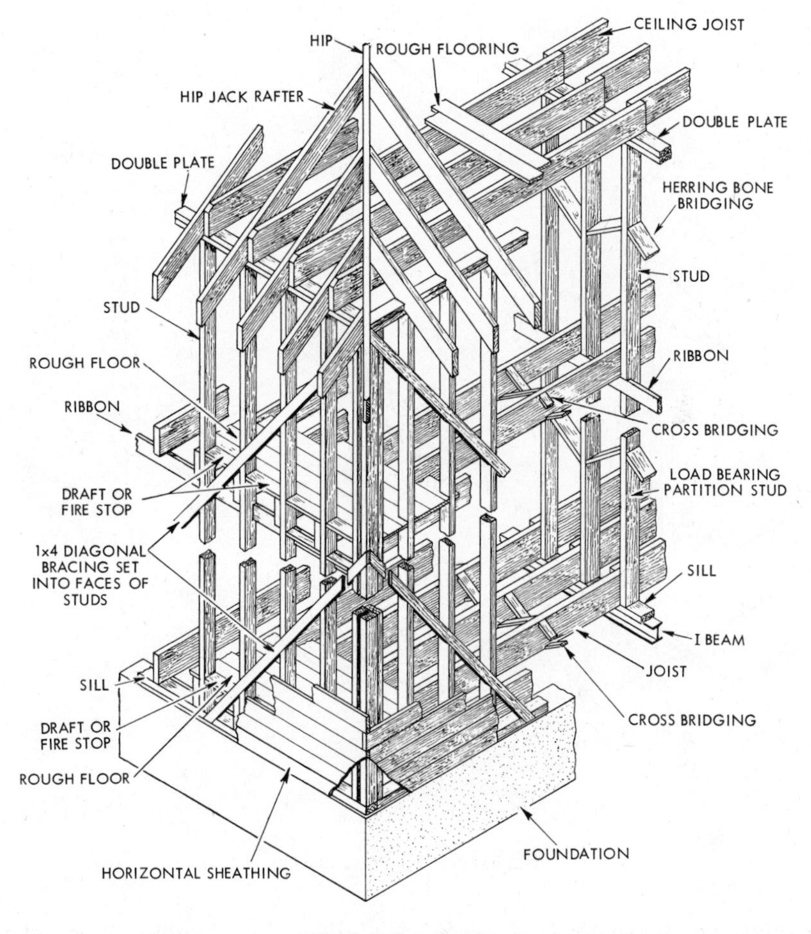

Fig. 5. Balloon framing.

is, load-bearing) members and must be well-seasoned high-grade lumber. Very often these planks will be left exposed and serve as the ceiling for the rooms below. Note that No. 1 common grade lumber is usually the minimum acceptable quality. Carefully check the drawings and specifications for the lumber requirements.

Sheathing. After the frame is constructed it is covered with sheathing. The sheathing helps to insulate the building while providing a base for siding, brick, and exterior trim. See Fig. 7.

Lumber (usually tongued-and-grooved or matched), fiber insulation board, and plywood are commonly used as material for sheathing.

The most common size of insulation board and plywood is 4′ × 8′ panels. Insulation board is about ¾″ thick

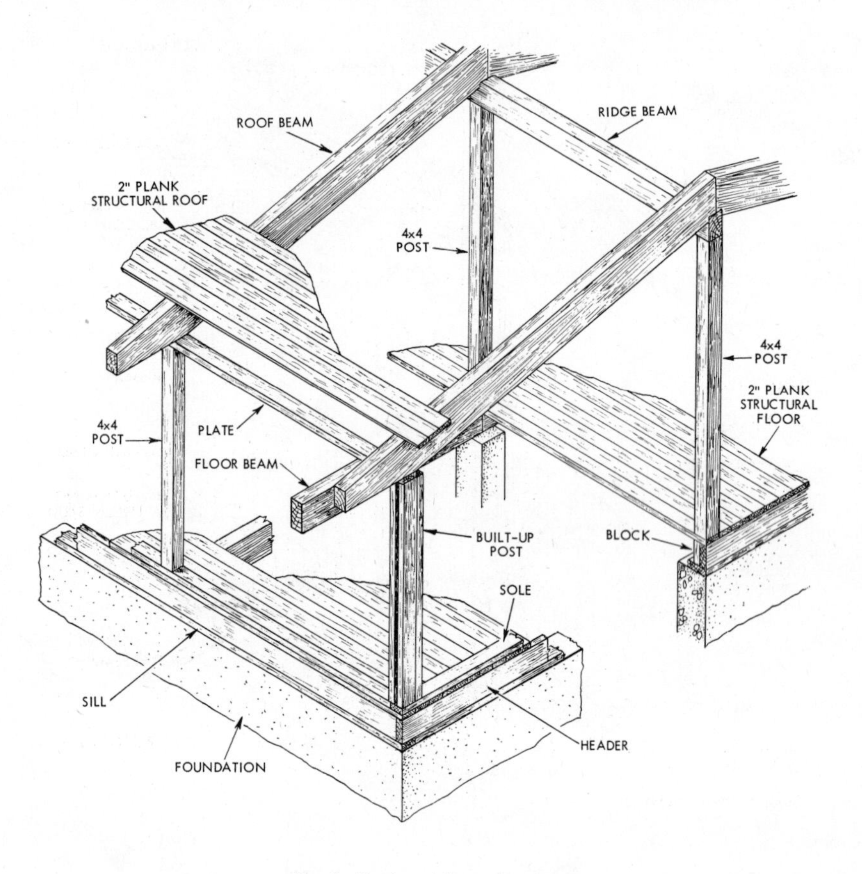

Fig. 6. Plank and beam framing.

and plywood is usually 5⁄16" or ⅜" thick, depending on the stud spacing and load to be supported.

In both the western and balloon frames, the sheathing is nailed directly to the outside edge of the stud. Either boards or panels may be used depending on the specifications and local building codes.

Building paper or *felt paper* is customarily nailed to the sheathing and the exterior finish (such as shingles or siding) is nailed over the paper, Fig. 7.

Siding. Fig. 7 shows three different types of siding. Wood boards, Fig. 7-*A*, are often used for siding but in recent years metal siding has become more popular. Sizes vary and for this reason the specs should give the exact requirements.

Clapboards are large siding material approximately 6" wide, ½" thick at the butt, and about ⅛" thick at the opposite edge, Fig. 7-*A*.

Wood shingles are shown in Fig. 7-*B*. These are from 20" to 24" long, ½" to 9⁄16" thick at the butt, and of

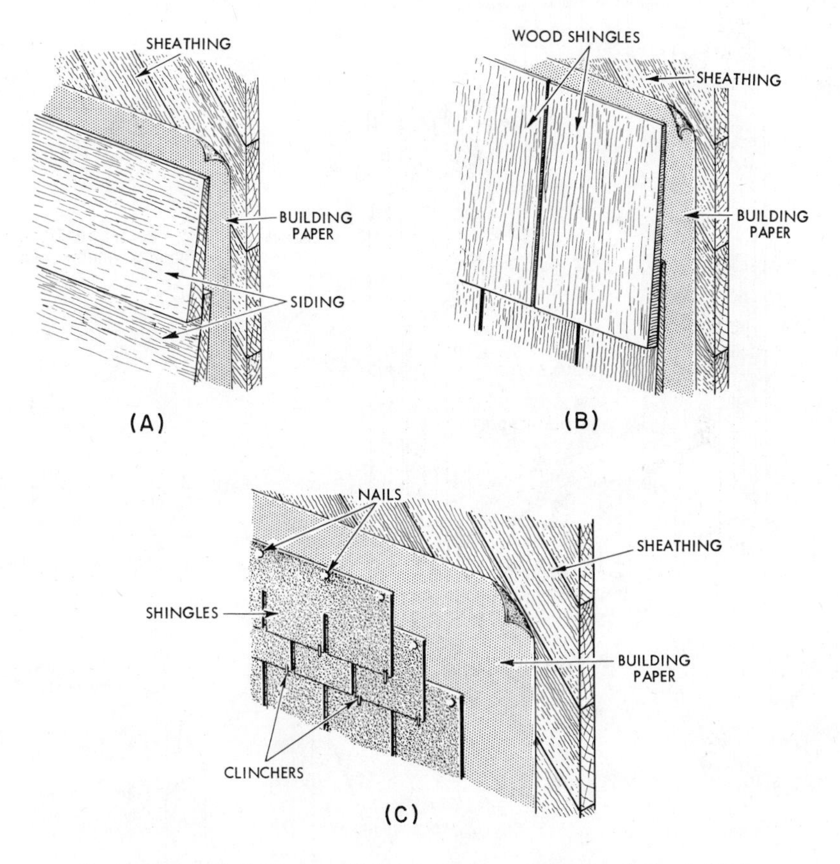

Fig. 7. Siding: (A) Clapboards, (B) Wood shingles, and (C) Manufactured shingles.

negligible thickness at the opposite edge. Some roofing shingles are of random widths varying from 2½″ to 14″ or 16″. Dimension shingles (for walls) are sawed to uniform lengths of 4″, 5″, or 6″.

Asbestos, cork, slate, and other varieties of manufactured shingles, Fig. 7-*C,* vary in size. The specifications should either state the size or give enough information so that the size can be accurately determined. This is important for the labor estimate—the sizes of the shingles will affect the cost.

Bracing. Specifications may require bracing for some frames. Figs. 4 and 5 show one kind of bracing. These are 1 × 4's which are nailed between the studs after being cut and beveled. Other stud bracing can consist of horizontal pieces of 2 × 4 between studs. Still other forms require the studs to be notched to permit 1 × 6 boards to be placed diagonally.

Walls and Partitions. Figs. 8 and 9 show details of outside wall framing. Figs. 10 and 11 show details of interior wall

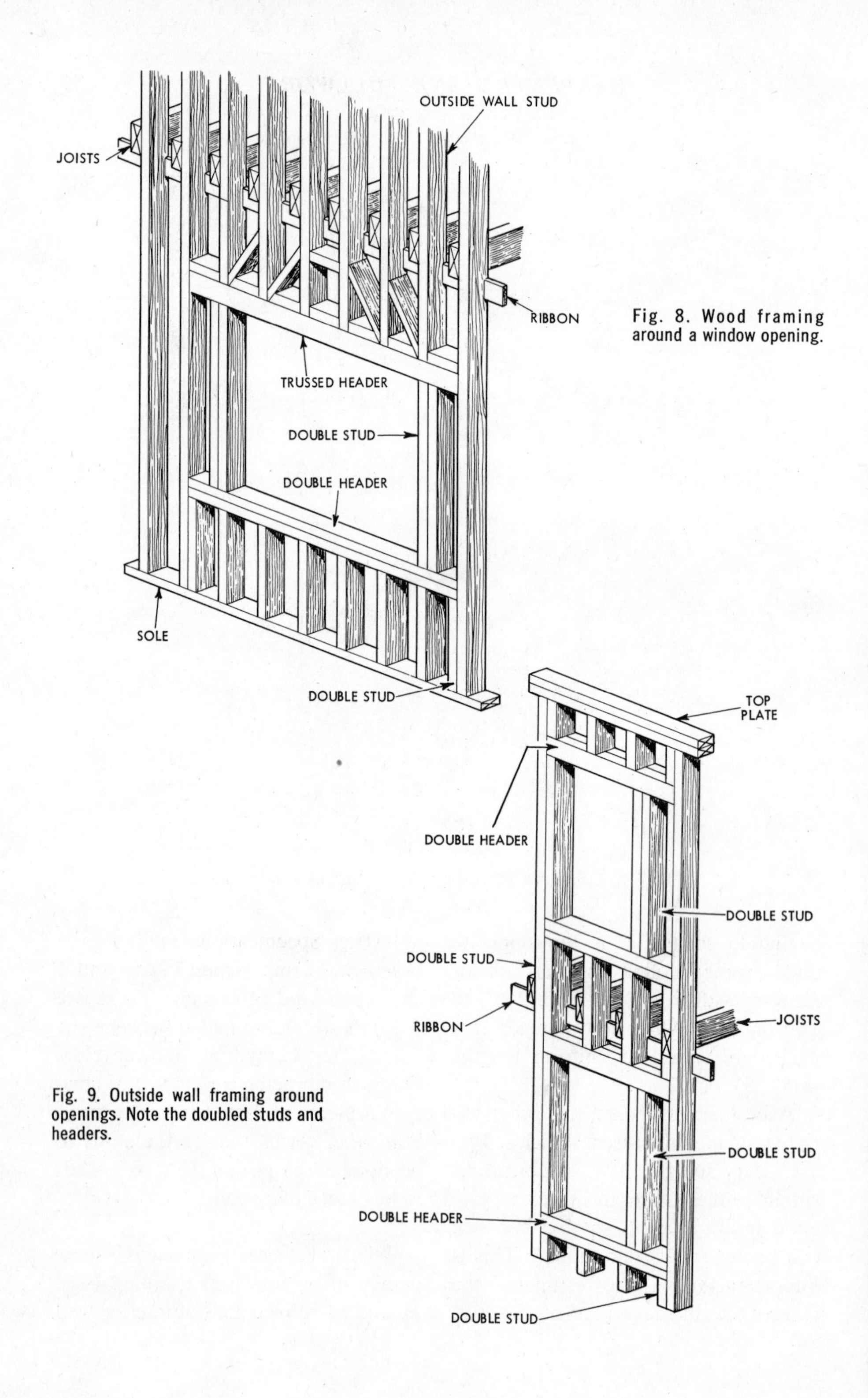

Fig. 8. Wood framing around a window opening.

Fig. 9. Outside wall framing around openings. Note the doubled studs and headers.

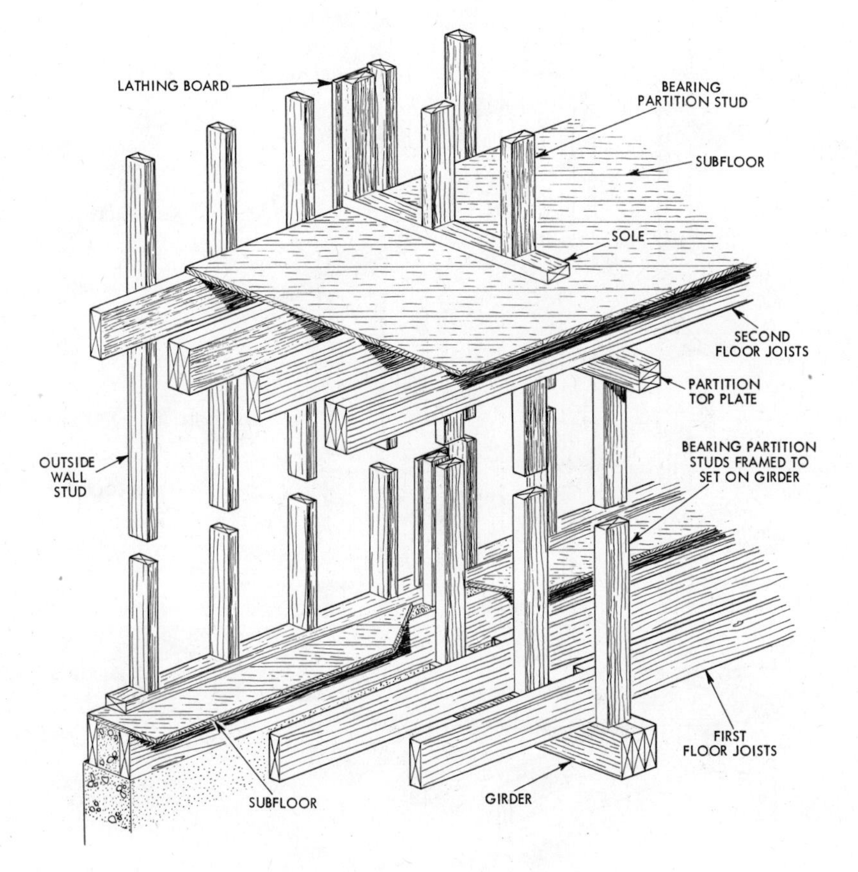

Fig. 10. Interior wall and partition framing.

framing. The soles are nailed to the rough floor and the studs are toe-nailed to the sole.

Interior Framing. In frame buildings the interior walls or partitions are framed using 2 × 4 vertical studs, usually spaced 16″ o.c. (on centers). Lathing and plaster, or dry wall covering is then applied.

Sometimes an interior wall is framed with 2 × 6 studs in order to have more space between the walls. (See Fig. 21-*B*.) This is required for walls which have soil pipes or ducts in them, or where extra wall strength is desired. In any case, a wall should have a sole plate or else the stud should extend down to the girder as in Fig. 10. If no girder comes under a bearing wall or partition, the joists directly under the wall or partition are doubled, as shown in Fig. 11.

Insulated Walls. Insulation is necessary for both comfort and to keep

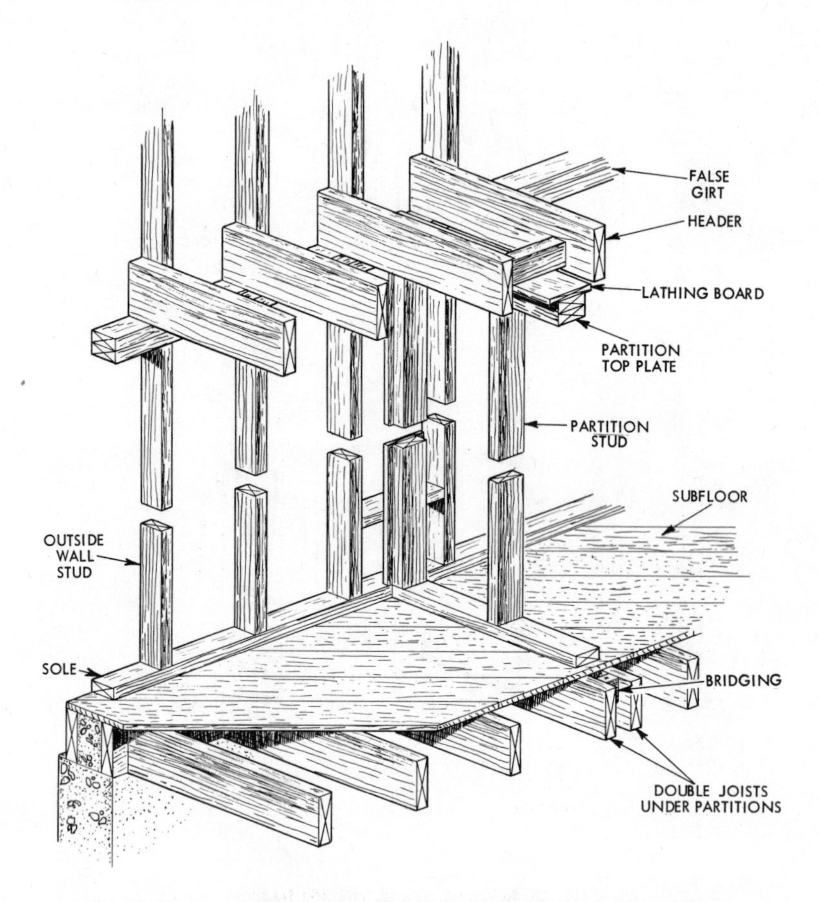

Fig. 11. Interior partition framing. The joists are doubled under the partition.

fuel costs reasonable. Many insulation materials are also used as structural members.

Fig. 12 illustrates common insulating practice. *Rigid* insulation is substituted for sheathing and lathing. The insulation is delivered to the job in sheets of various sizes. These sheets are nailed directly to the studs and provide the same rigidity as wood sheathing. Plaster is applied directly to the insulation. Sometimes only sheathing insulation is specified, other times only plaster backing. The sheets can be installed rapidly and are easily handled. The air space between the bricks and sheathing also helps to insulate the house.

The undersides of the first-floor joists (Fig. 12) may have rigid insulation applied to them. This produces a desirable basement ceiling and reduces heat transmission between the basement and the first floor.

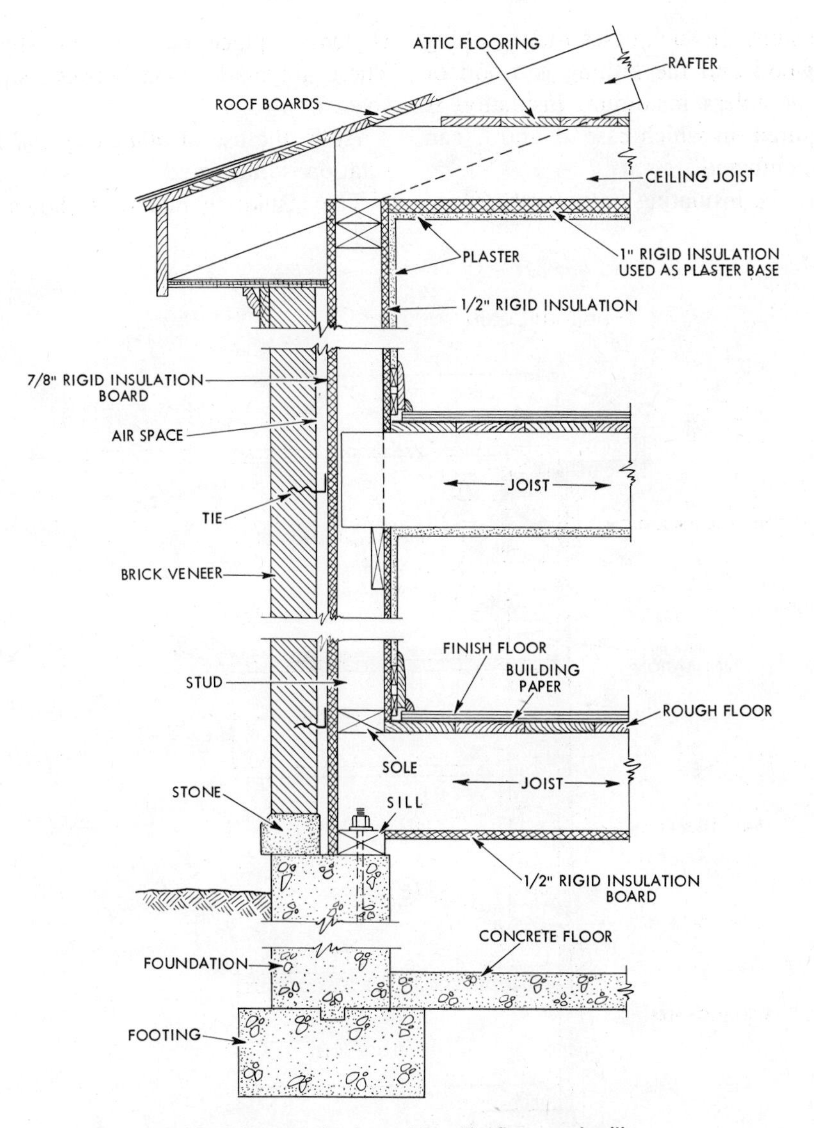

Fig. 12. Rigid insulation is used in walls, floors, and ceilings.

A full frame wall, such as Fig. 13, can be insulated in many ways. At *A* is the same type of rigid insulation as in Fig. 12. In this case the exterior sheathing is nailed directly to the sheet insulation.

At *B,* between attic joists, *loose fill* insulation is used as a method of protecting the second-floor rooms from the varying summer and winter temperatures of the attic.

At *C,* loose fill is shown between

the studs. In such cases the sheathing is wood and the lathing is wood or metal unless maximum insulation is required, in which case *A* and *C* can be combined.

At *D*, insulating *batts* (or *bats*) are shown in place between the studs. These are made to fit between studs spaced 16″ o.c.

At *E*, the use of *aluminum foil* insulation is illustrated.

The insulating materials shown in

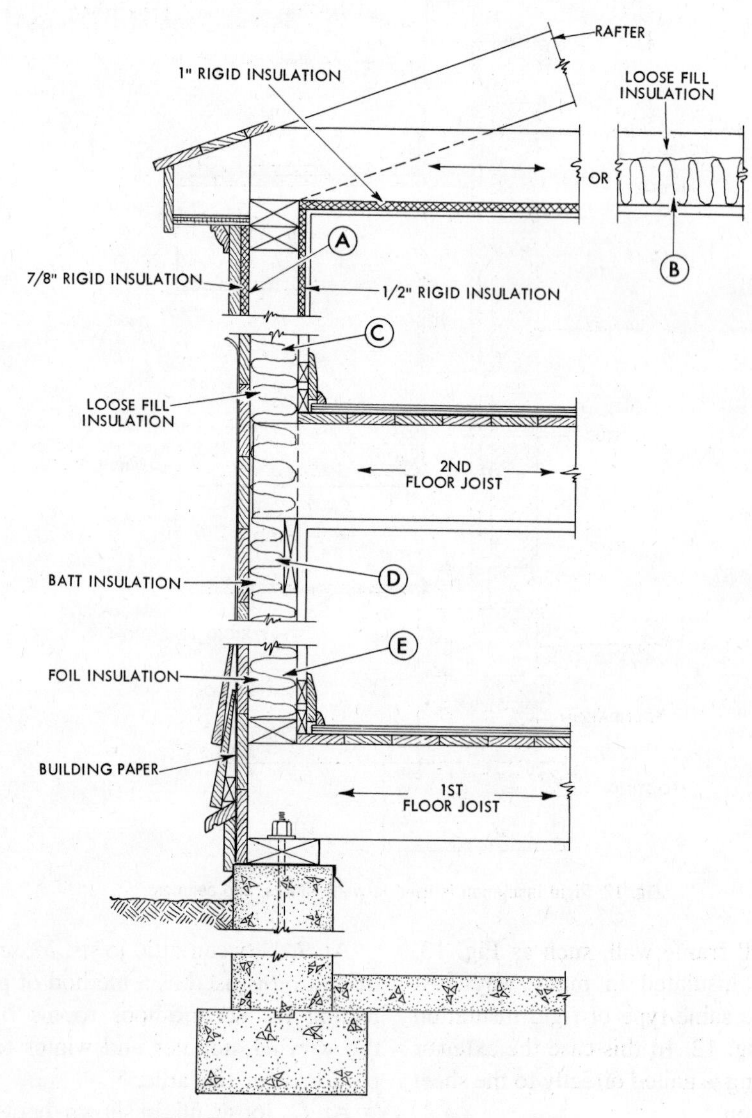

Fig. 13. Various types of insulation in a frame wall.

Fig. 13 can be applied to stucco and stone veneer walls as well as brick veneer and all frame walls. Many different combinations of insulating materials are possible.

Solid brick walls, Fig. 14, can be insulated by using plaster backing as shown at *A*, by plastic insulation placed in the air space as shown at *F*, or by various blanket forms of insulation as shown at *E*. The insulation shown at *B* serves to insulate the floor. At *C*, the insulation provides a finished surface for the foundation wall and lessens the transmission of heat between the basement and the concrete. At *D*, aluminum foil is shown between joists.

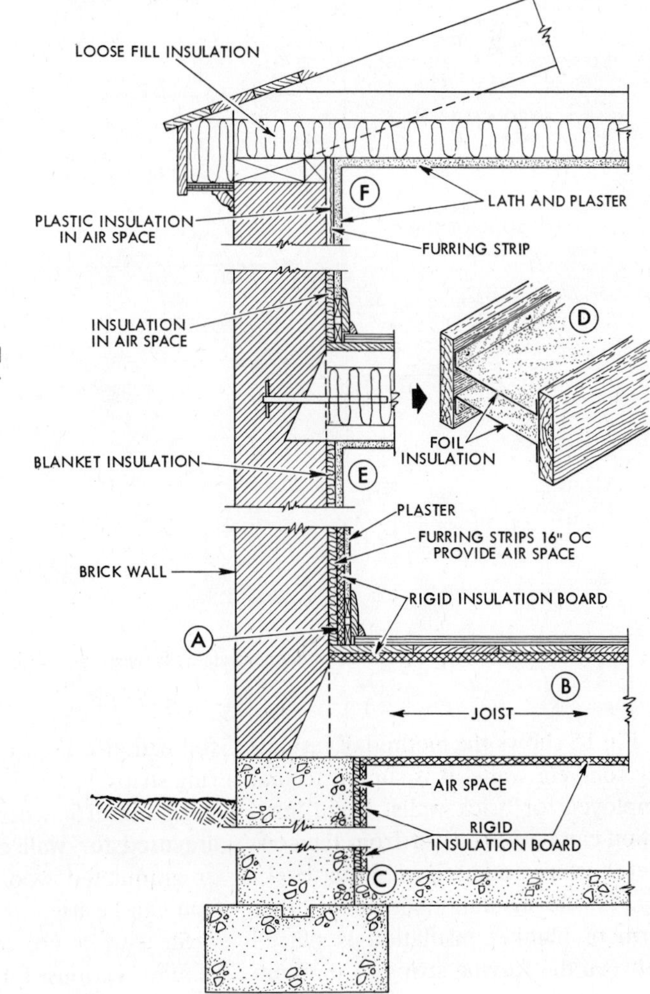

Fig. 14. A solid brick wall with various kinds of insulation.

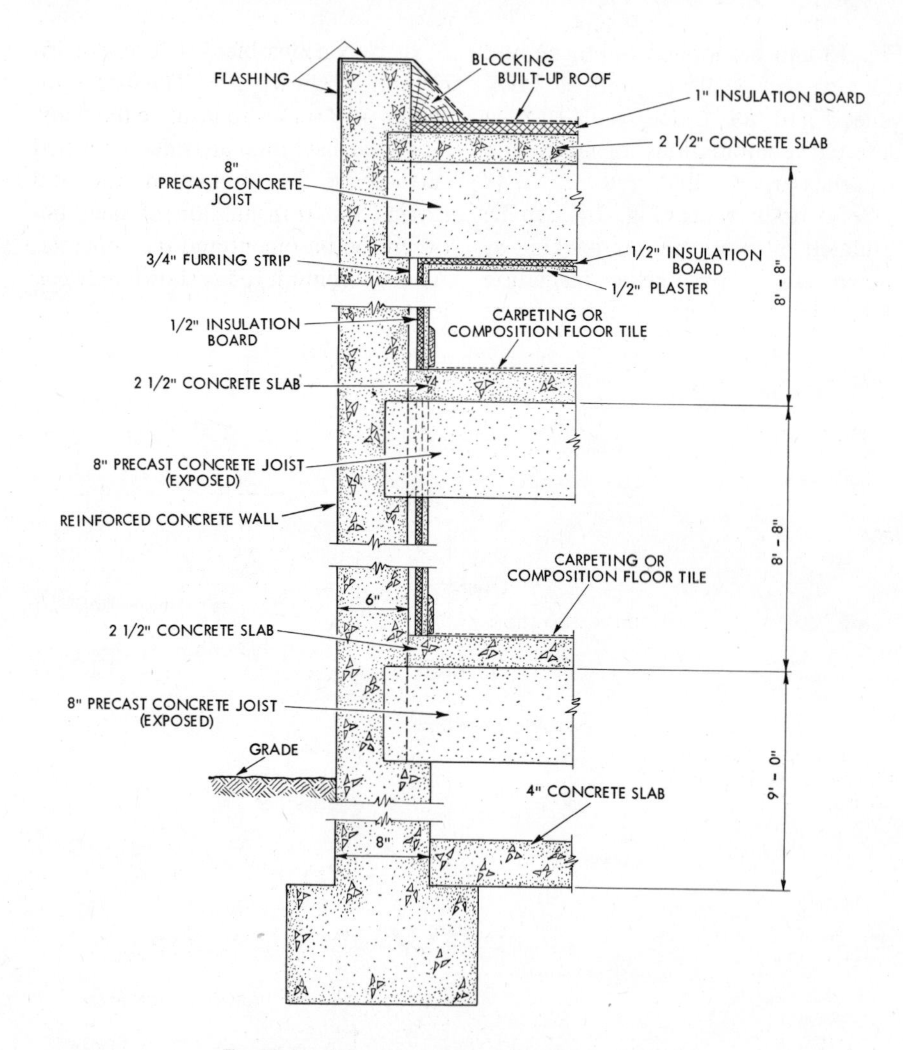

Fig. 15. Insulation for concrete walls, joists, and slabs.

Fig 15 shows the method of insulating concrete walls. It is similar to that employed for brick walls. Rigid insulation can be *furred* out from the concrete surface and used as plaster backing. Plastic insulation or some other form of blanket insulation is placed between the furring strips. Aluminum foil can also be used between the furring strips.

In Fig. 16, where concrete blocks are used for wall construction, loose or granulated wool and rigid insulation can be used as shown.

Steel walls are of many types and require various forms of insulation.

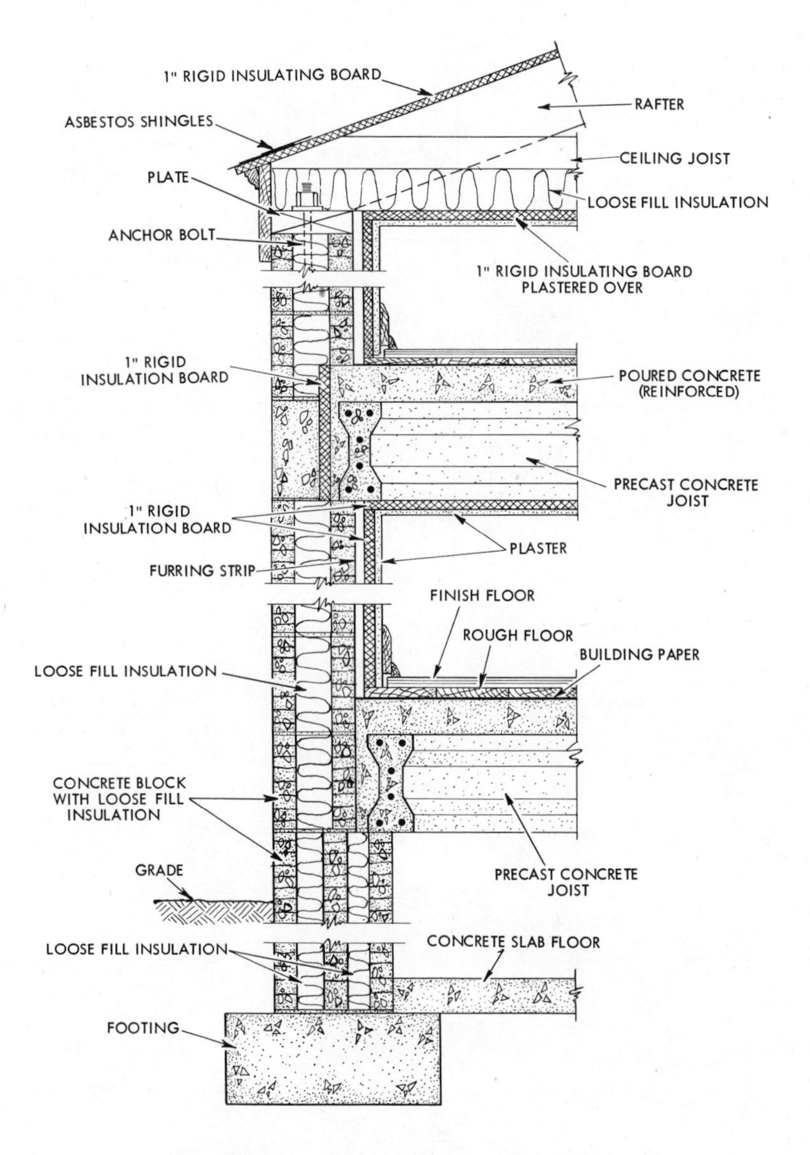

Fig. 16. Insulation for concrete block walls.

The wall in Fig. 17 is relatively thick, permitting the use of rock wool or similar types of insulating material. Rigid insulation may be necessary for other types of steel walls. The estimator should therefore study the wall section drawings to better visualize the areas or volumes which the insulation will occupy.

Basement Partitions. These are gen-

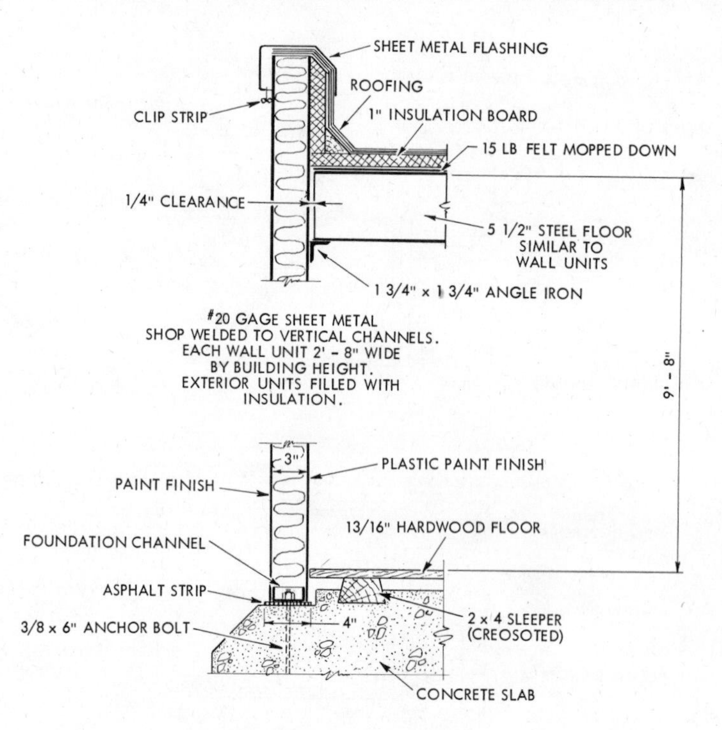

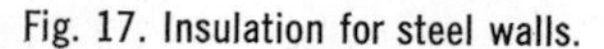
Fig. 17. Insulation for steel walls.

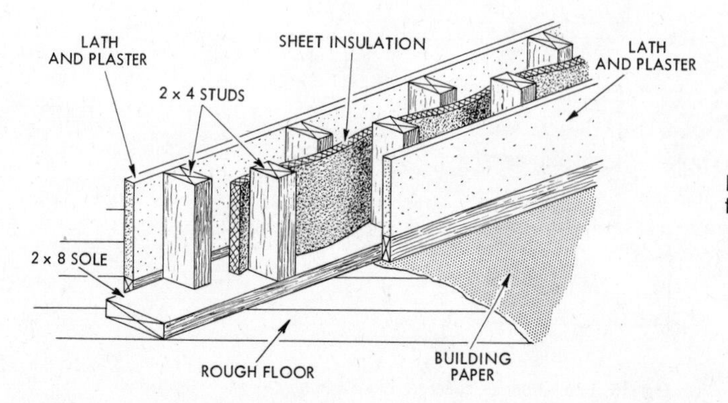

Fig. 18. Soundproofing a frame wall.

erally of wood or brick with wood being the most popular. Where partitions must serve as a support for first-floor loads, the partitions can be constructed economically of 2 × 4's and sheathing (or matched boards).

For basement recreation rooms the partitions are built of 2 × 4's and then plastered or covered with plasterboard on one or both sides. Furring is used if plaster is to be applied to foundation walls. Concrete, concrete blocks, and tile are often used depending on the other structural features of the building. In any case, the symbols or written instructions should plainly indicate wall specifications and details.

Soundproofing. Soundproofing of the type shown in Fig. 18 requires double the usual number of studs, with sheet insulation between rows, and almost double the usual amount of labor is required.

Floor Framing. In frame buildings, the floor joists are wood members specified as 2 × 8's, 2 × 10's, 2 × 12's, etc., spaced at intervals of 12″, 16″, or 24″ o.c. The lumber is sold in lengths which are multiples of two—such as 6′, 8′, 10′, and 12′. A good designer always plans in such a way as to use as many short length joists as possible. The length of lumber is an important item to the estimator.

Fig. 19 shows a girder being used as a joist support. Girders are made of concrete, I-beams, solid timber, or several 2 × 8's, 2 × 10's, etc., spiked together. The joists can rest directly upon the girder, or they can bear on a 2 × 6 as shown in the illustration.

Joists can be fitted between the flanges of an I-beam girder as shown in Fig. 20, in which case the estimator must consider such additional items as *iron dogs* and *bolts*. The added amounts of material and labor of the I-beam method over the method shown in Fig. 19 must also be considered.

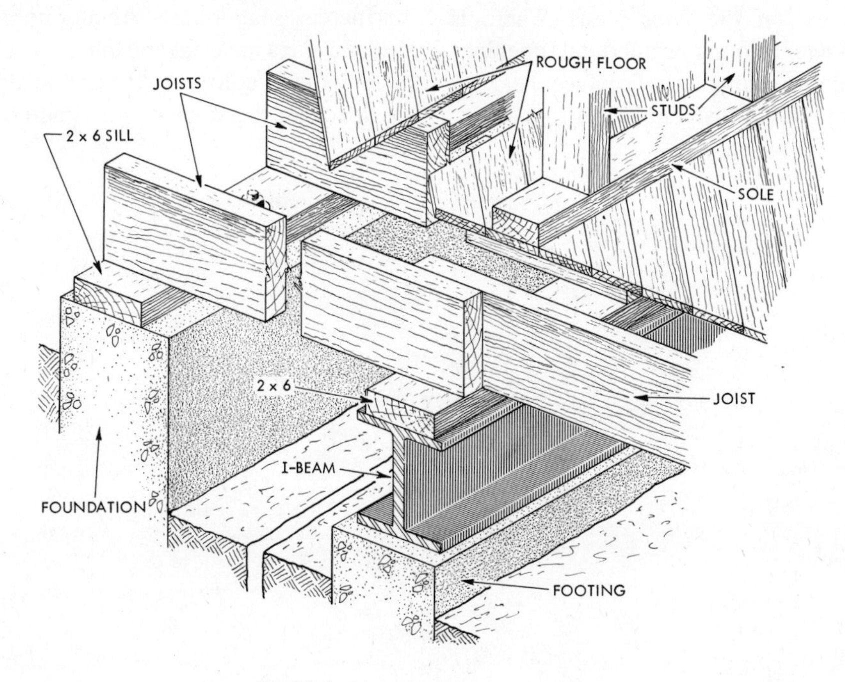

Fig. 19. An I-beam used to support joists.

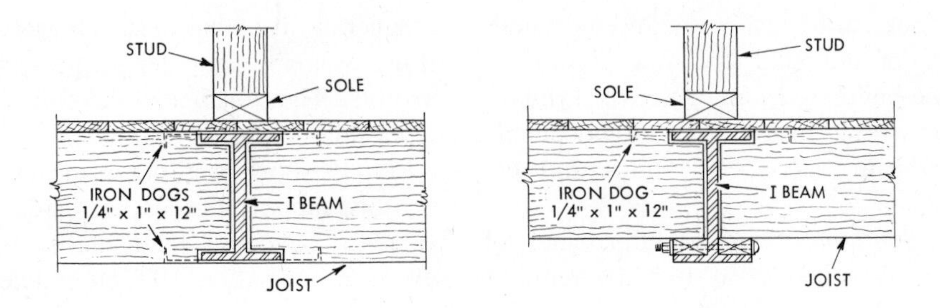

Fig. 20. Wood joists fitted between the flanges of an I-beam.

Sometimes the architect will supply framing plans, in which case the estimator can easily count the number of joists and note the lengths, sizes, and spacing. If framing plans are not supplied, the estimator must determine the sizes and spaces from the specifications and then calculate the amount needed according to the dimensions shown on the floor plans. There is *double* framing around all openings such as chimneys, fireplace hearths, and stair wells, Fig. 21-*A*.

Under bathrooms the joists are usually doubled or spaced much closer together because of the weight of the bathroom fixtures. Wherever a first- or second-floor wall occurs, the joists under it are doubled. Joists under a bathroom are shown in Fig. 21-*B*.

Where the outside walls are brick, the joists must be beveled at the ends; this increases labor costs. Anchor bolts also add extra material and labor costs.

Sometimes concrete or steel joists are specified together with a concrete

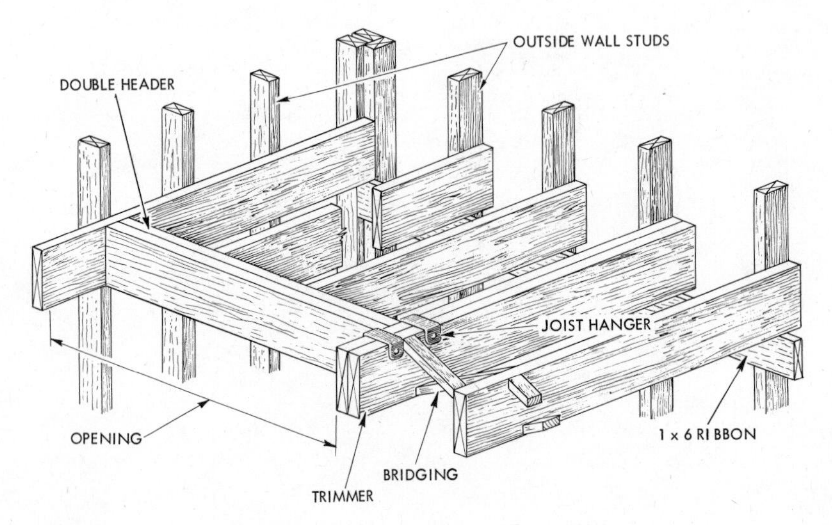

Fig. 21-A. Double framing around a floor opening.

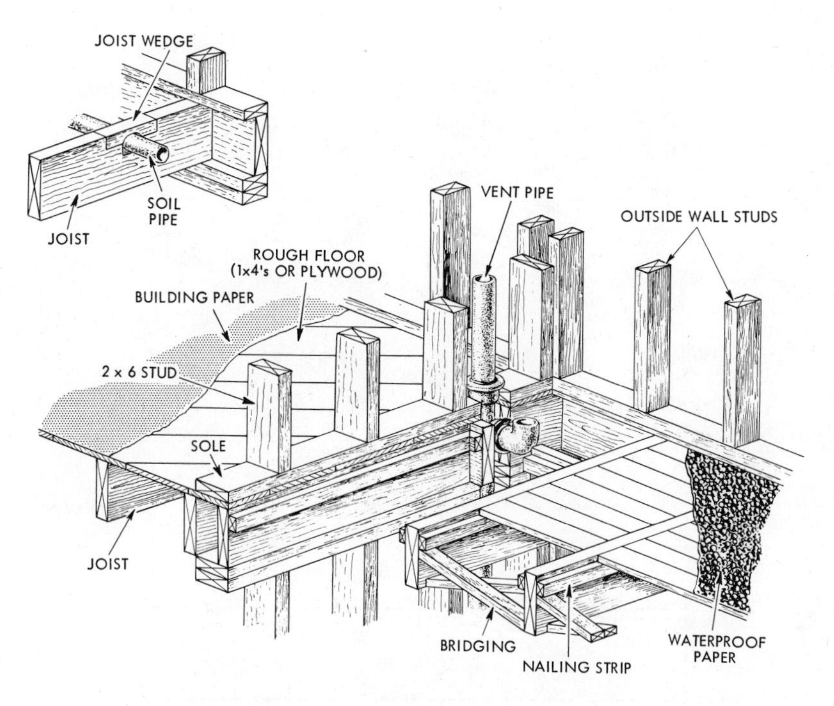

Fig. 21-B. Bathroom joists and wall framing.

floor. In such a case, material and form costs must be studied carefully. If framing plans are supplied, they are used for quantity take-offs.

Wood Floors. The rough flooring is laid over the joists. This flooring generally consists of cheap boards ranging from ¾″ to 1″ thick and from 4″ to 6″ wide. The boards are sometimes tongued-and-grooved. Plywood panels are also used as subflooring.

Fig. 22 is a detailed drawing of joists, bridging, rough floor, and finish floor. Rough flooring is sometimes put on diagonally and nailed directly to the joists.

The finish floors are not laid until the building is practically complete. Finish flooring is laid over the rough flooring after it has been overlaid with building paper. The laying of finish flooring involves considerably more time than rough flooring because more careful matching and fitting is required.

Different kinds and qualities of lumber are used for the finish flooring. Some rooms, such as a kitchen where linoleum is to be laid, are floored with pine or other less expensive lumber. Generally the rooms are floored with oak, maple, or similar wood. If wall-to-wall carpeting is to be laid, the floor can be of pine. The specifications should contain the exact kind and quality of flooring.

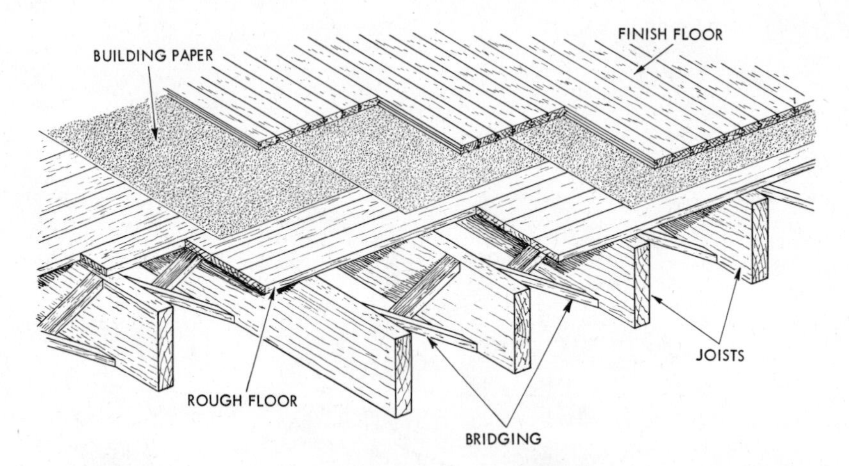

Fig. 22. Typical wood floor construction.

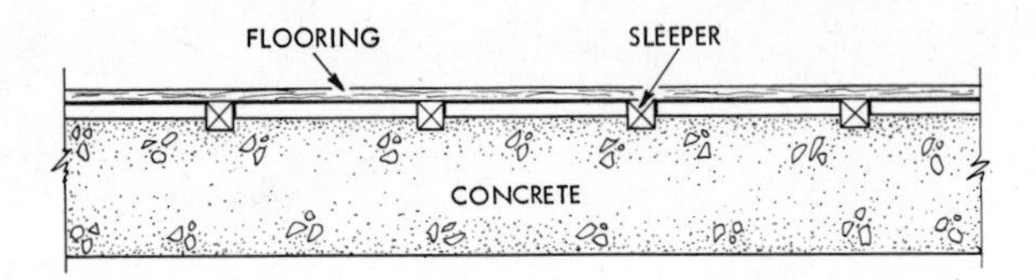

Fig. 23. Wood floors can be secured to concrete slabs by placing sleepers in the wet concrete.

Where concrete floors are indicated, wood *sleepers* must be embedded in the concrete before it sets. See Fig. 23. The sleepers are *creosoted* or otherwise treated to resist deterioration. They extend above the concrete a specified distance and provide an insulating air space.

If wall-to-wall carpeting is laid over concrete floors, wood flooring is not always necessary. The surface of the concrete is ground smooth and a heavy layer of felt is cemented to it. The carpet is laid over the felt.

Sound Insulation. This item adds greatly to both material and labor costs, so it is an important item to the estimator. Fig. 24-*A* is a cross section of an ordinary ceiling-floor without soundproofing. Various methods of soundproofing are shown at *B* through *E*. The most expensive method shown is at *D*; here double the usual number of joists are required.

Doors. If one stud must be eliminated for a door opening in a load-bearing partition, the framing is done as in Fig. 25. Two 2 × 4's are doubled and set on edge over the door opening. They in turn bear on an extra set of 2 × 4's added to the vertical framing. In this

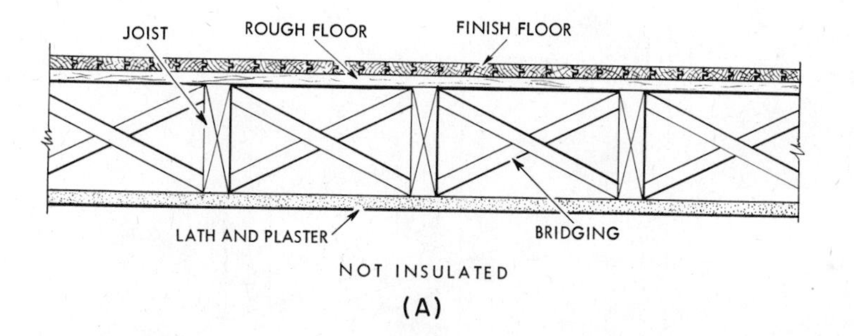

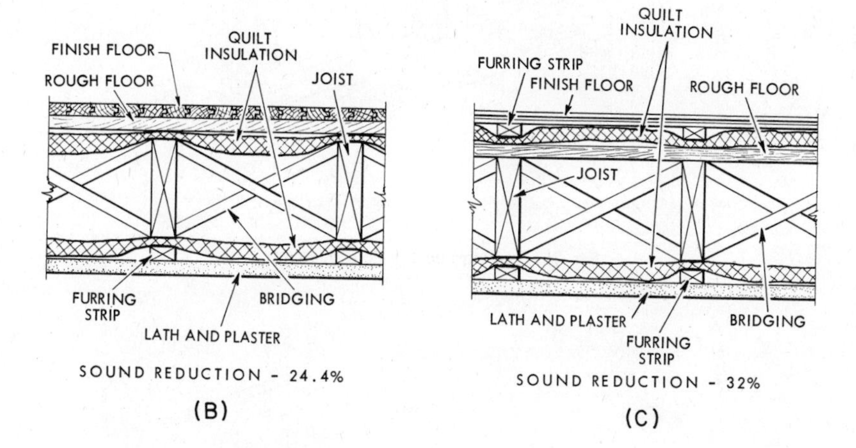

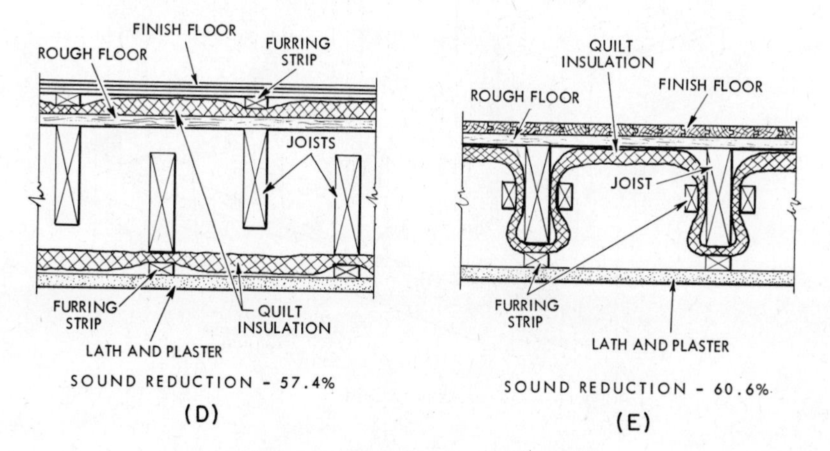

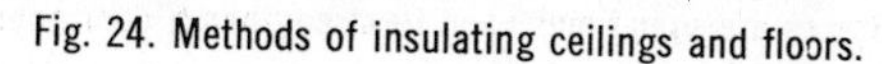
Fig. 24. Methods of insulating ceilings and floors.

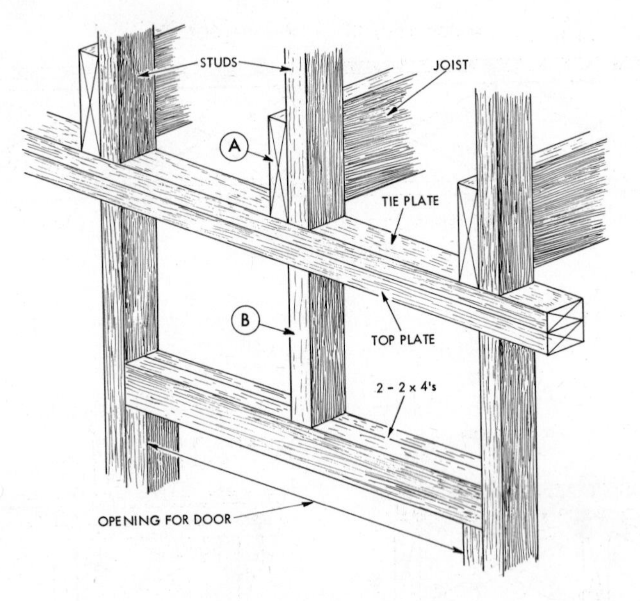

Fig. 25. Framing above a door opening.

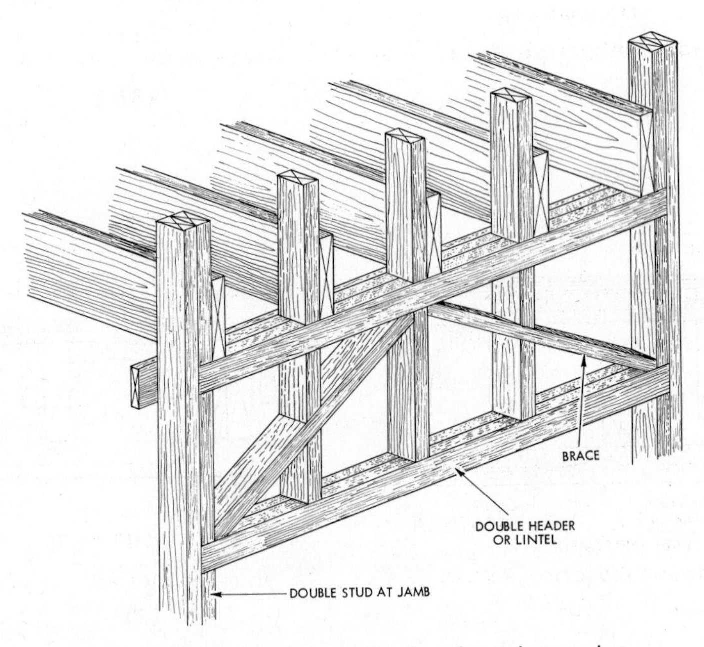

Fig. 26. Braces are sometimes used to strengthen a door opening.

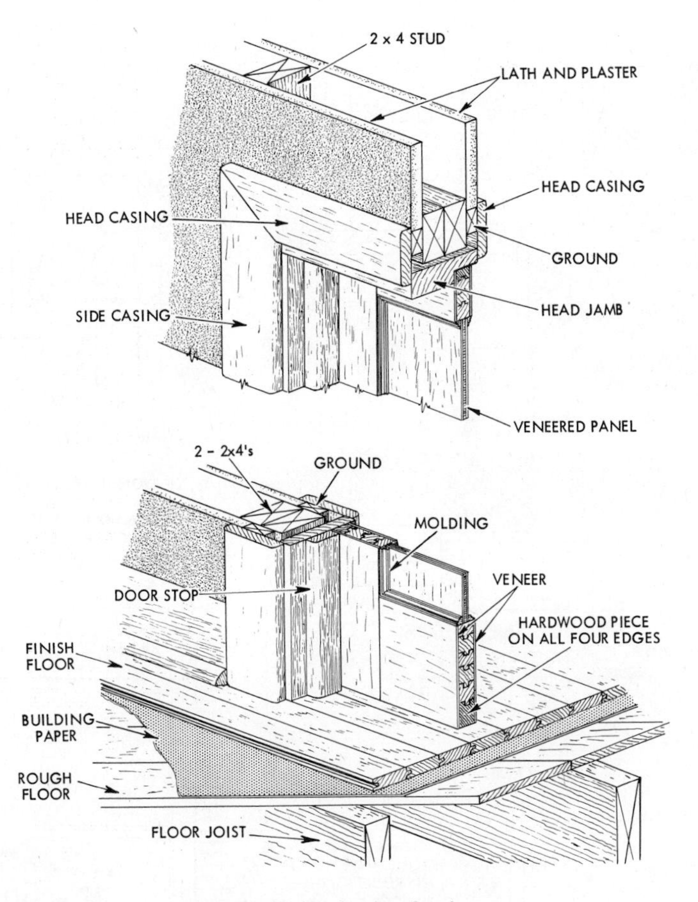

Fig. 27. Interior door framing.

manner the joist *A* and stud *B* have a solid bearing.

Fig. 26 is another method of framing a door opening. The braces strengthen the framework.

Door openings in partitions which are not load-bearing need only one 2 × 4 on either side and one across the top of the opening.

The framing around a door is done at the same time as the wall and partition framing. Fig. 27 shows a typical interior door. The jambs and casings are generally made on the job and fitted into place from straight lumber, cut and shaped for that purpose.

The door itself is delivered to the job ready to be installed. It is therefore considered as millwork. Hinges and locks are separate items and are usually listed as hardware.

Windows. Fig. 28 shows wall framing details around windows. There are many methods of window framing so the estimator should be sure that he

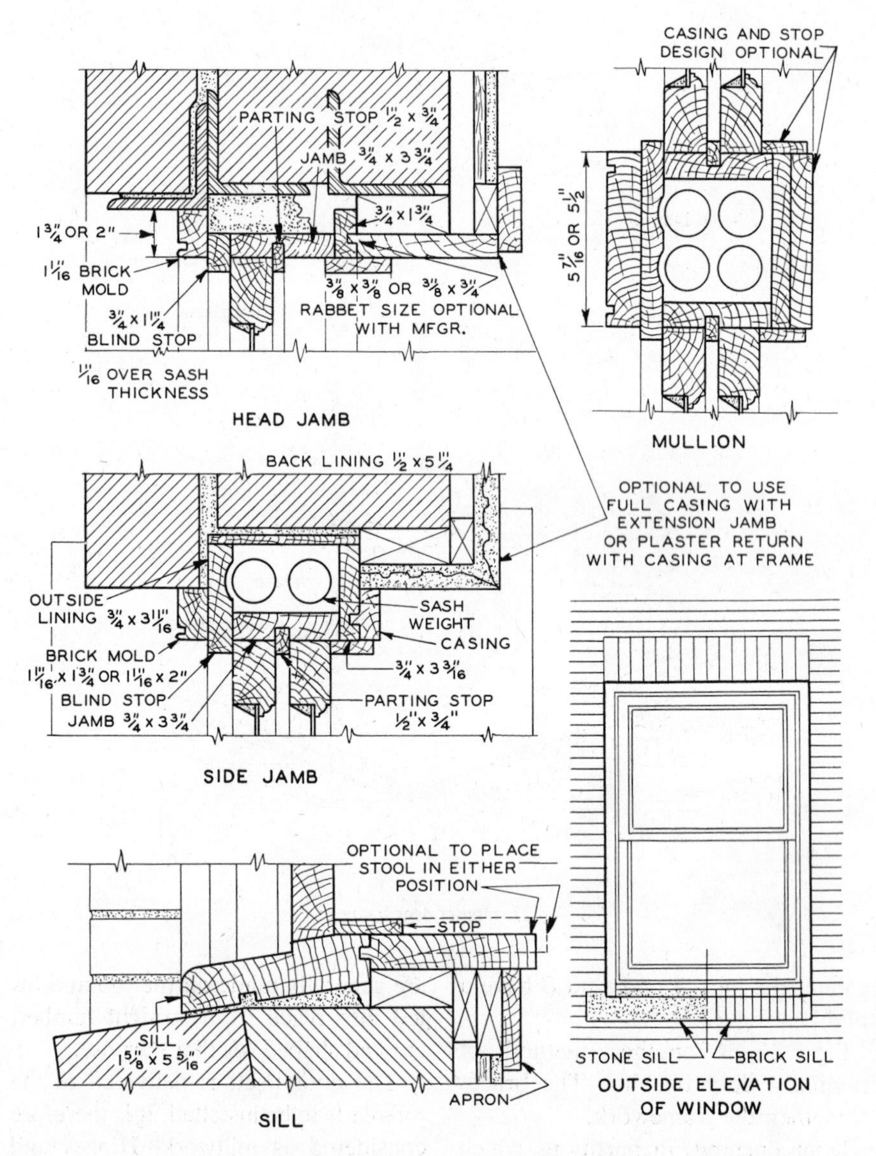

Fig. 28. Window framing details.

knows the method which will be used on the particular job he is estimating. Knowing the various methods will make labor estimates much easier.

Lintels. For both windows and doors in masonry walls, some form of lintel is necessary. Sometimes timbers are used, but generally angle irons set back-to-back are used.

Window frames are rarely made on

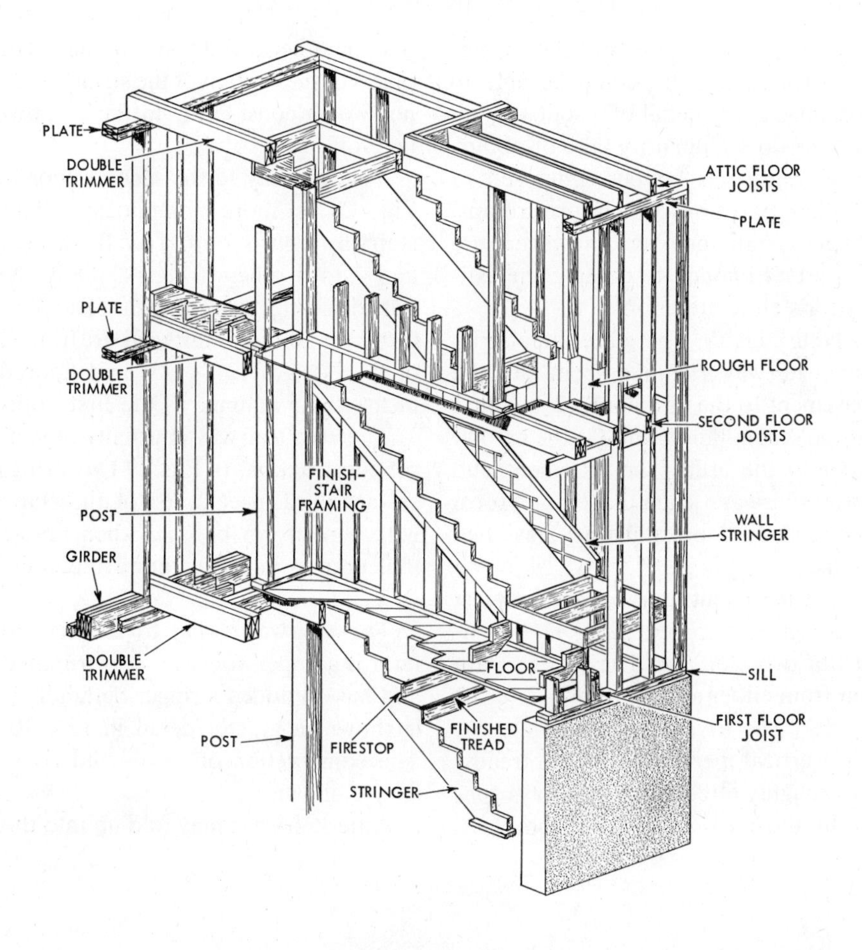

Fig. 29. Stairway framing.

the job. They are purchased complete from mills and delivered partially or completely assembled. The labor cost occurs in the placing of the frame in position in the wall. Window trim is purchased separately and cut and fitted on the job.

Any pulleys, ropes, sashweights, etc., which are to be installed, are usually estimated as hardware. (Note: These items have been generally replaced by friction holding devices.)

Window frames for masonry walls are slightly different in construction, but like windows for frame walls, they are delivered to the job ready to be installed. The sash and glass are delivered to the job ready to be installed in the frame. Quantity take-offs for glass are listed separately.

Stairs. This is a difficult item of construction for the inexperienced estimator to visualize, unless he has prior experience with stairs and their framing. The estimator should be able to visualize every detail of a stairway before he does a quantity take-off. Stairway framing is largely composed of walls or partitions which form rooms. Typical stair members, however, are important enough to deserve the estimator's close attention.

Note Fig. 29. There are three separate stairways leading from (1) the basement to the first floor, (2) the first to the second floor, and (3) the second floor to the attic. The basement and attic stairways each have one turn; the second-floor stairway has two turns.

The basement stairs are the simplest in design and construction. They consist of two *stringers*. The stringers are cut from either a 2 × 10 or a 2 × 12 timber. The *treads* (rests) and the *risers* (vertical members between treads) are roughly fitted. Such stairways generally have a railing made from 2 × 4's or other available lumber. The plates at the bottom of the stringers are not worth considering in the quantity take-off.

The stairway to the second floor in Fig. 29 is more complicated. This stairway consists of a wall stringer and 2 to 4 other stringers. (Only the outside stringers are shown.) The platforms for all three stairways are framed of 2 × 4's or 2 × 8's and floored in the same manner as the first floor.

The attic stairway is usually of simple construction. In Fig. 29 two stringers are used in each leg of the stairs with a platform between them. Note that double joists surround the stairwell.

The stairway going to the second floor is an open stairway; it is trimmed and has an added stringer detail. This is shown in greater detail in Fig. 30. The construction of treads and risers is also shown.

Attic stairways may fold up into the

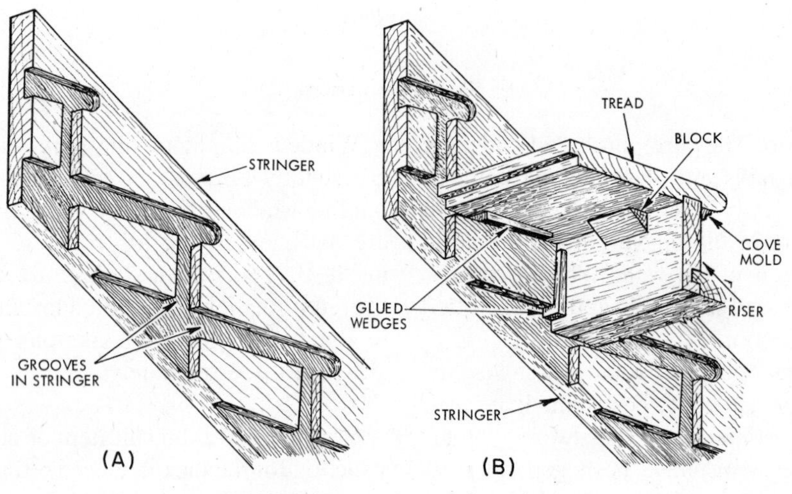

Fig. 30. Stringers and parts of a stairway.

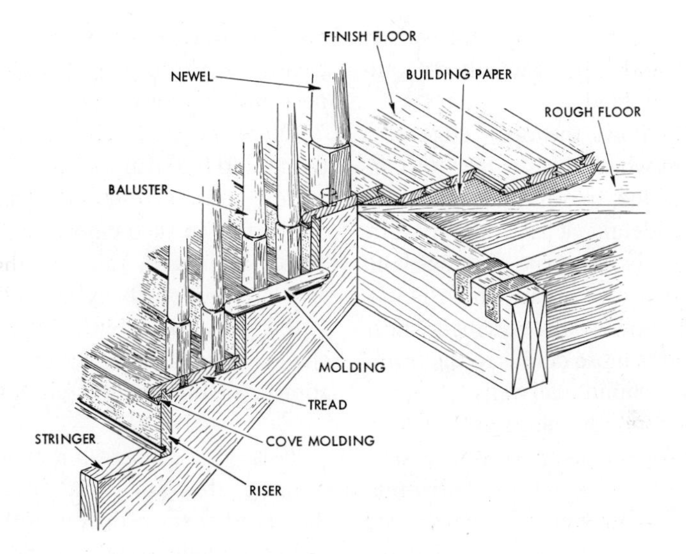

Fig. 31. The open edge of an inside stairway.

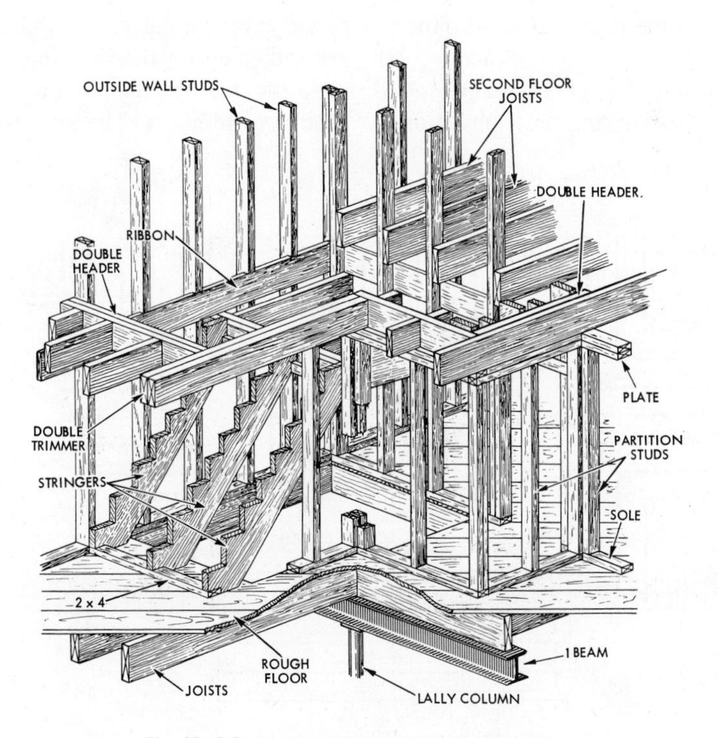

Fig. 32. A 3-stringer stairway and stairwell details.

attic floor. Such stairways are usually mill made and either partially or completely assembled.

Fig. 31 shows the open edge of a stairway in which the stringer (or carriage) and the treads and risers are shown in detail. The trim for the stringer is also shown. Railings and balusters can vary greatly in design.

Fig. 32 shows a three-stringer stairway and adds more details to the framing in and around stairwells.

The working drawings will contain timber dimensions—length and size. The *run* of a stairway is the horizontal projection. Knowing this length and the *rise* (distance between landings or height of each leg), the stringer lengths can be approximated. The timber size depends on the riser and tread dimensions.

Dimensions for platforms are also given on the working drawings. Lumber for risers and basement and attic treads is generally specified as to thickness and kind of wood.

Often the contractor does no more than build the stringers (for main stairways) and surrounding frame and platforms. In such cases, the stairs together with trim, railings, and balusters are supplied by a mill, ready to be installed. Generally, only the cheaper residential designs call for contractor-built main stairways.

Roofs. Fig. 33 shows a typical roof framing diagram. The *rafters* are shown in their proper positions and from such a diagram the estimator can easily count the number of rafters, *ridge pieces,* etc. The lengths of the pieces must be calculated unless they are indicated on the drawing. Generally, the size of the rafters and the spacings are given. The specifications

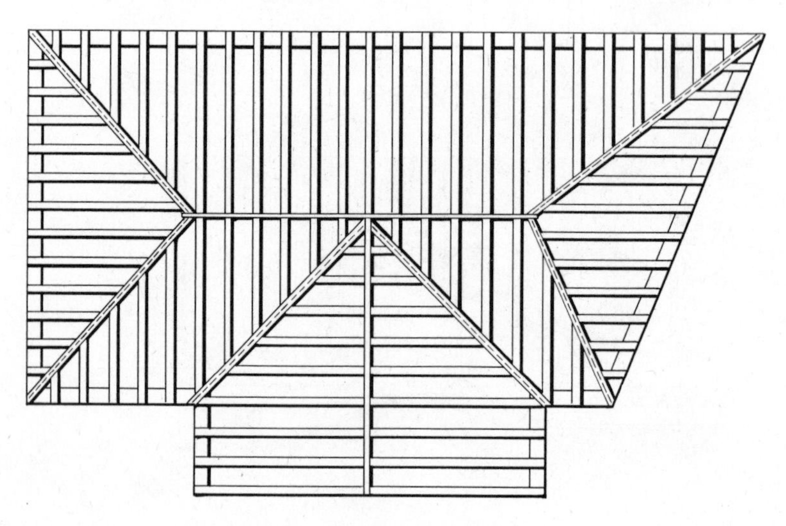

Fig. 33. A typical roof framing diagram.

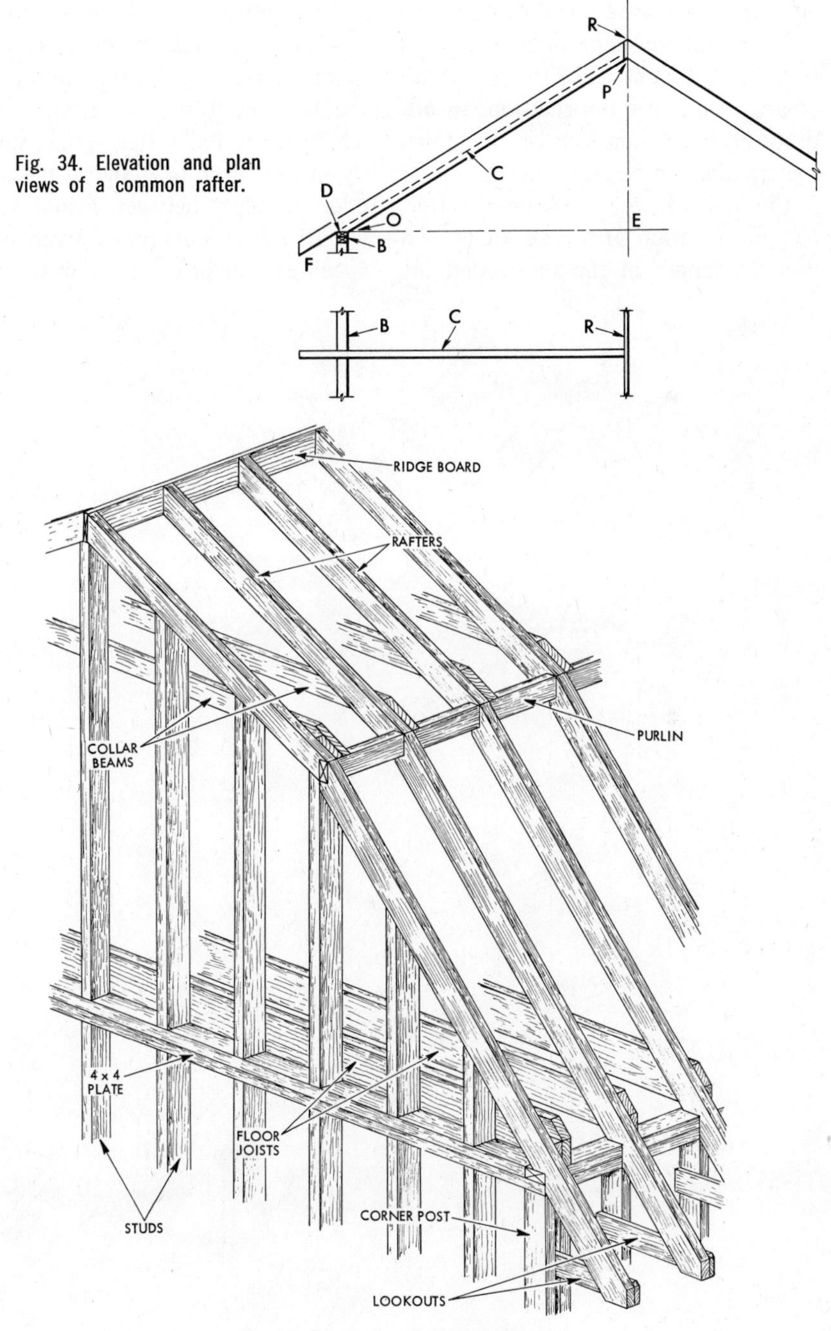

Fig. 34. Elevation and plan views of a common rafter.

Fig. 35. Gambrel roof construction.

or details should indicate the *pitch* or the rise and run. The pitch is the ratio of roof height (rise) to roof width (span). With this information, in addition to the dimensions on the plans, the lengths can be computed.

Fig. 34 shows a common rafter, *FP,* in elevation and plan views. To find the length of timber needed for such a rafter, the distances from *F* to *D* and from *D* to *P* must be calculated. The run of the rafter *FP* is the horizontal distance from *F* to *E*. The rise of the rafter is the vertical distance from *E* to *P,* that is the difference in height between *F* and *P*. The length *FP* is sometimes given in the specifications or on the plan or eleva-

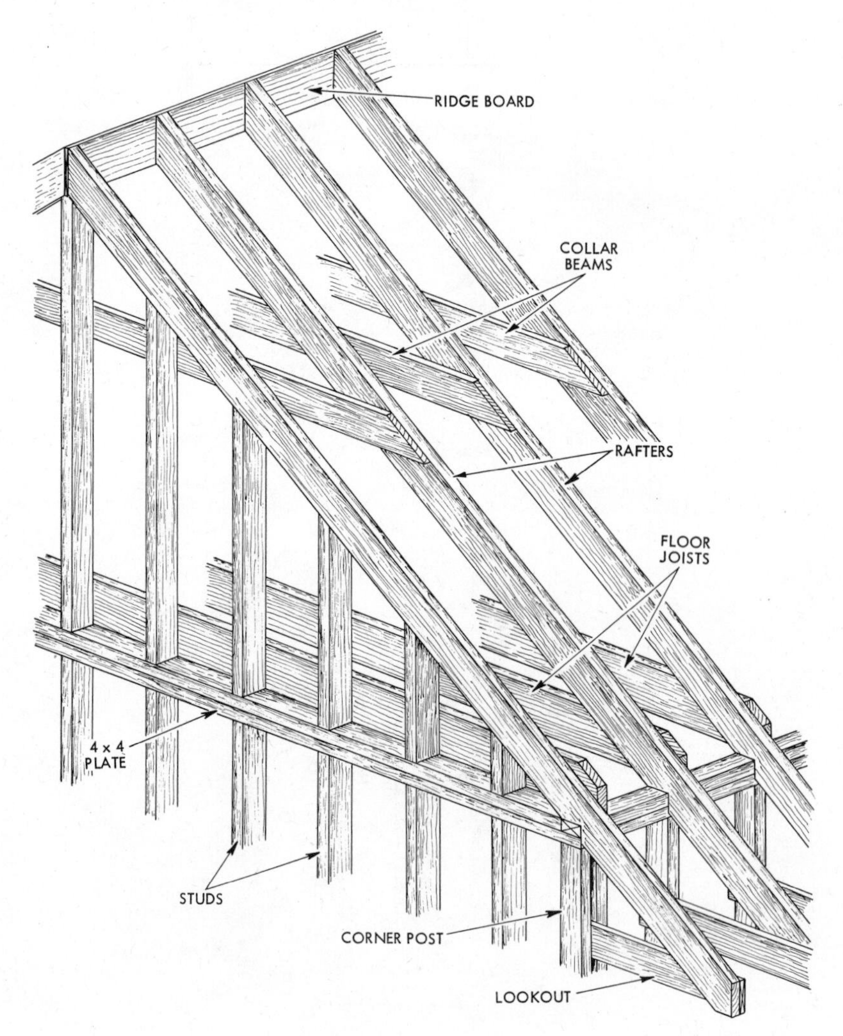

Fig. 36. Gable roof construction.

tion drawings. Timbers or sticks should be somewhat longer than the actual calculated length to allow for the bevels such as at *RP*.

Figs. 35 and 36 show two common roof types. From these illustrations you can visualize the necessary roof construction materials.

For a gambrel roof (Fig. 35) two lengths of rafters must be calculated.

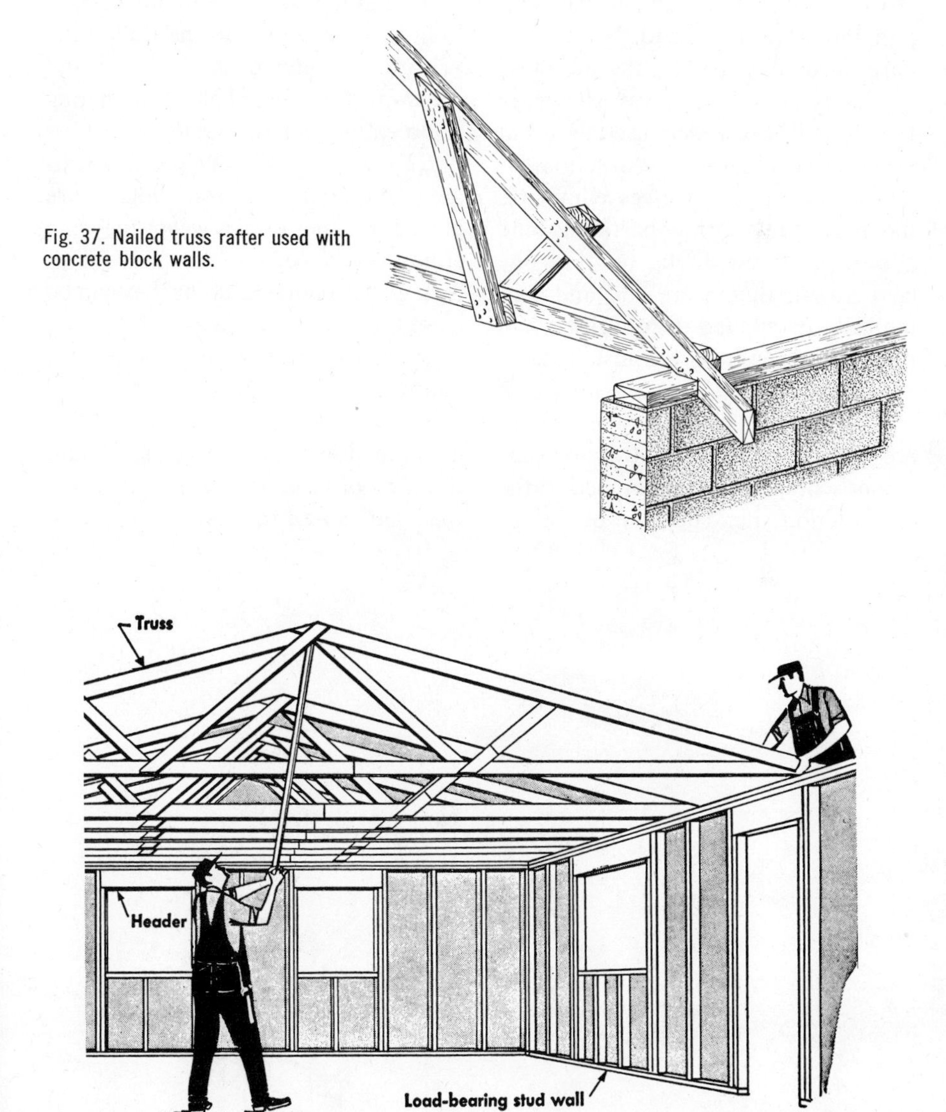

Fig. 37. Nailed truss rafter used with concrete block walls.

Fig. 38. Prefabricated trusses are swung into place, aligned, and fastened quickly. University of Illinois, Small Homes Council.

The *ridge pieces* and *purlins* may be 1 × 6 or 1 × 8 boards or they may be 2 × 6 or 2 × 8 timbers.

Fig. 37 shows nailed truss rafters used as roof framing for a concrete block-wall house. The block walls support the entire roof load. No bearing partitions are needed for the structure.

The type of roof truss shown in Fig. 37 has been widely used in recent years for small house construction.

The use of roof trusses eliminates the need for bearing partitions and allows the house to be built as one large room. Floors are laid and ceiling hung or applied from wall to wall before the partitions are installed.

Fig. 38 shows the ease with which trusses can be installed. The trusses are usually prefabricated at the carpenter's shop and then shipped to the construction site. They can be assembled in several ways. The simplest are nailed together; the larger, heavier trusses are bolted together. Many use a combination of nails and glue while others are simply glued.

Special hardware is often necessary to attach the trusses to the wall plate. A careful study of all drawings and specs must be made before estimating a job calling for roof trusses.

Roof Boards. Fig. 39 shows a simple gable roof. The roof boards are placed in either of two ways: they can either be spaced as shown, with at least 2″ between boards, or they can be close together as in Fig. 40. The open or spaced roof board method is generally used in connection with wood shingles, although some other roofing methods use spaced boards. Where shingles of asbestos, cork, etc., are specified, closed roof boards are used.

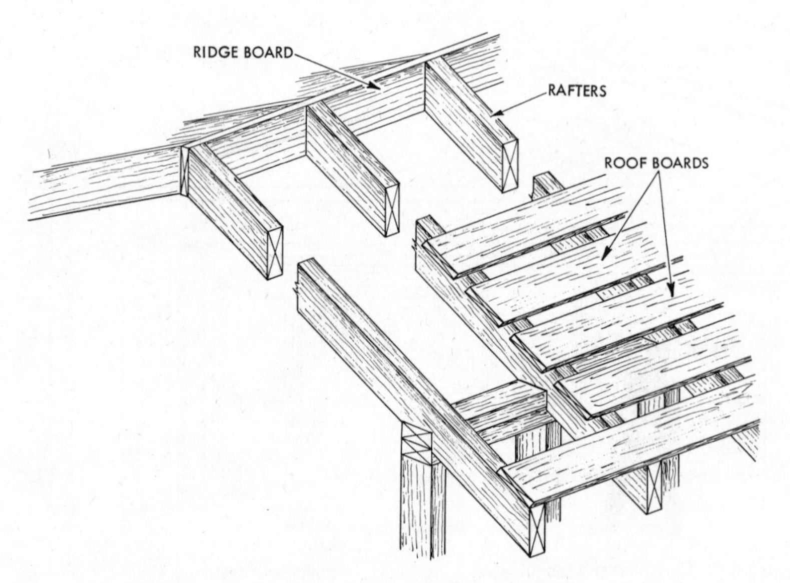

Fig. 39. Open-spaced roof boards are generally covered with wood shingles.

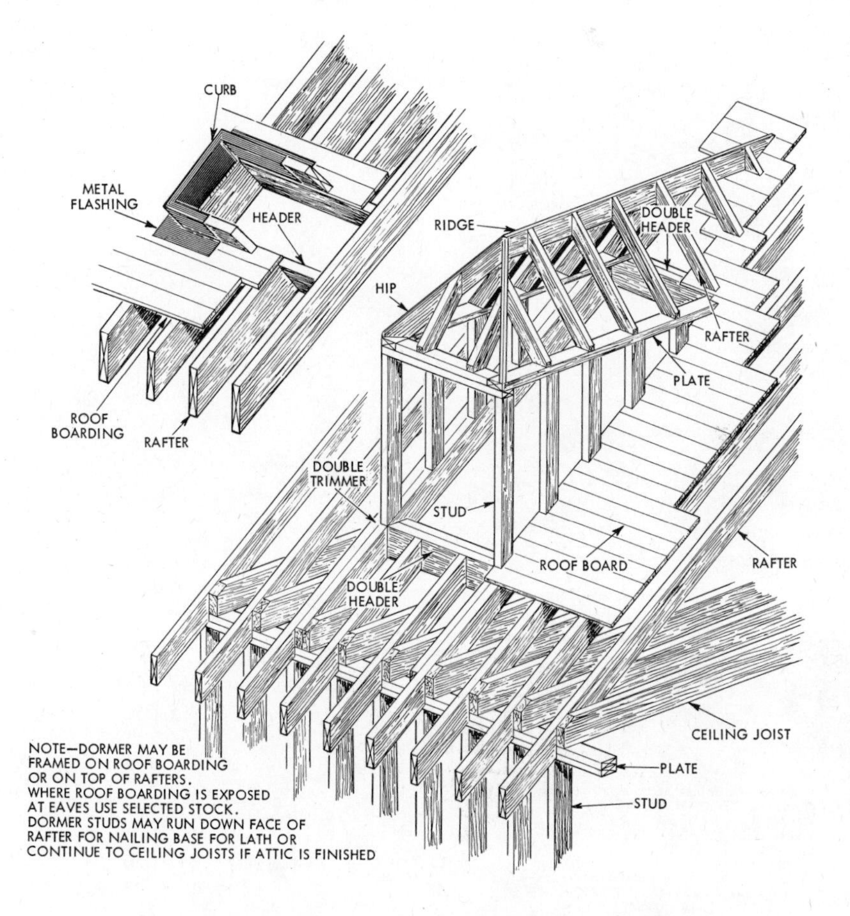

Fig. 40. Framing for dormer and roof opening on a gable roof.

There is a considerable difference between the two in both material and labor.

Roof boards vary in thickness from ⅝″ to 1″. Their widths may be from 4″ to 6″.

Dormers. Figs. 40 and 41 show typical dormers. These drawings should be carefully studied to obtain a clear concept of each of the members used in dormer construction. For example, in Fig. 41 the common rafters are discontinued and held in position by a double header. The header also serves as a support for the rafters over the dormer. The ceiling of the dormer is supported by a second set of timbers placed beneath the dormer rafters. The corner posts of the dormer are 4 × 4's or two 2 × 4's spiked together. Sometimes one or more special details are specified, such as the tying pieces, shown by the detail drawing in Fig. 41. (Fig. 40 has a different assembly for

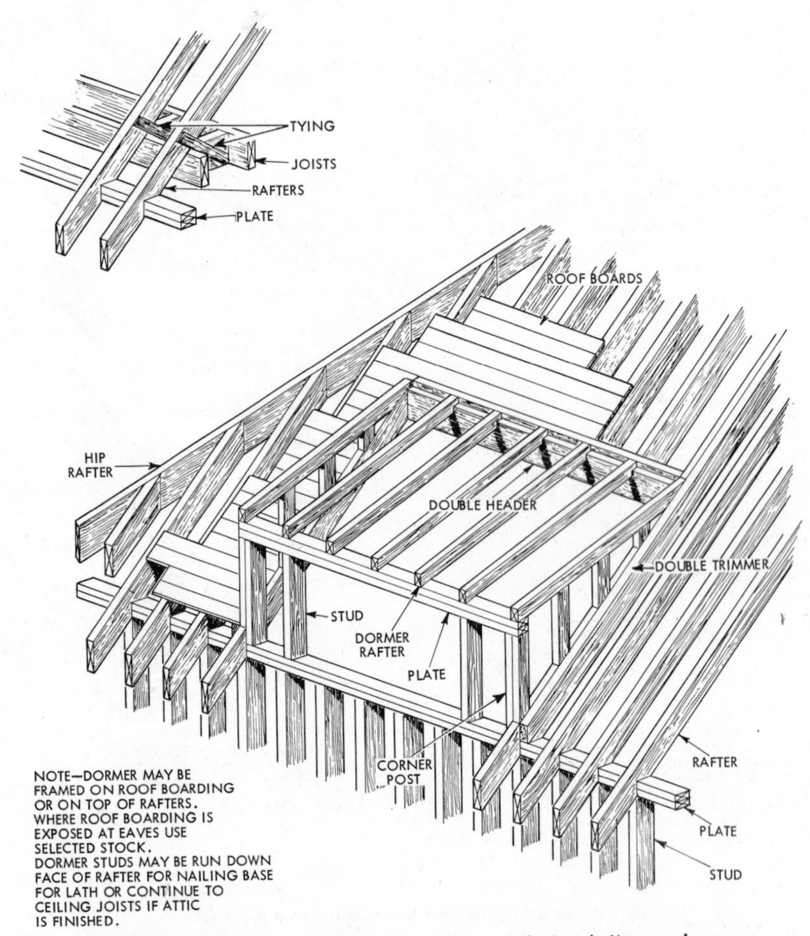

Fig. 41. Dormers require doubled studs and plates at the top, bottom, and corners.

the dormer, but you can still see all the parts.)

The lengths of the various members of the dormer assembly must be calculated or estimated from the working drawings. Enough dimensions can be obtained from the drawings (from actual dimensions or by scaling) for close calculation of the various lengths.

The sides of dormers are covered with regular sheathing; the roofs are covered with the same types of roof boarding used on the main roof area. Dormer construction is costly because of the labor involved in cutting, beveling, and fitting the large number of members.

Cornices and Sills. Five typical cornices for frame and brick walls are shown in Fig. 42. At *A, B,* and *C* the cornices are the kinds used in frame wall construction. The top plate is

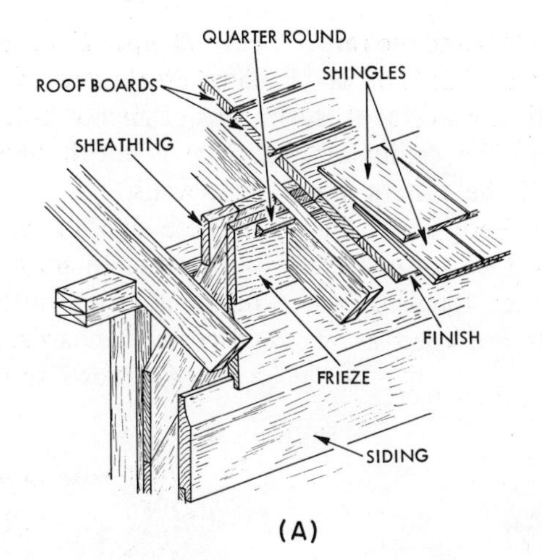

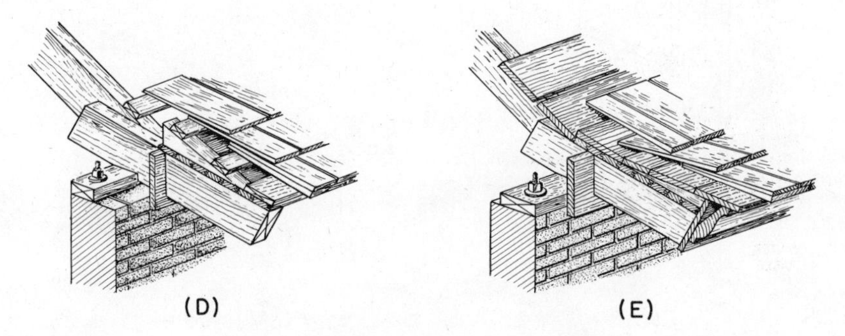

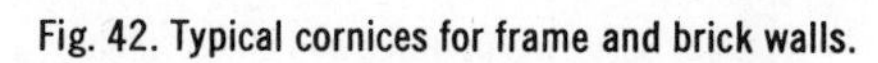
Fig. 42. Typical cornices for frame and brick walls.

composed of two 2 × 4's and the rafters are notched to fit over it. The ornamental work on the cornice at *C* is an important item to the estimator; he must determine if the rafters are delivered to the job ready to be installed or if they must be formed on the job. In some cases tongued-and grooved roof boards are used as at *D* and *E*.

At *D* and *E* of Fig. 42, the top plates are bolted to the masonry by means of anchor bolts. Otherwise, the cornice is much like those used on frame walls.

Where slate or tile roofing is used, added strength must be provided by additional larger rafters and tongued-and-grooved boards.

Fig. 43 shows four sills for frame

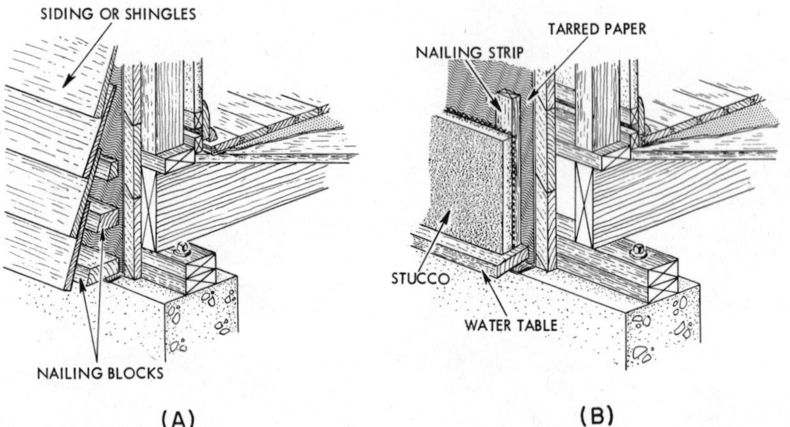

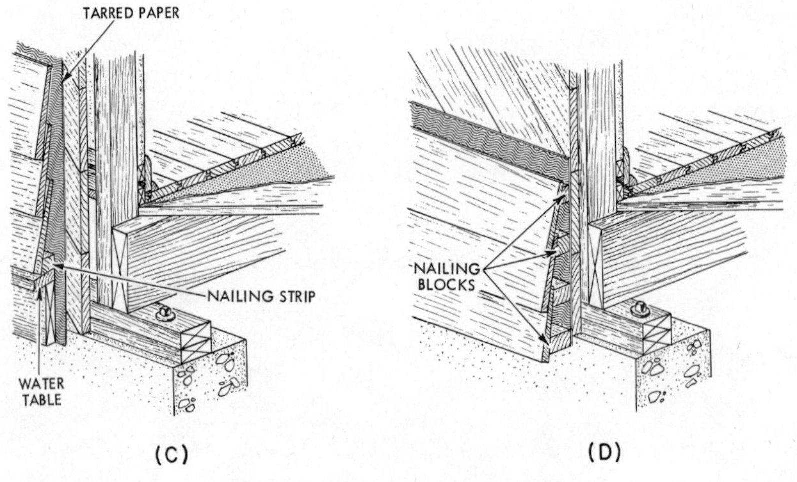

Fig. 43. Sills used in frame and stucco walls.

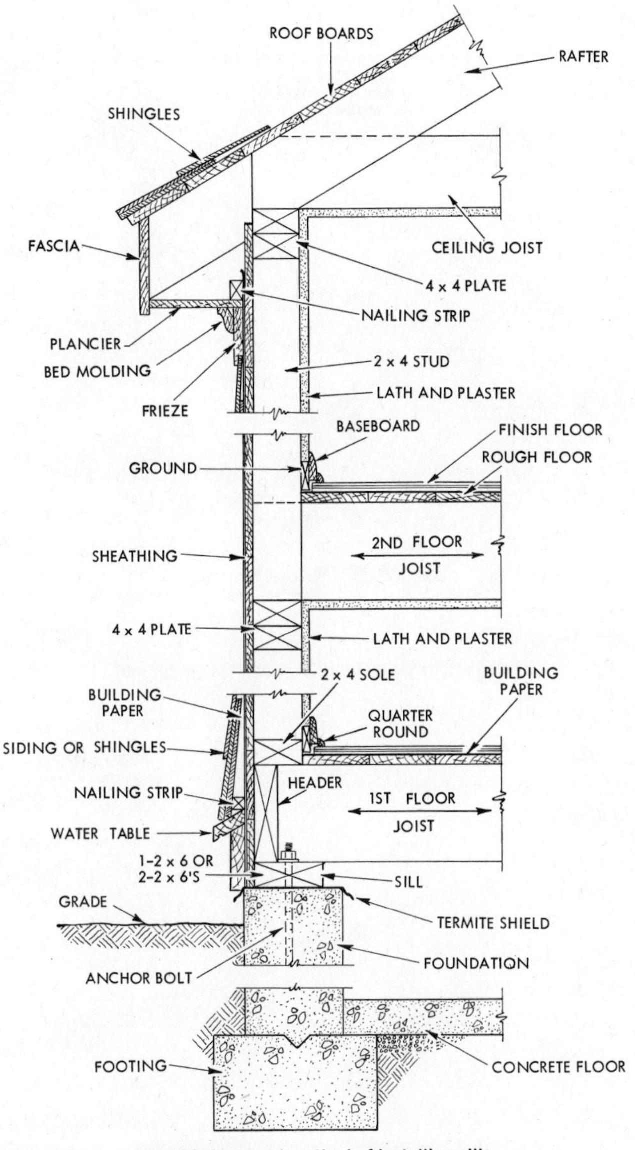

Fig. 44. A general method of installing sills.

and stucco walls. These sills plus the more general kind shown in Fig. 44 are only a few of the many types in use. Most plans indicate the type of sill to be used. If there is no indication of the type of sills, the contractor must be consulted.

Fig. 45 shows one recommended method for installing *flashings* on a flat roof connected to a brick wall. The

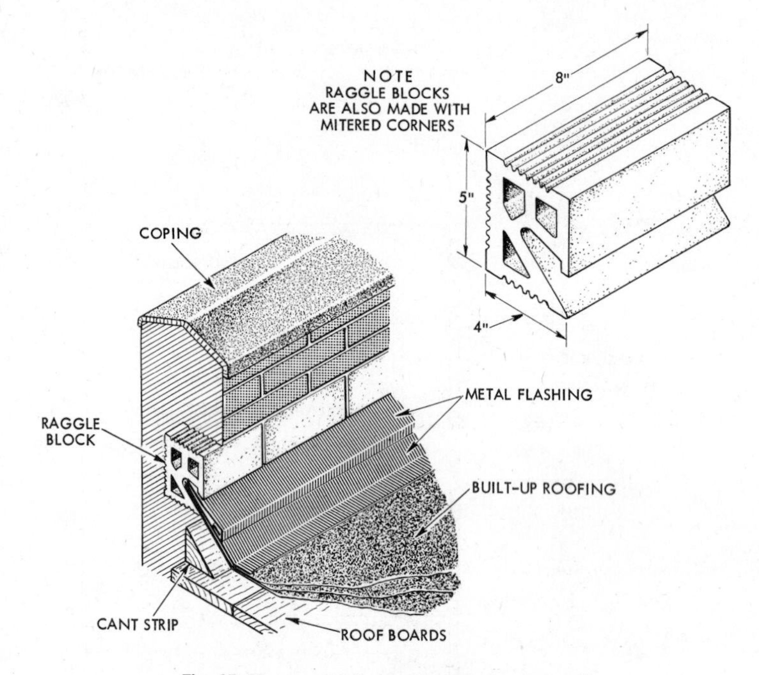

Fig. 45. Sheet metal flashing installed on flat roof.

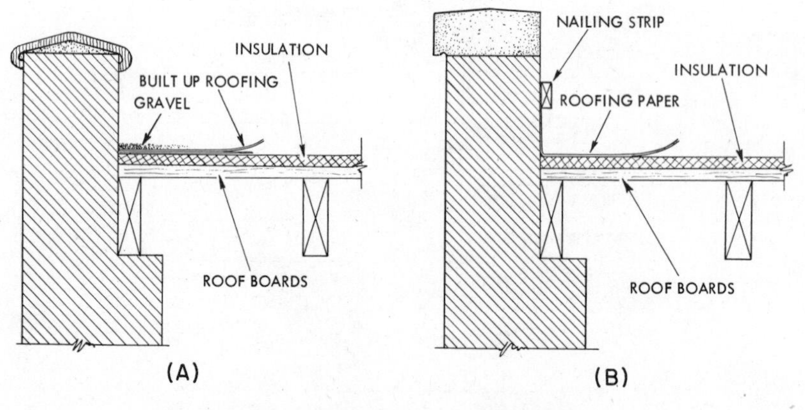

Fig. 46. An older method of installing roofing and flashing.

important part of this detail, a *raggle block* with a *mitered* corner, is purchased specifically for this purpose.

An older method of roof topping (and flashing) is shown in Fig. 46. In *(B)* the building paper (or flashing) is nailed to the masonry using a 1″ × ¼″ nailing strip.

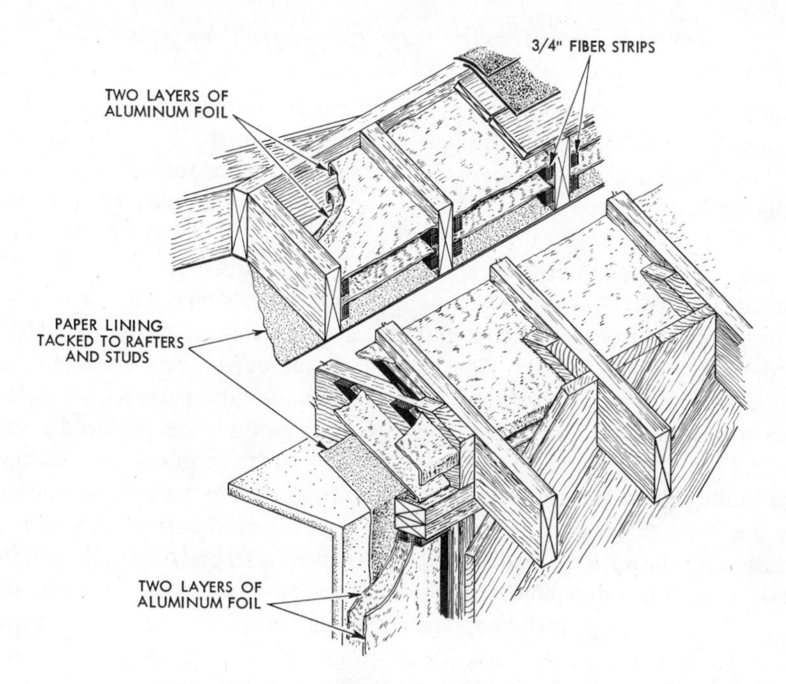

Fig. 47. Aluminum foil used as roof insulation.

Methods of constructing and flashing flat roofs on masonry and concrete walls vary widely. If a concrete slab roof is poured, the insulation and roofing are applied as explained for frame roofs, and flashing is installed in a similar manner.

Roof Insulation. Aluminum foil is commonly used as insulating material in frame roofs and is well adapted here because of its thinness and ability to reflect rather than hold heat. Fig. 47 shows a typical installation.

Rigid insulation is commonly used

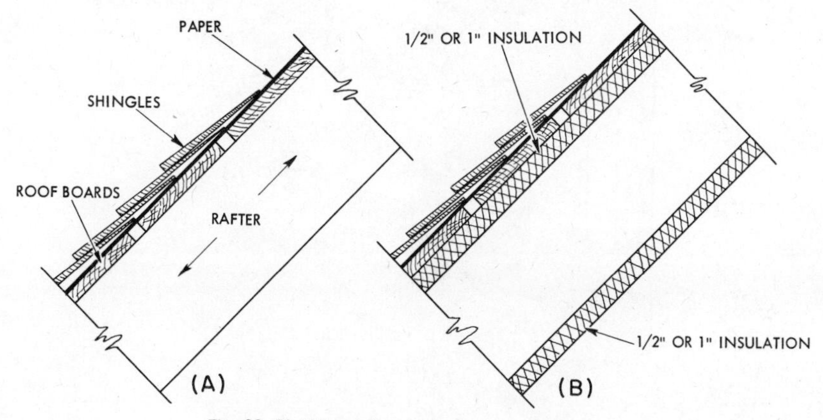

Fig. 48. Rigid insulation is also used on frame roofs.

in frame roofs as shown in Fig. 48. Other types of insulation such as blankets, felts, wools, etc., are used for various kinds of roofs.

Grounds. In Fig. 44 there are small rectangular strips or grounds shown next to the 2 × 4 sole at the first and second floors. (Grounds may also be required around windows.) These are pieces of wood about ¾″ thick and from ¾″ to 1½″ wide. They are nailed to the sole or the studs and run along the floor all around the rooms. The lath and plaster is placed on top of and flush with the grounds. When baseboards are being installed, the grounds form a nailing surface. An enlarged view is shown in Fig. 49.

If *chair rails* (wooden molding at chair-back height) are to be installed on one or more walls, grounds should be applied to the 2 × 4's under the designated place for the rail.

Fig. 50 shows various places where grounds are used. At *(A)* two pieces of 1 × 2 are used where an inside partition meets an outside wall exactly at a stud location. At *(B)* the partition meets the outside wall between studs. (Where a partition meets floor joists, at *(C)*, no grounds are required.) Sometimes a partition intersects a ceiling as shown at *(D)*, in which case one piece of ground is required. Grounds are also used as shown at *(E)*, and here either one or two pieces may be used. Ceiling cornices may also require grounds.

Grounds can require significant material and labor, so this item should be examined carefully.

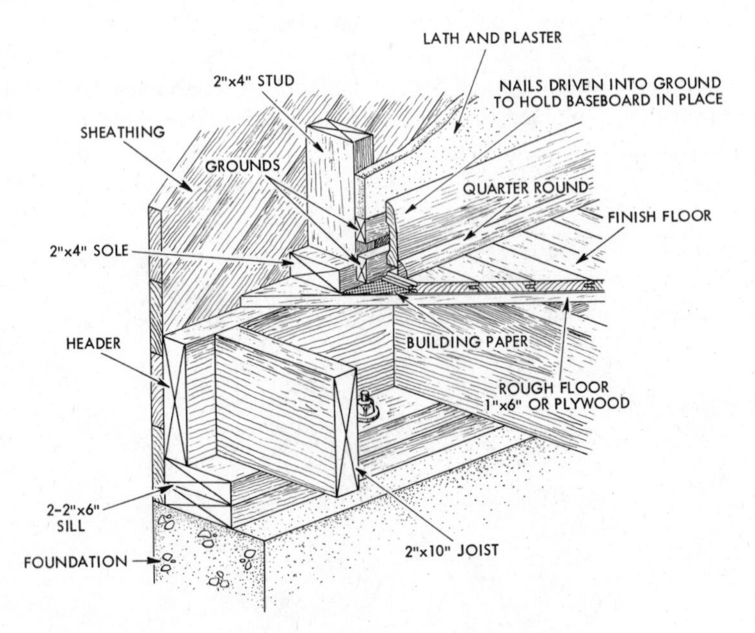

Fig. 49. Grounds installed in a wood frame building.

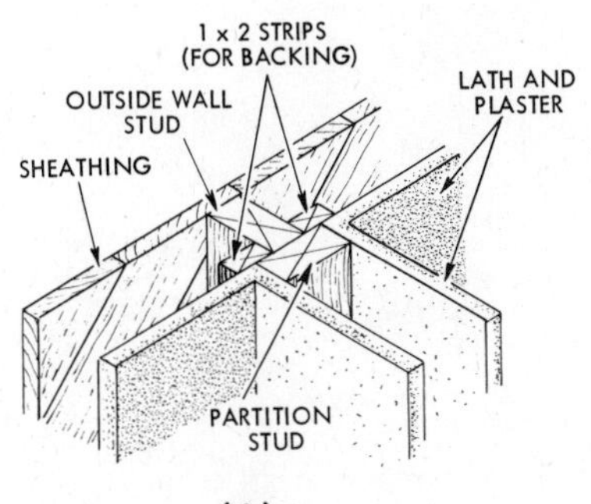

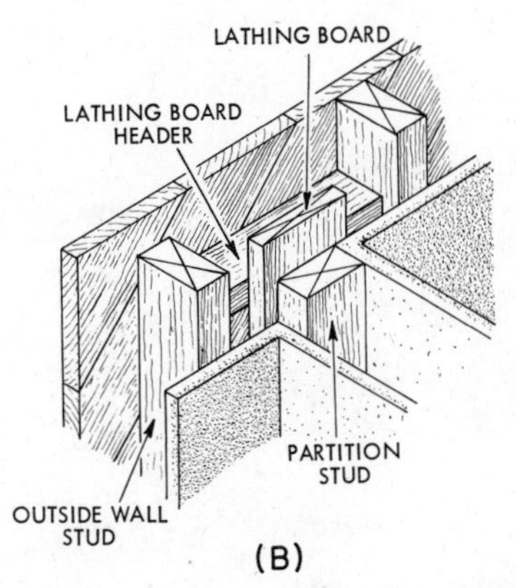

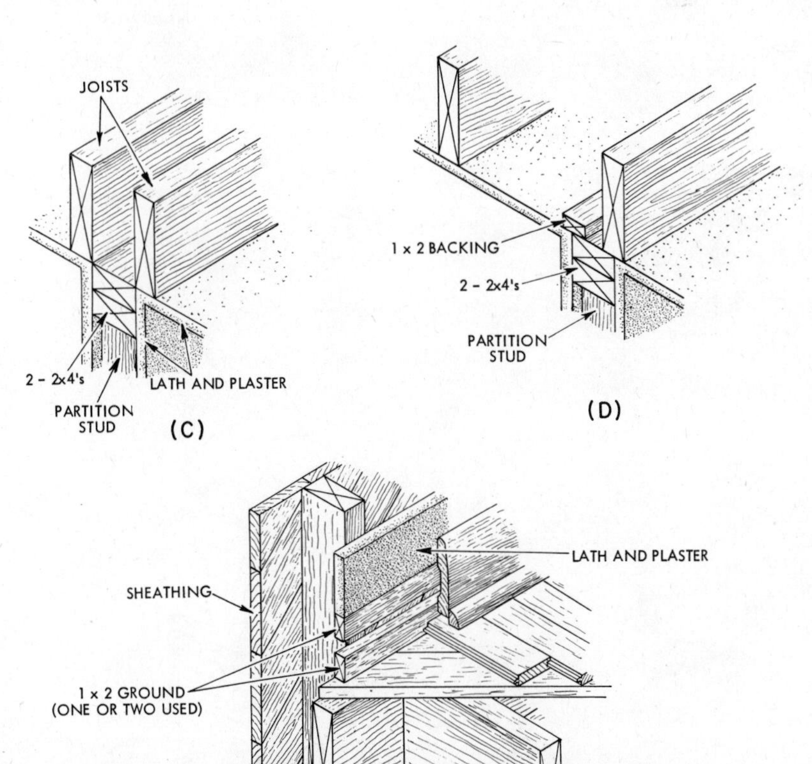

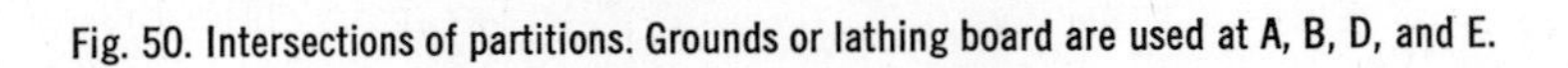
Fig. 50. Intersections of partitions. Grounds or lathing board are used at A, B, D, and E.

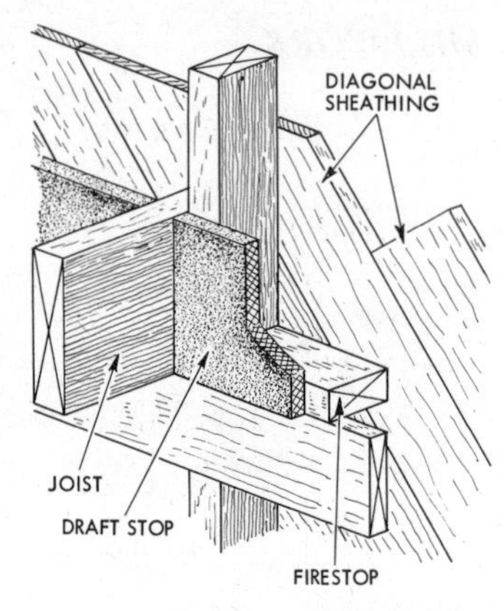

FIRESTOP-BALLOON FRAME

JOIST

DRAFT STOP MAY BE USED
AS FIRESTOP ON ALL SILLS

SILL

FIRESTOP-WESTERN FRAME

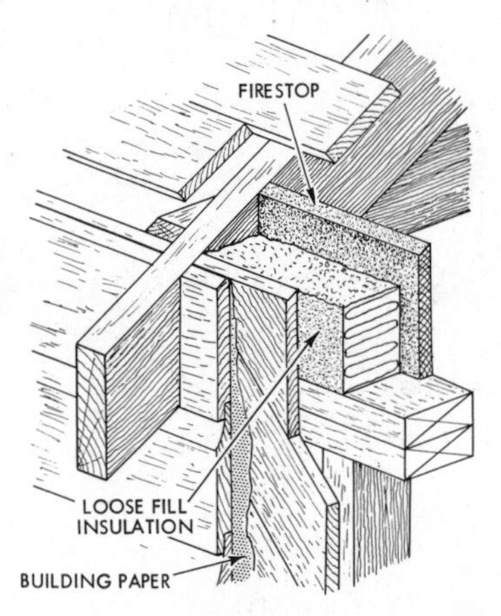

FIRESTOP AT CORNICE

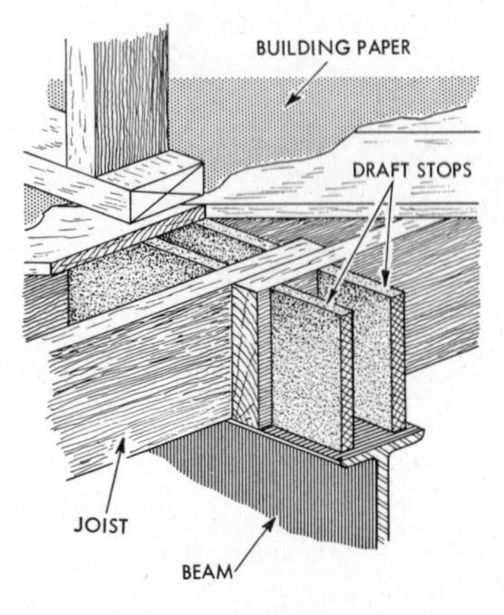

FIRESTOP AT GIRDER

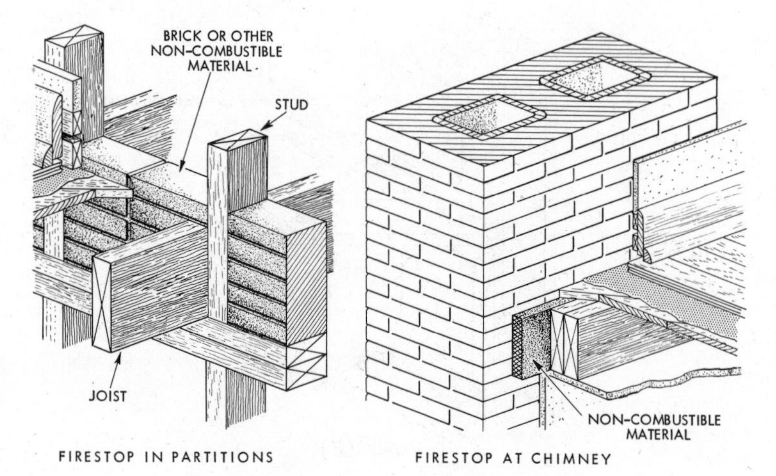

FIRESTOP IN PARTITIONS

FIRESTOP AT CHIMNEY

Fig. 51. Firestopping can be built into wood walls.

Fireproofing. Where buildings are constructed of flammable materials, such as wood, fireproofing is impossible unless some form of fire-resisting paint or chemical is applied to all wood parts. *Firestopping,* however, can be built into the walls as shown in Fig. 51.

The main purpose of firestopping is to prevent the spread of fire. For example, if fire originates in the basement of a frame residence, the hot smoke, gases, and flame have a tendency to escape and spread up through the walls. The walls form a chimney unless they are built to obstruct the sweep of fire, smoke, and gases.

Chimneys themselves can be fire hazards unless some fireproofing material separates them from adjoining combustible surfaces.

Many insulating materials are designated fireproof or *slow-burning.* For example, most of the wool and blanket types are made of non-combustible materials or they are treated to make them fireproof. The rigid types burn slowly; they smolder but do not burst into flame. Thus a well-insulated structure is also well fireproofed, or at least made fire retarding.

Concrete structures are naturally fireproof. Steel structures generally contain much combustible woodwork which is not fireproof. Light steel members used in residential construction will not long withstand fire and intense heat, so many designers specify large amounts of non-inflammable material to serve the dual purpose of insulating and firestopping.

Special Details. There are many special

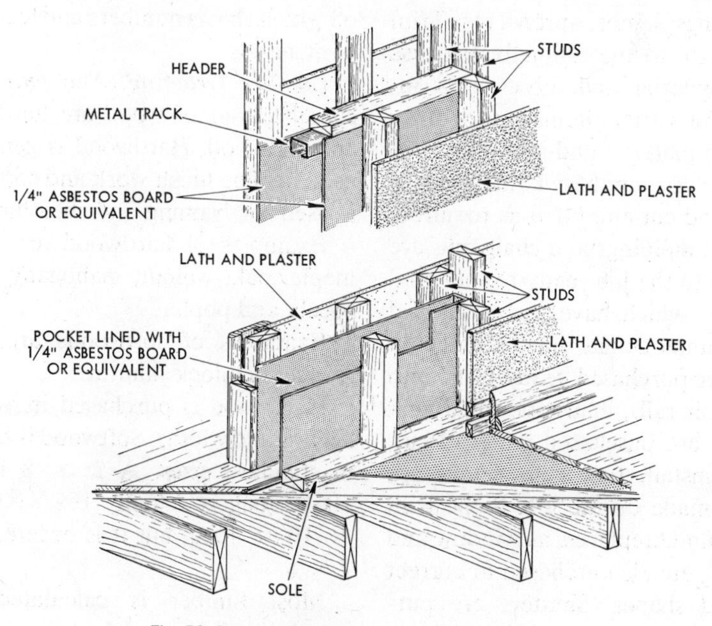

Fig. 52. Details of framing around a sliding door.

details encountered in wood construction. One such item is shown in Fig. 52. The framing around a sliding door as shown in the illustration is typical and the estimator can easily do a quantity take-off.

Cabinets. Kitchen cabinets and special cabinets of all kinds are delivered to the job ready to install in most cases. Shelves for linen closets, pantry shelves, etc., are usually cut and fitted on the job.

Storage Areas. Storage spaces, especially in attics, are not generally finished. The specifications should cover any special work to be done.

Closet Accessories. The coat rods and shelves in clothes closets are cut and fitted on the job.

Trim. This includes such items as chair rails, baseboards, picture moldings, casings, jambs, aprons, etc. Trim is important to the estimator because of the material and labor costs involved. The various items of trim differ greatly in material and size. All of it, however, is purchased in straight lengths and cut and fitted as required.

Picture moldings and chair rails are delivered to the job ready-formed and in lengths which have to be cut and fitted. Mantel details may be made on the job or purchased ready to fit into place. Stair rails, balusters, tread nosing, etc., are purchased ready to cut, fit, and install. Few trim details are actually made on the job. Stone trim used around fireplaces, as window and door sills, etc., is purchased in correct sizes and shapes. Shutters are purchased complete and ready to hang.

Estimating material

Wood and Lumber. The lumber take-off should show each piece of material, location, number of pieces, board-feet, and linear footage. The take-off is then given to the contractor, who reviews it and then forwards it to the lumber yard. At the lumber yard, the wood is selected and fabricated in accordance with the take-off, and then delivered to the building site as needed, over a period of time. (This is done to lessen the possibility of theft, destruction, and the improper use of lumber which is not immediately required by the carpenters.)

Copies of the final estimate are given to the carpenter and contractor who, along with the plans, can locate each piece of framing material on the job. You can readily understand the importance of a correct quantity take-off which shows numbers and locations of materials.

Lumber Ordering. The two main classifications of wood are hardwood and softwood. Hardwood is generally used for fine finish work and softwood is used for framing and sheathing.

Examples of hardwood are birch, maple, oak, walnut, mahogany, gum, beech, and poplar.

Examples of softwoods are pine, spruce, hemlock, and fir.

Hardwood is purchased in various lengths and widths. Softwood is bought in standard sizes. A 2 × 8 board, 10′-0″ long is actually 1⅝″ × 7½″ × 10′-0″ long, but it is ordered as a 2 × 8.

Most lumber is calculated and bought by the board-foot.

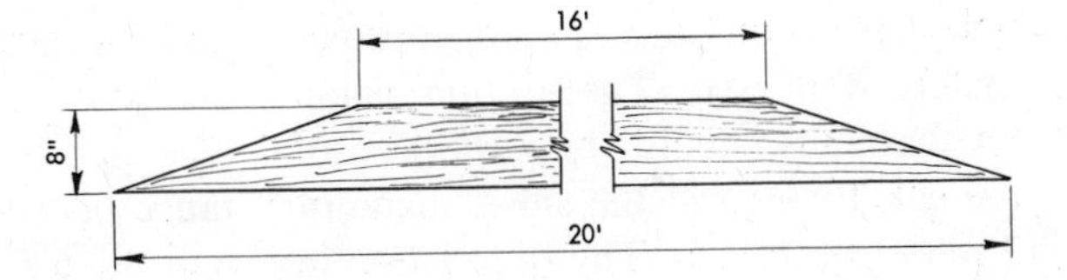

Fig. 53. A board beveled at both ends.

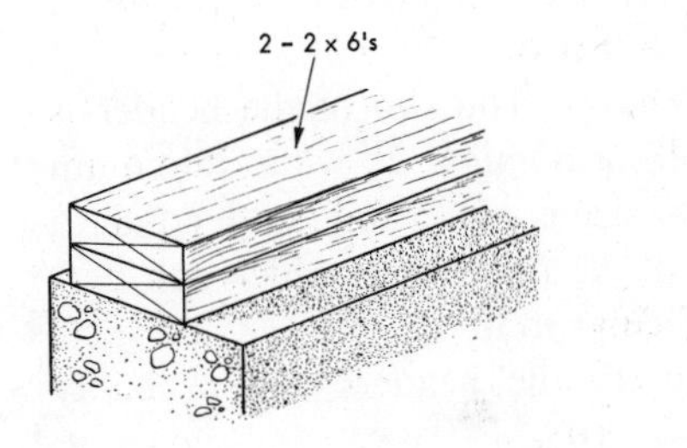

Fig. 54. A typical sill on top of a foundation wall.

Dressed-and-Matched Lumber. D & M lumber is calculated as the area to be covered plus 20%. Deduct all openings.

Wood Paneling. In interior dry wall construction (walls which are not plastered or cemented), the finish material is often wood paneling, plywood, or wallboard.

Wood paneling is figured as the area to be covered plus 10% for waste. All calculations are in square feet. Paneling is ordinarily ordered in lengths to fit the space indicated, or in sheets 4′ × 7′, 4′ × 8′, or 4′ × 10′.

Plywood. This material is also estimated in square feet and ordered in either square feet or by the panel. The contractor should order the sizes which can be used with a minimum of waste. Careful planning is necessary in order to specify which sizes cut to the best advantage.

Irregularly Shaped Boards. Sometimes boards may taper or even be curved. To find the board measure of an irregularly shaped board, the procedure is to find the average length and the average width.

Example 1. Find the board measure content of a beveled board 1″ thick, whose parallel sides are 16′ and 20′ long and 8″ wide. Fig. 53 illustrates this board. (NOTE: The board is not drawn to scale.)

Solution: Add together the two parallel sides and divide by 2; then multiply this average length by the width (in inches) and divide by 12.

$$16 + 20 = 36 \qquad \frac{36}{2} = 18$$

$$18 \times 8 = 144$$

$$\frac{144}{12} = 12 \text{ feet board measure}$$

Wall and Partition Framing. This section is concerned mainly with the lumber and insulating material.

Sills are usually secured to the concrete foundation wall with anchor bolts. The bolts are spaced approximately 8′-0″ o.c. The section detail on the plans should show the size of the sill. In Fig. 54 the sill is made of 2 pieces of 2 × 6. The foundation in Fig. 55 needs approximately 240′ of 2 × 6 board.

Headers. The size of the header is usually a nominal 2″ wide (minimum actual size is 1⅝″) and depth equal to that of the floor joists which run perpendicular to it. Thus if 2 × 10 joists are used, the headers will probably be 2 × 10's.

Outside Wall Studs. Calculate the perimeter of the building, that is, the lengths of the outside walls. The procedure is then to find the lengths of the studs, the distances between studs, and the number of studs; then calculate the required board feet.

For a balloon frame residence, you can usually assume one stud for every foot. This takes care of double studs and wastage. Thus if the outside walls are 200′ long, you will need 200 studs. The stud lengths are equal to the distance between the bottom plate (sole) and the top plate.

For a western frame residence, the procedure is similar. The western frame, however, has 2 sets of studs—one for the first floor and one for the second floor. (If the house is to be only one story high, only one set is necessary.) If the distance around the residence is 150′, you would assume one stud per foot, or 150 studs. If 2 sets are required, you would assume 300 studs. The lengths of each set are equal to the distance between the various plates.

Wall Plates. Plates are figured separately from the studs, even though they are usually the same size — 2 × 4. Measure the linear footage or calculate the lengths from the dimensions on the working drawings. Some plates

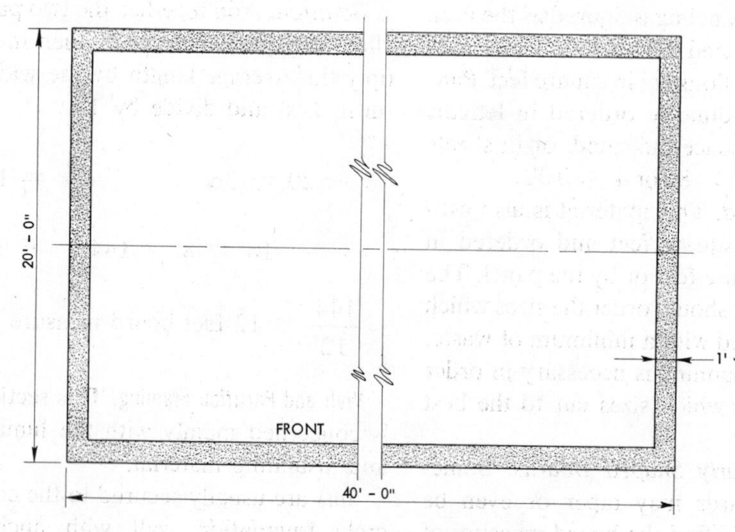

Fig. 55. Plan view of a foundation wall.

are single 2 × 4's and others are double 2 × 4's. Where double, simply multiply by two.

The western frame requires three different plates (soles) in the wall in addition to the foundation plate (sill). For this type of frame, the linear footage of one plate is multiplied by 3; this gives the total linear footage of the sole plates.

For the balloon frame, the linear footage is multiplied by 2 because there is only the upper plate (a 4 × 4 or two 2 × 4's) to be considered. (See Fig. 5.)

The balloon frame requires a 1 × 4 *ribbon* at the second-floor level. (In Fig. 5 the ribbon is just below the second-floor joists.) To find the length of ribbon, measure the linear footage (or use the given dimensions) of the outside walls.

Sheathing. Measure the length of walls requiring sheathing and multiply by the height to which the sheathing will be applied. Then deduct the areas of all openings such as windows and doors. Where boards are applied diagonally, add 30% for waste. Sometimes the combined areas requiring sheathing must be figured individually because of irregular shapes. The lengths of all areas can either be scaled or found from the dimensions. The heights should be found in a similar manner.

Insulating rigid sheathing is calculated by square feet. Allow about 6% additional for waste. Deduct all openings such as windows and doors. A carpenter can help to cut wastage by figuring the most economical sheets to order. The sheets are 4′-0″ wide and come in lengths of 6′, 8′, 10′, and 12′.

Insulation. Batt insulation for frame walls is figured in square feet. Calculate the square footage of surface area to be insulated and deduct all openings such as windows and doors. The material comes 4″ thick in sizes 8″ × 15″ and 16″ × 48″. No wastage need be considered.

The calculation of side wall areas is simple and is done in the same manner as for sheathing.

Quilt insulation is also figured by the square foot. Calculate the area to be insulated and deduct all openings. No wastage need be considered. The material is in rolls 48″ wide and ½″ or 1″ thick. Area calculations are the same as batt insulation. (For brick walls the process is the same as for frame walls.)

Loose wool insulation is bought in bags. Each bag covers 15 square feet, 4″ thick. Therefore, calculate the area to be insulated and divide by 15 to determine the number of bags. Deduct all openings. Areas are calculated as explained for batt insulation.

Wall and Partition Boarding. Boards for walls and partitions are calculated by finding the total surface area and deducting the openings. Then add 20% for wastage.

Wall Studs. The lengths of interior wall studs depend on the room height. These figures can be obtained from the elevation drawings. Studs (2 × 4's) are purchased in lengths which are multiples of two and also in 9′ lengths. Be careful to order the most economical lengths. Sometimes two or more

TABLE II. BEVELED SIDING

SIZE (INCHES)	EXPOSED INCHES	TO AREA TO BE COVERED ADD	SIZE (INCHES)	EXPOSED INCHES	TO AREA TO BE COVERED ADD
1/2 x 4	3 1/4	1/4	1/2 x 5	3 3/4	1/3
1/2 x 4	3	1/3	1/2 x 6	5 1/4	1/5
1/2 x 4	2 3/4	1/2	1/2 x 6	5	9/40
1/2 x 5	4 1/4	1/5	1/2 x 6	4 3/4	1/4
1/2 x 5	4	1/4	......	....	....

short studs are obtained from one long stud.

When finding the stud dimensions, do not include the floor and joist thicknesses or heights.

After the heights have been determined, the linear footage of each partition height can be scaled, allowing one stud per foot (when the spacing is 16″ o.c.).

Studs in inside walls of dormers are counted with interior studs. Exterior dormer wall studs are counted with exterior studs.

Siding and Felt. The amount of siding is calculated by combining the area of all surfaces requiring siding, and adding 25%. The additional amount takes care of lapping. This applies to ½″ × 6″ and ½″ × 8″ beveled lap siding. An allowance of 20% should be made for wastage. Deduct for all openings. Table II can be used for other sizes and exposures.

The amount of felt is determined by adding 10% to the required area. This felt is laid between the sheathing and the siding. Felt for side walls usually comes in rolls containing 500 square feet.

Floors. The main areas of concern under this heading are joists, bridging, subflooring, finish flooring, and floor felt.

Joists. The positions of all floor joists should be indicated on the plans. For example, HOUSE PLAN A has double-headed arrows. These arrows indicate floor joists and are generally accompanied by the size of the floor beams and the spacing of the beams. The direction of the arrows indicate the direction the floor joists are to run. If no arrow is shown, the estimator must decide for himself (or ask the architect) which direction the joists will run. Remember that the joists must seat or bear at least 4″ at each bearing point. Thus if the length of a room is given as 12′-8″, and the arrow indicates that the joists are running the long way, you must add 4″ for bearing at each end. This gives 12′-8″ + 4″ + 4″, or 13′-4″ as the required joist length. Since joists are bought in lengths corresponding to multiples of two, you would have to buy floor joists 14′-0″ long.

Sometimes architects include several framing plans with their drawings. The joists for all floors are indicated on these plans by dash- or double lines in their proper positions. In such cases the estimator need only count the number of joists and determine the various lengths according to standard lengths which are multiples of 2′-0″.

If no framing plans are included, there are several methods which can be used. The rule for estimating joists spaced 16″ o.c. is shown in the following example.

Example 2. Find the number of joists required along the rear of Fig. 55. The joist spacing is 16″ o.c.

Solution: Divide the length (40′-0″) by 1.33 (or 4/3) and add 1.

$$\frac{40.0'}{1.33'} = 30$$

$$30 + 1 = 31 \text{ joists}$$

(If each of the joists is 10′ long, then 310′ of board is needed.)

For estimating joists spaced 12″ o.c., the length is divided by 12″, or 1′. Then add 1.

For estimating joists spaced 24″ o.c., you divide the length by 2′ (or 24″) and then add 1.

Bridging. Each joist normally requires two pieces of bridging. To determine the total linear footage of required bridging, you must determine the length of an average piece of bridging. Each piece of bridging forms the hypotenuse of a right triangle with the height of the joist as one leg and the distance between joists as the other leg. Thus if 2 × 10 joists are used, the height of the joist is 10″. If the joists are spaced 12″ o.c., the other leg of the triangle is 10″. (Remember that

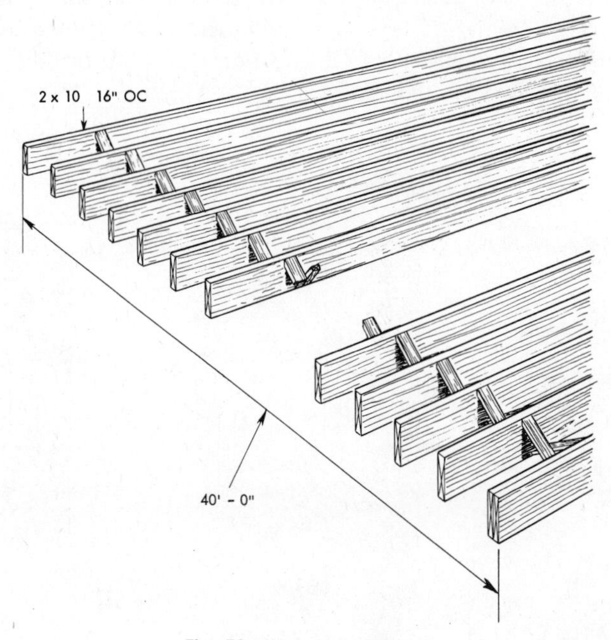

Fig. 56. Joists and bridging.

the o.c. distance is greater than the actual distance between joists.)

Example 3. Find the length of board needed for bridging of the house foundation shown in Fig. 55. The joists are 2 × 10's spaced 16″ o.c. There are two sets of bridging per run. (See Fig. 56.)

Solution: First find the length of an average piece of bridging.

$$10^2 + 14^2 = 296$$

$$\sqrt{296} = 17.2\text{, or about } 17'' \text{ each}$$

$$\text{No. sets} = \frac{40.0'}{1.33'}$$

$$= 30 \text{ sets per run}$$

$$30 \times 2 = 60 \text{ pieces}$$

Since there are 2 runs, 120 pieces, each 17″ long are required.

$$120 \times 17'' = 2{,}040''\text{, or } 170.0'$$

If metal bridging is to be used, only the number of required sets need be found.

Subflooring. The rough floor is calculated as the area of the entire floor minus such large openings as stairways. The openings are figured roughly, being sure to allow too little rather than too much area for them. The flooring, as shown by the typical wall section in Fig. 57, does not always extend below sheathing or even to the sheathing, Fig. 57*(B)*. Thus the dimensions to the outside of the building cannot always be used. For irregularly shaped floors it is a good idea to divide the surface into rectangles, squares, etc., to simplify area calculations.

If the subflooring is specified as diagonally applied, add 30% to the calculated area. If not diagonally applied, add 20% to the area.

Finish Flooring. For flooring which is ¾″ × 2″ or ¾″ × 2¼″ in thickness and width: Find area of each floor surface requiring finish flooring, then add together these areas; add 33%.

For 1″ × 4″ fir or pine add 20% to the total area.

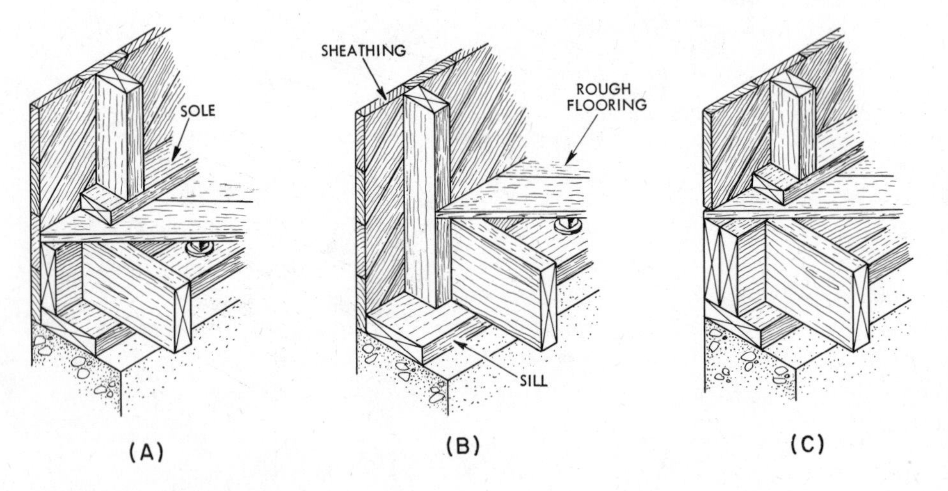

Fig. 57. Rough flooring may reach up to the studs or under the sheathing.

Drop Flooring for Tile. Drop flooring requires additional material. This material is figured the same as for ordinary subflooring. Add 20% for waste. The 1 × 2 or 2 × 2 cleats are measured by assuming two per joist. They are figured in linear feet and ordered in lengths which can be used with a minimum of waste.

Floor Felt. This item is calculated in square feet. Find the area of the floor, deduct the areas of openings, and add 10%. One roll of felt usually contains 500 square feet.

Doors. These are designated *exterior* or *interior* because they are usually two distinct types. The exterior doors are usually thicker. Doors are listed in a *Door Schedule* as in the following:

EXTERIOR DOORS

Size	Description	Amount
3' - 0" x 6' - 8" x 1 3/4"	Flush Solid Core	1
2' - 8" x 6' - 8" x 1 3/4"	3 Lite 2 Panel	1

Screen doors are generally mill made and purchased complete.

Door Casings. These are bought by sides which include two side casings and one head casing for each member. The same amount of molding is used regardless of whether it is one member or two. The material list in the door schedule or quantity take-off should show the sizes of doors and sides.

Door Stops. Stops are bought by the set which consists of two side pieces and one head piece.

Windows. If the windows are not listed in a *Window Schedule,* you must list them as in the following:

WINDOWS

Size	Description	Amount
3' - 0" x 4' - 0"	6 over 1 DH	1
2 (3' - 0" x 4' - 0")	gang of 2 DH 6 over 1	1
2 (3' - 0" x 4' - 0") + (4' - 0" x 4' - 0")	gang of 3 DH 6 over 1	1

Windows ordered in this manner are usually delivered complete with sash, frame, trim, casing, mullions, aprons, and stools.

Chapter 12 explains in detail how glass is estimated.

Screens and Storm Windows. Screens are usually made just half the full window size; full-sized screens are not very common.

The screening material may be copper, galvanized metal, aluminum, etc. The screening is purchased in standard widths and in rolls. For odd-sized windows there is usually some waste of material.

Screen frames are generally 1″ × 2″ stock. This is ordered in linear feet. The screen molding goes completely around the screen.

Screen moldings are joined in several ways, but for estimating the necessary amount of 1 × 2, the width and height are just doubled for regular half window screens.

Example 4. Two windows with identical dimensions need screens. What is the amount of framing needed if the windows are each 3′-0″ long and require screens which are each 3′-6″ high?

Solution: First find the perimeter of one screen, then double this figure to obtain the total length.

$(2 \times 3.0') + (2 \times 3.5') =$
13.0′ for each screen
$2 \times 13.0' = 27.0'$ needed for
both screens

Although the required molding is slightly less than the framing (because it is nailed to the inside edge of the frame), most contractors assume that it is the same length as the framing.

Storm windows are generally mill made. No quantity take-off is necessary except for the windows themselves—each counting as a unit.

Storm sash can either be bought as a stock item or made at the mill.

Window Weights. Most modern homes have double-hung windows installed with spring-loaded balances or friction holding devices. Thus window weights and sash cord are not usually necessary.

Where window weights are required, it is not difficult to calculate the required amount. If the glass in a double-hung window (where only two large panes are used) is of identical size, measure one pane, calculate its area, and multiply by two to find the total glass area. Allow 1 pound of window weight for each 100 square inches of glass.

Example 5. Assume a double-hung window has two large panes of glass, each 36″ × 24″. Find the amount of window weight needed.

Solution: 36″ × 24″ = 864 sq. in.

$\frac{864}{100} = 8.64$, or approximately 9 lbs. per pane—18 lbs. for the entire window.

(NOTE: Each sash will then have about 4½ lbs. of weights.)

If the window panes are unequal in size, the weights are figured separately for each pane.

Sash Cord. Each sash requires approximately 10 feet of cord if window weights are used. This is about 5 feet for each side. Thus 20 feet of cord is required for an ordinary double-hung window which uses sash cords.

Sash cord is purchased by the *hank.* A single hank contains 100 feet of cord which is ordinarily sufficient for 5 ordinary-size windows.

Window Trim. Trim is usually bought in sets consisting of stools, aprons, casings, and stops. Linings are required as part of the set when the walls exceed 8″ thickness.

Stairs. Finished stairs are made in the mill or stair shops where all parts including stringers, risers, treads, railings, etc., are included. The stair is shipped to the construction site in parts. A main stairway is listed as "one flight, main stairs."

Basement and Attic Stairs. Basement stairs are not always shown in detail on the plans; the estimator must therefore do some preliminary calculations before doing a quantity take-off. Fig. 58 shows how a typical stairway is planned. Without the necessary details and elevations, the estimator (or contractor) must calculate or measure the *rise* and the *run* of the stairs. Once these dimensions are known, the length of the stringers is found by the Right Triangle Law.

After the stringer length is determined, the 2 × 10 or 2 × 12 board is determined, keeping in mind that

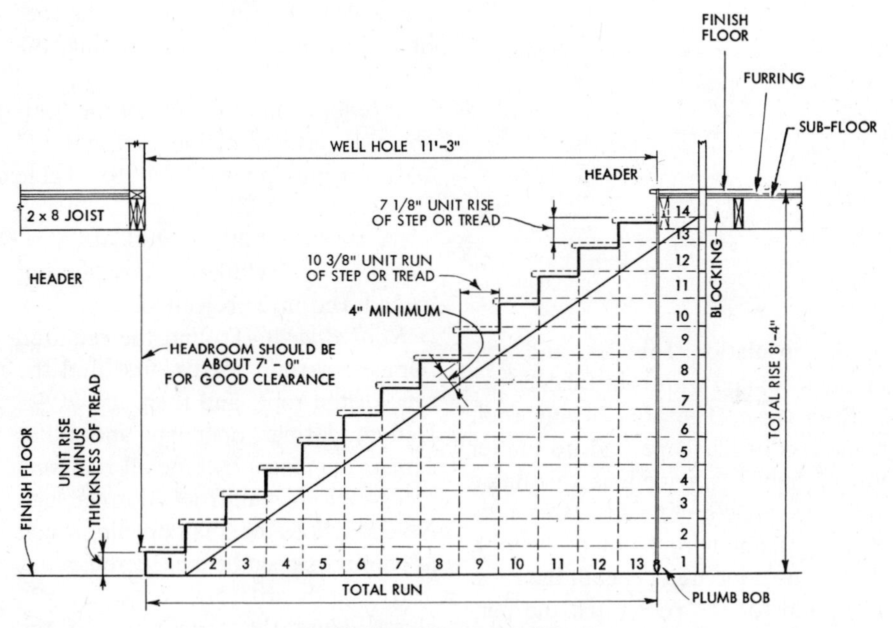

Fig. 58. How stairways are planned.

such lumber comes in lengths which are multiples of two. Two to five stringers are normally required. The tread and riser size depend on the riser height.

If a 7½″ riser is required, divide the height by 7½″; if the answer doesn't come out even, the height of each riser is increased or decreased by small amounts until it does come out even. The treads are always one less in number than the risers. The tread width is found by dividing the number of treads into the run of the stairway. (Sometimes there is only a limited distance available. Using that distance as a limit, the above process is used.)

Attic stairs are calculated in the same manner as the above.

The material for treads is sometimes 2″ × 10″ stock and sometimes thinner. One-inch stock is ample for risers. Knowing the width of the stairs, the total length of tread and riser material is determined and lengths of 2″ × 10″ and 1″ stock are ordered, thus minimizing waste.

Stair Railings. Stock for railings is figured in linear feet. Rough railings are composed of upper and lower 2 × 4 pieces having some millwork done on them and the pickets. The pickets can be 1 × 2, 1 × 3, 2 × 2, etc.; they are purchased in long lengths, as are the 2 × 4 pieces.

Roofs. In addition to the roof boards, there are various types of rafters,

TABLE III. ROOF AREAS OF PITCHED ROOFS

Pitch	Increase of Area Over Flat Roof	Multiplication Factor
1/4	12%	1.12
1/3	20%	1.20
3/8	25%	1.25
1/2	42%	1.42
5/8	60%	1.60
3/4	80%	1.80
7/8	101%	2.01

beams, insulation, shingles, and cornices associated with roof take-offs.

Roof Area. The area of any roof, regardless of its shape and no matter how it may be divided, is calculated as follows: Determine the exact area of the horizontal projection of the roof, that is, the floor area beneath the roof. Then add for the roof pitch the percentage specified in Table III. To the results thus obtained, add the cornice projection. This is a sufficiently accurate method of finding roof areas for most purposes.

If the length of a common rafter is known, the roof area can be accurately determined by multiplying the rafter length by the total length of the roof which the rafter spans over.

Example 6. Find the area of the roof of a 1/3 pitch hip-roof building, 30′ by 30′.

Solution: Find the area of the horizontal projection of the roof and add 20% (or multiply by 1.20). See Table III.

$$A = 30' \times 30' = 900 \text{ sq. ft.}$$

(NOTE: This includes all dormers but excludes cornice projections.)

Roof Boards. To find the required number of roof boards, first find the area of the roof, and then add 20%. Deduct dormer openings and other similar openings. Include all roof *surfaces* such as main roof, dormer, etc. No deductions for small openings such as chimneys need be considered.

If roof boards are to be laid 2″ apart, figure the exact area of the surface to be boarded. Do not add anything for waste.

Roof Rafter Lengths. The required lengths of roof rafters are determined from the working drawings by either scaling or calculating from the given dimensions. Roof rafters 12′-0″ long must be ordered for a roof which requires 10′-9″ rafters because rafters,

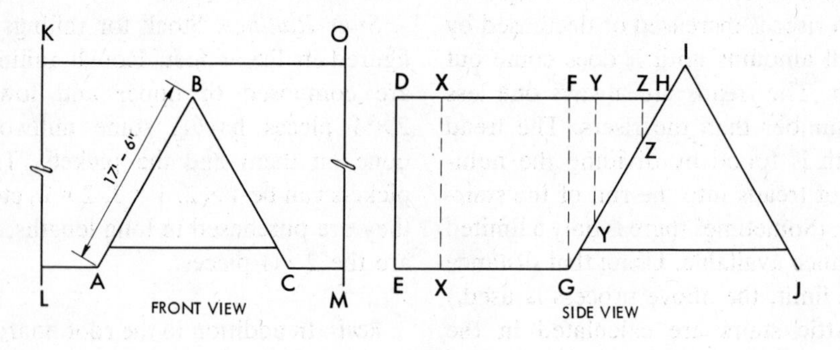

Fig. 59. A two-section sloped roof.

like other dimension lumber, come in lengths which are multiples of two.

Fig. 59 shows a front view of roof section *ABC* intersecting roof section *KOML*. In the side view, section *ABC* becomes *DHGE*. To determine the rafter lengths required for roof section *ABC* you must scale the true length of the longest rafters, which are the *common* rafters. These rafters are used in the area *DFGE* in the side view. The true length of rafters such as *XX* cannot be scaled from the side view. You must use the front view and scale the line *AB* or *BC*. *BC* is the same length as *XX*. The rafters in the area *DFGE* are therefore the same length as *BC*.

Example 7. Find the length of the common rafters in roof section *DFGE* in Fig. 59.

Solution: The lengths are equal to *BC*. Since the length of *BC* is the same as *AB* and the length of *AB* is 17′-6″, the rafters in section *DFGE* are 17′-6″. The closest available length is 18′-0″.

For the roof area *FHG* in Fig. 59 the same 18′-0″ lengths are ordered even though the rafters gradually become shorter. The amount sawed off

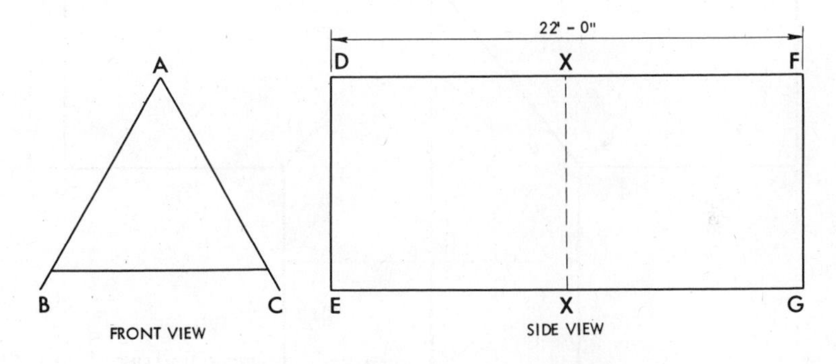

Fig. 60. A common sloped roof.

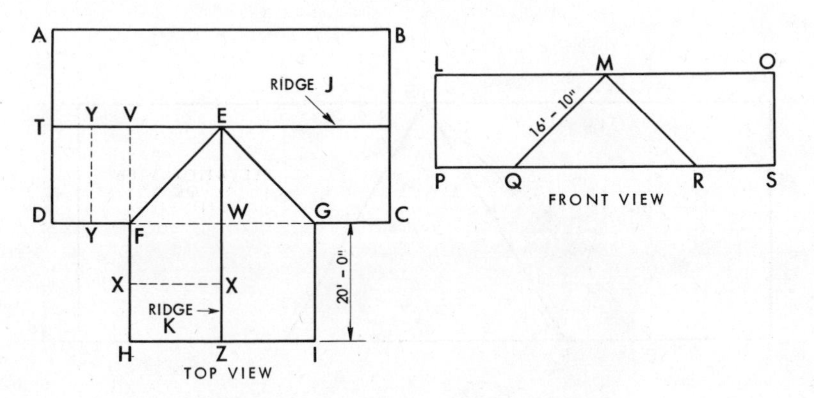

Fig. 61. Determining rafter lengths from roof drawings.

an 18′-0″ rafter to make rafter *YY* leaves enough for rafter *ZZ* so that there is little waste.

Fig. 60 shows a common sloped roof. This is the type of roof which is often used on a rectangular-shaped house. To find the length of rafter *XX* (side view), you measure its length on the front view, because this is the only place where the true length can be measured. Thus if either *AB* or *BC* (front view) measures 15′-0″ you know the length of rafter *XX* is 15′-0″. (You would order 16′-0″ lengths.)

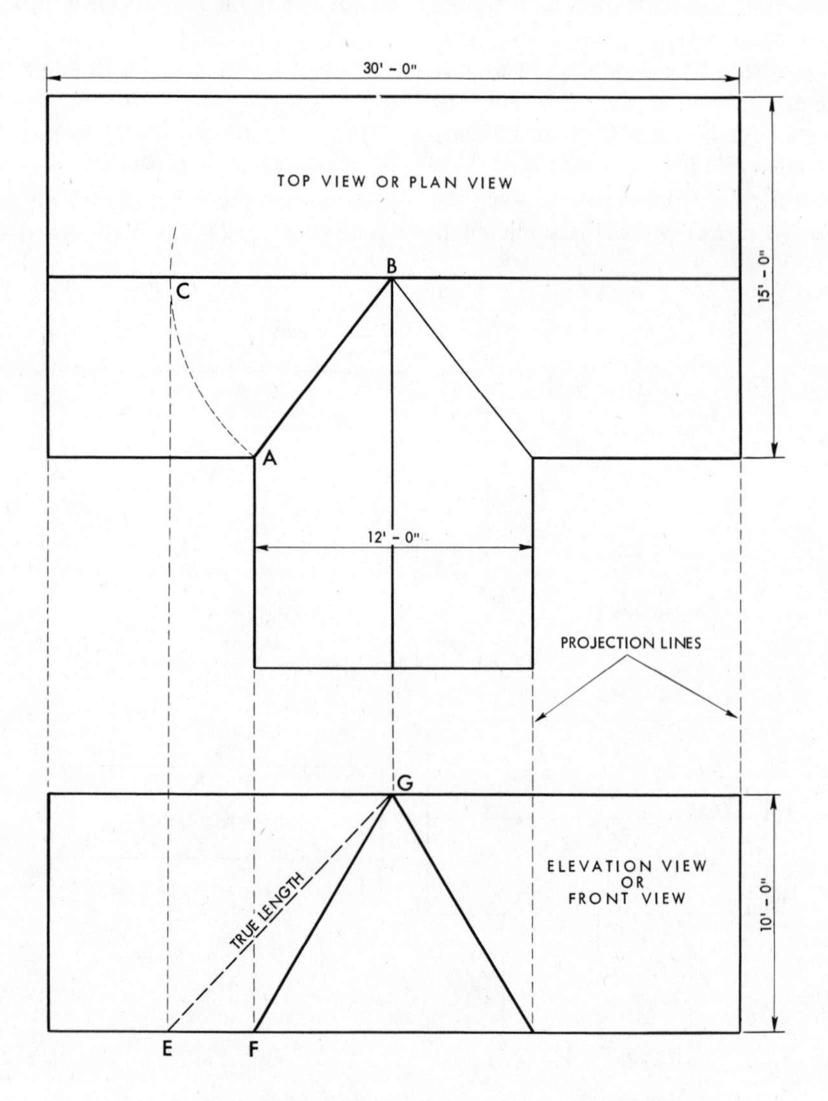

Fig. 62. A method for finding true lengths of valley rafters.

Fig. 61 shows another common roof type. The top view shows the ridges *J* and *K*. Suppose that you desire to find the length of rafter *XX*. To do so you must find the length of *MQ* on the front view. *MQ* is the same length as *MR* and is found to be 16′-10″. Therefore, 18′-0″ rafters will be ordered for *XX* and all common rafters in the areas *HFWZ* and *ZWGI*. For areas *FWE* and *EWG* you would also order 18′-0″ lengths which will be cut into shorter lengths.

To determine the length of rafter *YY* in Fig. 61 you will need an end view of roof section *DABC*. Since the working drawings usually show all four outside wall dimensions, all true lengths can be measured.

Valley Rafters. Another common problem is finding the lengths of valley rafters such as *FE* and *EG* in Fig. 61. Valley rafters are longer than common rafters. There is no place where the true lengths of these rafters can be scaled quickly.

Fig. 62 illustrates a method for finding the true lengths of the valley rafters. The top view (plan) is drawn directly above a front view (elevation). Using a radius equal to *AB*, draw the arc *AC*. (The center of the arc is at *B*.) From point *C* drop line *CE* to the elevation view. Then connect *E* to *G*. The line *EG* is the true length of the valley rafter. The difference between lines *FG* and *EG* is the difference between the true and apparent lengths.

Fig. 63 illustrates a roof which has two roof sections of unequal widths

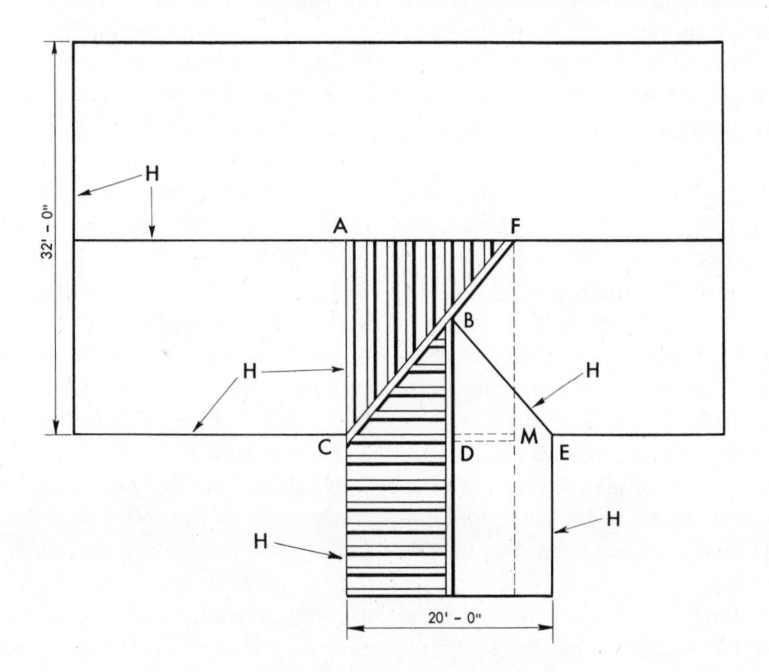

Fig. 63. Two roof sections with the same pitch.

but identical pitch. Since the two roofs have the same pitch, the valley rafter *CF* intersects the ridge of the main roof section at the point *F*, which is apparently just as far from the point *A* as is the point *C*. In other words, distance *AF* is equal to the distance *AC* when measured on a horizontal plane. The distance *AC*, however, is the *run* or horizontal projection of the common rafter in the main roof. Since the main roof is 32′ wide, the distance *AC* is 16′ and the distance *AF* is also 16′. If the two roof sections had been of identical width, the dotted line *FM* would have been the ridge of the smaller roof section; but since this roof is only 20′ wide, the line *BD* is the ridge of the narrower roof and intersects the valley rafter at *B*. The valley rafter is continued up to the main ridge at *F* and is supported there by the main ridge, thus giving support to the inner end, *B*, of the secondary ridge.

The point *F* is the same distance measured vertically above the level of the top of the wall plate as is the point *A*, and this distance is the *rise* of the common rafter *AC*. Now suppose that the slope of the main roof is 10″ to the foot. (For every foot in the run of rafter *AC*, this rafter and the main roof surface rise 10″.) Since the span of the main roof is 32′ and the half-span is 16′, the total rise is 16 × 10″, or 160″. This is also the rise of the valley rafter *CF*.

The two roof sections are of the same pitch so a run of 16 × 12″ along *AC* (from *C* to *A*) corresponds to a run of 16 × 12″ along *AF* (from *A* to *F*) which also corresponds to a run of 16 × ? along the valley rafter from *C* to *F*. Since the two distances *AC* and *AF* are equal (measured horizontally), it appears that *CF* is equal to the length of the hypotenuse of a right triangle, the other two sides each being 12″.

$$12^2 + 12^2 = CF^2$$

$$CF = \sqrt{288} = 17''$$

Thus the question mark in 16 × ? is replaced by 17″ and 16 × 17″, or 272″ is the run of the valley rafter.

Knowing the rise (160″) and the run of the valley rafter enables you to calculate its length. The length is the hypotenuse of the right triangle with legs equal to 160″ and 272″.

$$\sqrt{(160)^2 + (272)^2} = 315.5'', \text{ or } 26.3'$$

Using the steel square, Fig. 64, a quick method of finding the lengths of valley rafters is as follows: Find the 12″ mark on the tongue and 12″ mark on the blade; with a rule or another steel square measure the distance between these marks as shown in Fig. 64. The distance is almost exactly 17″. Then the run of the valley rafter is 16 × 17″. At *F*, a point 17″ from the point *C*, the rise is 10″. Thus the slope of the valley rafter is 10″ in 17″. Fig. 64 shows that the length of this portion (1/16) of the valley rafter is 19⅔″. The actual length of the valley rafter is therefore 16 × 19⅔″ or 26.3′.

Quantity of Rafters. For simple roofs, such as Fig. 60, the process of finding the quantity of rafters is simple. If the rafters are spaced 16″ o.c., as they generally are, you *divide the roof length by 1.33′ (or 16″) and add 1—or multiply by 0.75′ and add 1.*

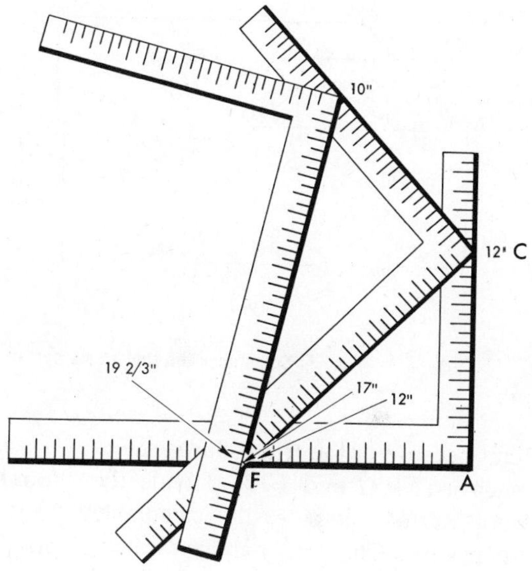

Fig. 64. A steel square can be used to find valley rafter lengths.

For rafters with 12″ spacing allow one rafter per foot (or 12″) and add 1.

Example 8. Find the number of rafters needed for the 22′-0″ roof length in Fig. 60. The spacing is 16″ o.c.

Solution:

$$\frac{22.0'}{1.33'} = 16.5 \text{ (call it 16)}$$

$$16 + 1 = 17$$

There are two sides to the roof in Fig. 60, so 2 × 17, or 34 rafters are required.

For roofs such as Fig. 61 the process is more laborious but just as simple. For example, the area *HZEF* is composed of common rafters between lines *HZ* and *FW*. The triangular section *FWE* is composed of *jack rafters*. For the area between *HZ* and *FW* the rule for 16″ o.c. rafters can be used.

$$\frac{20.0'}{1.33'} = 15$$

$$15 + 1 = 16 \text{ rafters}$$

For the *triangular* area *FWE*, another rule is used: For 16″ o.c. spacing, multiply the length by 0.5 and add 1. For 12″ o.c. spacing, multiply the length by 0.75 and add 1.

In a similar manner all other roof areas in Fig. 61 can be calculated for rafters. All dormers and other areas requiring rafters are computed in the same manner.

Extra rafters must be provided for valleys and hips; these usually are counted by studying the roof plans or elevations.

Collar beams are generally used on each pair of rafters—one per pair. The collar beams are counted and purchased in the nearest standard length.

Roof Insulation. The common roof type, Fig. 59, doesn't require special calculations when applying insulation because the area is easily determined.

For roofs such as that shown in Fig. 62 the area calculations are more diffi-

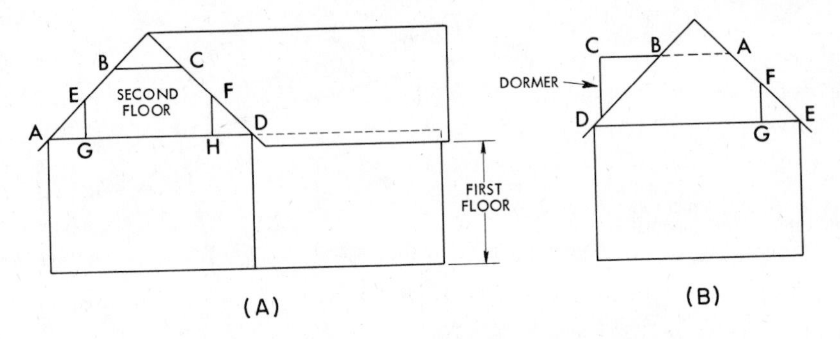

Fig. 65. Sometimes the roof forms part of the second floor ceiling.

cult because of the triangular surfaces such as *EWG* and *VFE*. Divide the roof surface into easily calculable shapes (see Chapter 4) and the calculation of areas will be simplified.

Sometimes roofs are encountered where the second-floor rooms have broken ceilings, Fig. 65-*A*. Here the roof forms part of the ceiling for the second floor. The lines *EB* and *CF* are parts of the roof. The line *BC* is the ceiling proper. In such roofs the insulation is usually specified for areas *GE, EB, BC, CF,* and *FH*. The areas *EG* and *FH* are considered side walls. To figure the areas of such roof parts, scale the width of *BC,* for example, and then scale its length and calculate the area.

Fig. 65-*B* shows a typical dormer. Here the ceiling of the room becomes *CA,* and *CD* and *FG* are side walls. Areas are calculated by scaling each part.

In full two-story houses the attic floor is sometimes insulated instead of the roof. It is then only a matter of calculating the attic floor area.

Shingles. These are usually bought by the *square* or in *bundles*. There are four bundles in one square (100 square feet). An 18″ shingle laid with 6″ exposed to the weather will cover at the rate of 1 square per 100 square feet of plain roof surface. Therefore, calculate the total roof area to be shingled and divide by 100 to find the required number of squares and bundles. Add 8% for waste on plain roofs.

Where there are hips and valleys as part of a roof, calculate the combined area and add 12% for waste. (Shingles around hips and valleys must be cut, resulting in considerable waste.)

For slate, tile, and composition shingles first calculate the area, then add 20% for slate, 18% for tile, and 5% for composition shingles.

Cornices. For cornices, you multiply the total linear footage by the width of frieze, fascia, and plancier (bottom board).

Example 9. A house has 60′ of cornice, and the plancier width is 12″, the frieze 8″, and the fascia 4″. Find the board measure necessary for this cornice.

Solution: 12″ + 8″ + 4″ = 24″,

or 2 ft. BM for every linear foot of cornice.

$$2 \times 60' = 120 \text{ board-feet}$$

Grounds and Backing. Both grounds and backing are measured in linear feet. Grounds are usually 1″×2″ strips.

Backing, generally 1″×1″, is used around door openings as a plaster guide and nailing strip. The average door requires about 35′ of backing.

The estimator or contractor must determine the places where backing is required or where grounds are necessary, and scale or calculate the linear footage. This is determined by a carefull study of the plans and all the details. Door openings are easily calculated and ground locations can be scaled.

Special Details. These details include case and cabinet work, shelf material, and small items which go into closets and storage areas.

Case and Cabinet Work. Ordinarily, cases and cabinets are mill made. The architect's designs are therefore sent to the mill. The cases or cabinets are delivered to the job ready to be installed. They are therefore listed as complete units in the quantity take-off.

If cases or cabinets are made by the contractor, the amounts and sizes of all materials must be calculated. This includes material for general enclosure, doors, drawers, glass, etc.

Shelf Material. Shelves are usually scaled to determine the required widths and lengths. The material is purchased in standard lengths and widths. Brackets for shelves are bought by the dozen. Wood strips, such as 1″×2″, are bought in standard lengths. Special mill made strips can be purchased for shelf supports.

Insulation Board for Storage Spaces. This item is found by determining the total area of the spaces or surfaces requiring insulation and then adding 6% for wastage.

Closets. For cedar closets you find the area and then add 20% for wastage.

The actual lengths of clothes poles are determined from the drawings. They are usually the width of the closet. Two brackets are required for each pole. The material is ordered in linear feet.

Hook strips are estimated by taking the length of the closet plus two times the width or depth of the closet.

Example 10. What length of hook strips is needed for a closet whose dimensions are 4′-6″ long and 2′-0″ deep?

Solution:

$$4.5' + (2 \times 2.0') = 8.5',$$

or approximately 9′

(The answer is rounded to the nearest highest foot to allow for wastage.)

Shutters. These are ordinarily mill made. Mills stock standard sizes or will make odd sizes. In any case, the shutters are listed in the quantity take-off as units.

Exterior Trim. Trim is ordered in linear feet. The necessary amounts of each kind or size are determined by scaling the drawings or adding the given

MATERIAL QUANTITY TAKE-OFF

FRAMING LUMBER

LOCATION		SIZE	PIECES/LENGTH	LIN. FT.	B.F.M.
Sill	(2)	2x6	-	350	350
Floor Beams Front Left		2x10	21/14	294	490
" Front Middle		2x10	5/10	50	84
" Front Right		3x10	13/20	260	650
" Rear Left		2x10	18/16	288	480
" Rear Middle		2x10	3/18	54	90
" Rear Middle		2x10	12/14	168	280
" Rear Right		2x10	5/12	60	100
" Right Middle		2x10	16/12	192	320
Ceiling Joists Front Left		2x8	21/14	294	392
" Front Middle		2x8	5/10	50	67
" Front Right		2x6	15/16	240	240
" Rear Left		2x8	18/16	288	384
" Rear Middle		2x8	3/18	54	72
" Rear Middle		2x8	12/14	168	224
" Rear Right		2x8	5/12	60	80
" Right Middle		2x8	16/12	192	256
" Covered Porch		2x4	9/8	72	48
" Attic		2x6	15/16	240	240
Bow Window Header	(2)	2x8	2/12	24	32
Porch Header	(2)	2x8	2/12	24	32
Porch Header	(2)	2x8	2/8	16	22
Kitchen Header	(2)	2x12	2/18	36	72
Dinette Window Header	(2)	2x8	2/8	16	22
Living Room Arch Header	(2)	2x8	2/8	16	22
Living Room Arch Header	(2)	2x10	2/8	16	22
Box Header		2x10	-	176	294
Bridging		1x3	-	660	165
Rafters House		2x8	60/20	1200	1600
Rafters Shed Dormer		2x6	15/12	180	180
Rafters Front Gable		2x6	26/14	364	364
Ridge at Gable		2x8	-	24	32
House Ridge		2x10	-	48	80
Collar Beams		2x4	16/12	192	128
Rafter Brackets		2x4	-	170	114
Post		4x4	1/8	8	11
Studs		2x4	650/8	5200	3467
Plates		2x4	-	1650	1100

NOTE: Under "Pieces" the first number denotes the number of pieces and the second number length of pieces, 21/14 is interpeted as 21 pieces, 14 feet long.

Fig. 66. House Plan A—Lumber take-off.

dimensions. When scaling drawings, care should be taken to scale only true lengths. For example, the trim for the cornice of the east and west ends of the main roof should be scaled using the east and west elevations.

Base Molding. This is figured as the perimeter of the room (in linear feet). Where many doors or a large opening exists, some deduction should be made. The perimeter of all rooms is added and the amount of base thus determined.

Base Board. This is calculated in

SHEATHING, FLOORING, SIDING, INSULATION, ETC.

LOCATION	DESCRIPTION	ACTUAL	ACTUAL + 10%
Sub floor	1x4 or plywood	2217	2439 sq. ft.
Finish floor	Oak	1932	2126 "
Roof Sheathing	1x6 or plywood	2216	2438 "
Roofing	235# asphalt	2216	2438 "
Side wall sheathing	Plywood	2253	2479 "
Siding	Flitch Pine siding	431	475 "
Wall paper	15# felt	2253	2479 "
Floor paper	15# felt	1932	2126 "
Roofing paper	15# felt	2216	2438 "
Wall insulation	2" batts	1200	1320 "
Ceiling insulation	4" batts	2100	2310 "
Kitchen underlayment	WP Plywood	174	192 "
Kitchen	Linoleum	174	192 "
Areaways	Cor. metal	5	5

NOTE: The waste factor is determined by the type of material used and the method of installation.

Fig. 67. House Plan A—Additional material.

EXTERIOR TRIM

Location	Description	Amount
Fascia	1x8	250 lin. ft.
Frieze	1x6	130 lin. ft.
Dove cote	Screened plywood......	1 unit
Gable Peak	Texture 1-11 plywood ..	1 unit
Shutters	Fixed wood	8 pieces
Soffit	WP plywood	200 sq. ft.

INTERIOR TRIM

Location	Description	Amount
Base	1x4	650 lin. ft.
Clothes pole	1 1/2" diam.	45 lin. ft.
Pole sockets	Wood	10 pair
Cleats	1x3	115 lin. ft.
Hook strip	1x4	85 lin. ft.
Shelving	1x12	200 lin. ft.

Fig. 68. House Plan A—Exterior and interior trim.

DOOR SCHEDULE

EXTERIOR DOORS

Size	Description	Amount
3'0" x 6'8" x 1 3/4"	9 lite	
	1 panel	1
2'8" x 6'8" x 1 3/4"	" "	2
	TOTAL	3

NOTE: All doors are to be ordered from the door schedule complete with doors, door frames, trim, casing, saddles, etc.

INTERIOR DOORS

Size	Description	Amount
2'6" x 6'8" x 1 3/8"	Flush H.C.	6
2'4" x 6'8" x 1 3/8"	Flush H.C.	1
2'0" x 6'8" x 1 3/8"	Flush H.C.	9
1'4" x 6'8" x 1 3/8"	Louvered	2
2'0" x 6'8"	Sliding	5
2'6" x 6'8"	Bi Fold	2 units
3'0" x 6'8"	Bi Fold	1 unit
	TOTAL	26

Fig. 69. House Plan A—Door schedule.

KITCHEN CABINETS

Location	Description	Amount
Counter	7'9" x 3'0" x 2'0"	1
Counter	4'6" x 3'0" x 2'0"	1
Counter	5'0" x 3'0" x 2'0"	1
Oven hanger	2'0" x 1'6" x 2'0"	1
Hanger	3'0" x 2'6" x 1'0"	1
Ref. hanger	3'0" x 1'6" x 1'0"	1
Sink hanger	3'0" x 1'6" x 1'0"	1
Hanger	2'6" x 2'6" x 1'0"	1
Hanger	2'0" x 2'6" x 1'0"	2
		TOTAL 10

NOTE: All counters are to be ordered complete with Formica top and back splash.

Fig. 70. House Plan A—Kitchen cabinets and accessories.

TABLE IV. CARPENTRY LABOR GUIDE

The time required for each phase of the construction of a small house.

ITEM	AMOUNT	TIME
Double sills with anchor bolts	100 lin. ft.	Man-Hours 1 1/2
Double sills without anchor bolts	100 lin. ft.	1
First floor beams	1,000 board feet	16
Sub flooring		
4'x8' plywood	1,000 sq. ft.	8
1"x6" board	"	16
1"x4" board	"	10
Exterior wall framing (partition)	100 lin. ft.	8
Interior wall framing (partition)	100 lin. ft.	8
Ceiling joists	100 board feet	2
Rafters and ridge boards	100 board feet	2 1/2
Bridging		
metal	1 set	3 minutes
wood	"	5 minutes
Wall sheathing		
1"x6" board	1,000 sq. ft.	18
4'x8' plywood	"	12
Roof sheathing		
1"x6" board	1,000 sq. ft.	22
4'x8' plywood	"	16

TABLE IV. CARPENTRY LABOR GUIDE (cont.)

ITEM	AMOUNT	TIME
Roof shingles -- 210# asphalt	100 sq. ft.	2
Wood Siding	1,000 sq. ft.	14
Exterior trim	100 lin. ft.	4
Finish floor (wood)	1,000 board feet	24
Door frames (wood)	1 frame	1
Window frames (wood)	1 frame	1
Doors (just hang)		
exterior	1 door	2
interior	"	1
Shutters	1 pair	1/2 to 2
Stairs		
main	1 stairway	3
main with open stringer, etc.	"	up to 16
basement	"	2 to 8
Door casing		
interior, one piece	1	1/2
exterior, ornamental	1	2
Window trim	1 DH window	1 1/2
Base (1"x4")	100 lin. ft.	1
Closets		
4' shelf		15 minutes
hook strip	1	10 "
clothes pole	8 lin. ft.	5 "
pole sockets	4 lin. ft.	5 "
misc. cutting & fitting	1 pair	25 "
		60 "
Door locks		
machine-installed in wood	1	1/2
hand-installed in wood	1	1
installed in metal door	1	15 min.
Window hardware (lifts and locks)	1 window (DH)	10 min.

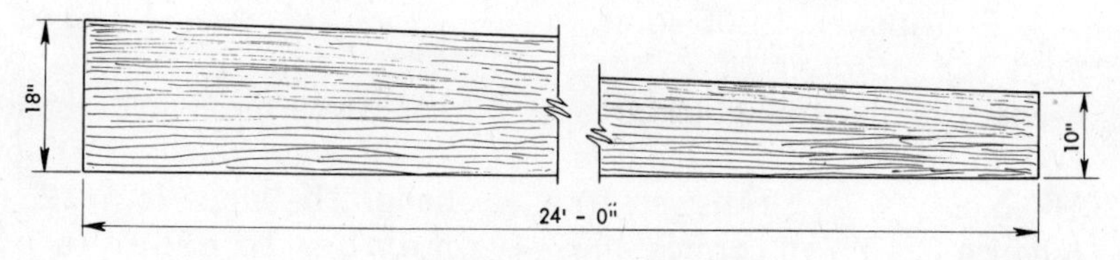

Fig. 71. A tapered board.

the same manner as base molding except that there is no deduction for door openings or arches. The widths of the door openings and arches is considered as the breakage or waste factor.

Quantity Take-Off—House Plan A. Figs. 66 through 70 are a quantity take-off for HOUSE PLAN A. The door schedule is frequently put on the working drawings (along with the window schedule —see Chapter 12). Sheathing, roofing, insulation, etc., are either on the working drawings or in the specifications.

Estimating labor

Table IV is a guide to the labor time involved in framing and millwork for a small home. The labor costs are calculated simply by multiplying the total man-hours by the hourly wages.

Review Problems

1. Find the board measure contents of a tapered board 1″ thick, 24′ long, whose ends are 10″ and 18″ wide, Fig. 71. (Answer: 28 fbm)
2. Find the number of joists required for the foundation shown in Fig. 55 if the joist spacing is 12″ o.c. (Answer: 41)
3. What is the length of molding required for a window screen which is 3′-0″ × 3′-6″? (Answer: 13.0′)
4. Find the area of a 3/8 pitch roof whose attic floor area is 60′-0″ × 20′-0″. (Disregard area of cornice.) (Answer: 1,500 sq. ft.)
5. Find the number of rafters needed for the triangular area **FWE** in Fig. 61. The spacing is 16″ o.c. (Answer: 9)
6. A house has 75′ of cornice; the plancier width is 12″, the frieze is 8″, and the fascia is 6″. Find the board measure (to the nearest foot) necessary for the cornice. (Answer: 163′ BM)
7. A room is 10′ × 13′ with a 3′-0″ wide door opening and a 6′-0″ wide arched opening. Find the amount of base molding needed for the room. (Answer: 37′)
8. How much ceiling molding is needed for the previous problem? (Answer: 46′)
9. Calculate the amount of sheathing needed for a sloped roof which is 40′-0″ long and has common rafters 18′-0″ long. (Answer: 1,440 sq. ft.)
10. What length of common rafter would you order for a building 20′-0″ wide with a 1′-0″ overhang? The slope is 5/12. (Remember to order in even lengths.) (Answer: 14′)

8

MASONRY

The masonry section of the specifications includes brick, concrete block, tile, and other structural clay or concrete products, plus the necessary mortar and anchors. The estimator should keep the complete set of plans together when he is making his estimate of the materials and labor involved in this section of the specifications.

Materials

Bricks and Blocks. The most common materials which the estimator will find in the specifications are brick, concrete block, and tile. Brick is usually one of four kinds: common brick, firebrick, face brick, and SCR. The kind of concrete block most commonly encountered is the three-hole variety. Bricks and blocks, of course, are available in dozens of varieties and sizes.

Common brick is the brick usually used in construction work. It is made of ordinary clays or shales, has no special color or surface texture, and is not specially marked or scored.

The size of the standard American common brick is 8″ × 3¾″ × 2¼″. However, because of the somewhat uneven temperature at different places in the usual brick kiln, bricks will vary somewhat in size. In quantity, the differences in size will tend to even out, however, so that for fairly large surfaces this size may be used for a quantity take-off. See Fig. 1.

Modular brick is usually 7½″ × 2⅙″ × 3½″. The size is based upon the module (usually a 4″ unit). The two dimensions, which are factors in the width of a building, will work out to modular units when the width of the mortar is included.

Firebrick is used in fireplaces, furnaces, and other places subjected to high temperatures. It is made of a special kind of clay which resists crumbling and cracking at high temperature.

The usual dimensions for firebrick (with mortar) are 8″ × 3½″ × 2¼″ or 9″ × 4½″ × 2½″ but there are innumerable sizes and shapes available.

Face brick is manufactured under special conditions. The temperature

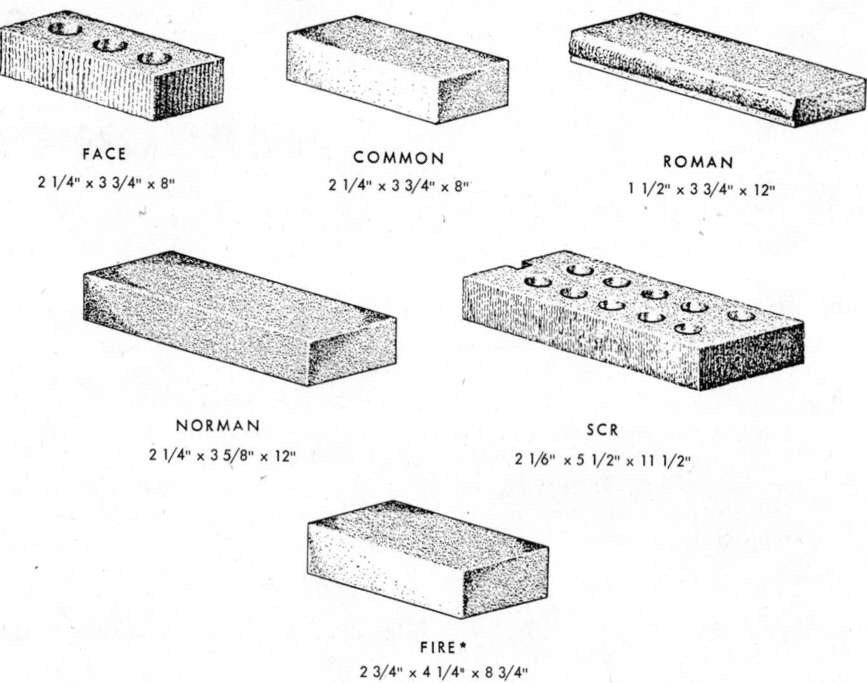

Fig. 1. Sizes of various kinds of bricks.

and materials are closely controlled, so that the color and the texture of the brick can be controlled. Hardness, uniform size, and strength are all of a high quality. Face brick is available in the same sizes as common brick. See Fig. 1.

SCR brick is characterized by a greater thickness than other brick. It is 11½" × 2⅙" × 5½". This thickness allows it to be used in single *wythe* construction. (A wythe is a single layer of bricks.) See Fig. 1.

Glass brick, often called glass block, is a hollow block made of glass. It has the advantage of being translucent, that is, it admits light but is not transparent. It is never used as a load-bearing wall, but it is often used for decorative effects.

Concrete block, as its name suggests, is a type of block made of sand and concrete. It is used in much the same manner as brick and, when it is to be used in a building, will be specified in both the drawings and specifications. The standard wall block is available with actual face dimensions of 15⅝" × 7⅝" and a variety of widths (7⅝" is common). Nominal dimensions of 16" and 8" are used for estimating purposes. See Fig. 2.

Cinder block, a lightweight concrete block, is also in common use.

Terra Cotta. This material is a mixture of high quality clays with cal-

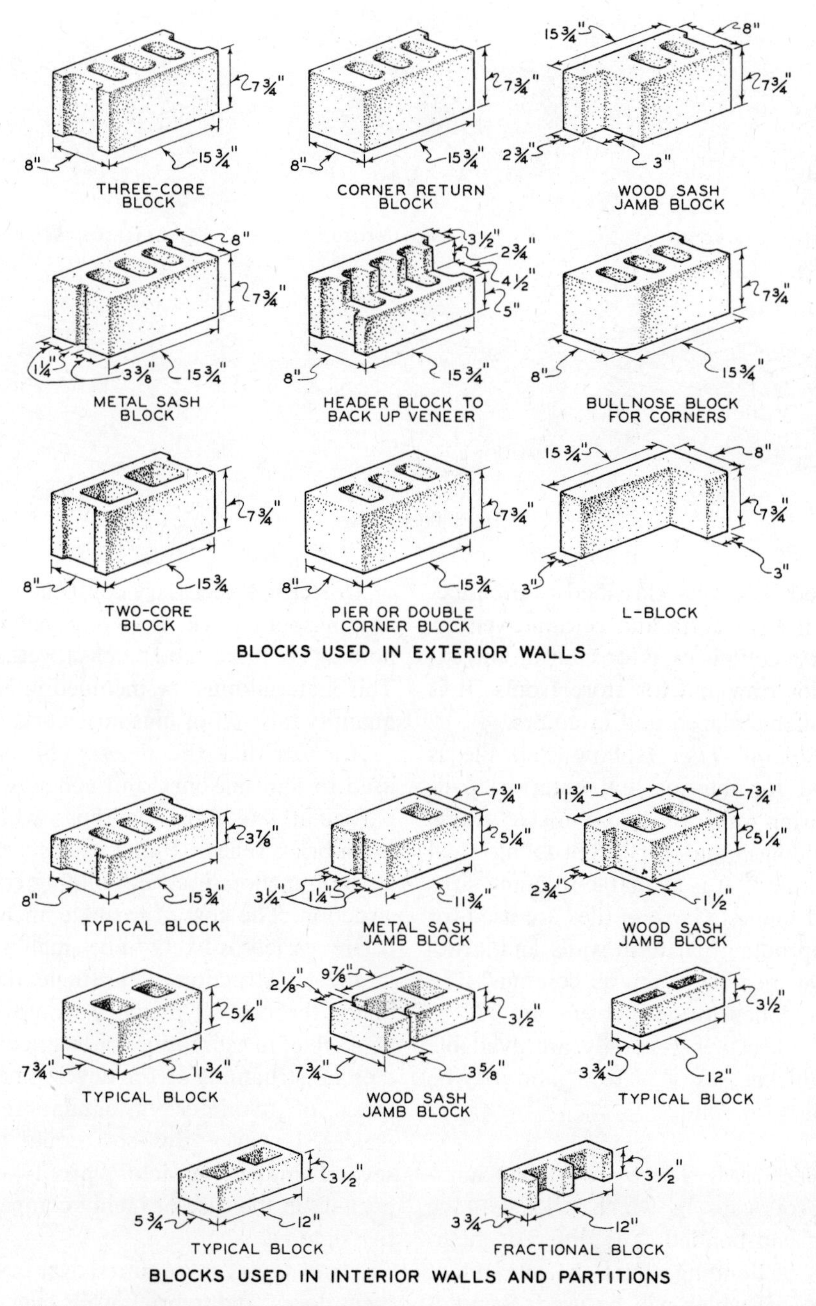

Fig. 2. Concrete blocks.

ENGLISH CROSS BOND

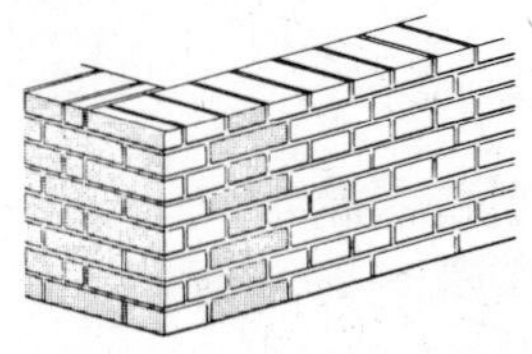

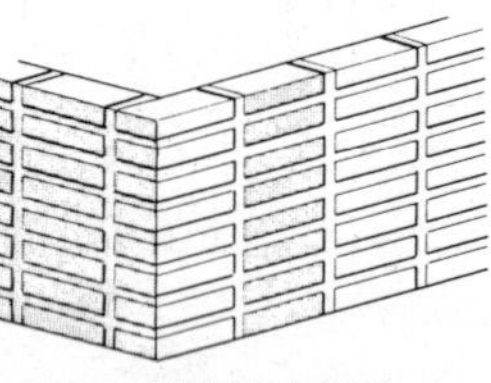

Fig. 3. Brick courses.

cined clay. It is classified as architectural terra cotta and ceramic veneer. Terra cotta is used for interior and exterior trim and for store fronts. It is available glazed and in colors.

Hollow Tile. Hollow clay tile is used for interior and exterior load-bearing walls, for floors in steel frame buildings, and for fireproofing steel members. It is available in many sizes and shapes. *Gypsum* tiles are used for fireproofing partition walls and structural members such as columns. The face dimensions of gypsum are 12″ × 30″. The tiles generally are available in thicknesses of 2″ to 6″ and may be hollow or solid.

Brick Bonds. There are a great number of ways in which brick can be laid and bonded. The different methods of bonding affect the amount of material which will be used. Some of the various bonds are illustrated in Fig. 3.

Mortar is a necessary constituent of any masonry work. It is required for holding the bricks and blocks together. This material must be included in any quantity take-off of masonry work.

Anchors, like the *sleeper clip,* are used to join masonry and concrete—particularly in those buildings which use a brick veneer. Fig. 4 shows a variety of anchors used in building construction. The cost of a single anchor of any variety is likely to be small and in the construction of a single residence the cost of anchors would be negligible. In estimating for large commercial structures or for development tracts of hundreds of similar residences, however, this cost could cut severely into a contractor's profit—especially in the case of highly competitive bidding.

Metal angles are required over basement doors, and in brick walls over all windows and doors. Angles also are used in fireplace construction. See Fig.

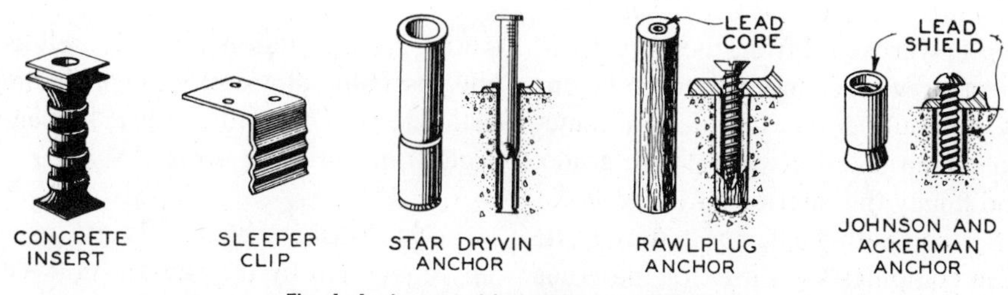

Fig. 4. Anchors used in building construction.

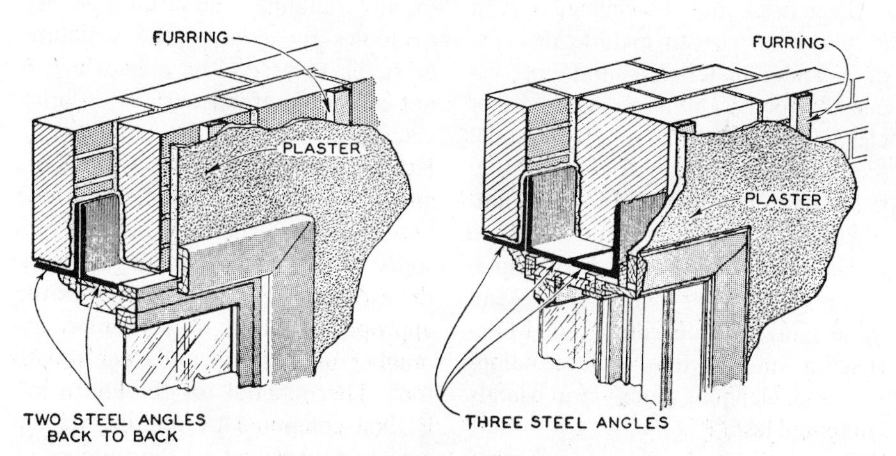

Fig. 5. Steel angles used above windows.

5. For example, a 3′-8″ long angle is required above windows in a basement for any 3′-0″ × 3′-0″ window opening.

Scaffolding. Many contractors use tubular scaffolding which can be readily assembled on the job. The cost of scaffolding materials is usually computed and spread out over the projected life span of the equipment. In cases where the contractor owns the scaffolding equipment, the estimator usually charges equipment rental to the job; this provides for replacement of the equipment when it becomes necessary to do so. However, scaffolding must sometimes be constructed on the job and, therefore, may be included with the cost of the masonry work.

Scaffolding may also be rented. Schedules of the cost of equipment rental are compiled yearly by the Associated Equipment Distributors.*

Estimating materials

When estimating the costs of materials, it is usual to work from the bot-

*For the cost of a copy of the current schedule, you should write to Associated Equipment Distributors, 30 E. Cedar, Chicago, Illinois 60611.

tom upwards and from the outside in. In other words, the estimator begins by computing the foundation materials, then exterior walls above grade, and finally the interior masonry work. After computing exterior materials, he then computes such interior materials as the fireplace veneer and hearth.

Since brick must be cleaned after it is laid, it is wise to include the cost of cleaning materials and labor, especially for *tuckpointing*. The cost of cleaning materials is relatively small (compared to brick, block, and tile) but it should be included in the quantity take-off. Since wood and metal are sometimes necessary for the completion of the masonry work, the cost of these must also be computed. Materials for lintels in the fireplace, dampers, and cleanout doors are usually computed last.

The preliminary steps in masonry work have been discussed in other sections of this book. Excavations, concrete work, footings, and foundations will not, therefore, be discussed here, although many estimators will consider them as part of the masonry cost of a building.

In general, when estimating masonry materials, the contractor should get the *delivered* cost of the materials. When taking a materials bid for a contract, he should assure himself that the quoted price includes delivery. In this way he can avoid the task of computing transportation costs, as well as the possibility of a sizeable error in the estimate which could result if he neglects transportation costs.

Bricks. Masonry materials are usually computed by the square foot. To estimate the number of masonry units in any building, the estimator first computes the area of the structure. Next, he subtracts the area which is not composed of masonry. (In a brick veneer residence, such as the one in HOUSE PLAN A, the areas of doors and windows are subtracted.) These two steps yield the total area of masonry material. After finding the area, the estimator uses one or more of the appropriate tables to determine the number of masonry units per square foot. The material for the entire job is then computed by multiplying the area in square feet by the number of masonry units per square foot.

Example 1. Calculate the total area of brick required for a 2-car, brick veneer garage. The floor plan dimensions are 25′-0″ × 25′-0″. The garage has a flat roof; the elevation is 10′-0″. There are two floor-to-ceiling doors, each 9′-0″ wide, in the front of the garage.

Solution: First compute the area of the exterior walls.

$$25.0' \times 10.0' = 250.0 \text{ sq. ft.}$$

Since there are 4 such walls,

$$4 \times 250 = 1{,}000 \text{ (sq. ft.)}$$

TABLE I. STANDARD SIZE FACE BRICK PER SQUARE FOOT

JOINT	1/8"	1/4"	3/8"	1/2"	5/8"	3/4"	1"
NUMBER OF BRICKS	7 1/2	7	6 1/2	6 1/8	5 3/4	5 1/2	5

TABLE II. PERCENTAGES ADDED FOR VARIOUS BONDS

	PER CENT	
COMMON (header course every 5th course)	20	(1/5)
COMMON (header course every 6th course)	16 2/3	(1/6)
COMMON (header course every 7th course)	14 1/3	(1/7)
ENGLISH OR ENGLISH CROSS (full headers every 6th course)	16 2/3	(1/6)
FLEMISH (full headers every 5th course)	6 2/3	(1/15)
FLEMISH (full headers every 6th course)	5 2/3	(1/18)
TWO STRETCHER GARDEN WALL (full headers every 5th course)	4	(1/25)
THREE STRETCHER GARDEN WALL (full headers every 5th course)	2 7/8	(1/35)
DOUBLE HEADERS (alternating with stretcher every 5th course)	10	(1/10)
DOUBLE HEADERS (alternating with stretcher every 6th course)	8 1/3	(1/12)

Now, find the areas of the openings.

$9.0' \times 10.0' = 90.0$ sq. ft.

$2 \times 90 = 180$ (sq. ft.)

Deduct the area of the openings.

$1{,}000 - 180 = 820$ (sq. ft.)

Example 2. The total area of brick work in the previous example was 820 sq. ft. Since the brick is unspecified, assume that common brick is to be used. Table I shows the number of common brick (or face brick) per sq. ft. (Since the mortar size is unspecified, assume a ½″ joint.) Six and one-eighth bricks are needed for every square foot of masonry area. Calculate the number of bricks needed for the garage.

Solution: Multiply the area by the number of bricks per unit of area.

$820 \times 6\frac{1}{8} = 5{,}022\frac{1}{2}$ bricks

As Table I gives the quantities for standard size brick laid in running bond, additional allowances must be made for the number of brick when bonds using headers are used. The percentages in Table II are added to the number of required bricks (as calculated by the use of Table I) when the face brick are laid with the bonds indicated.

Table III can be used to estimate the number of face bricks required for laying the more common bonds. With this table the number of face brick for any wall or area may be figured easily. To make such an estimate for face brick, calculate the total area of wall surface and deduct the area of all openings except those containing less than 10 square feet. Then from Table III the number of face brick required can be obtained for half-inch joints. If other sized joints are to be used, corrections can be figured using the data at the foot of the table.

Example 3. Assume a total wall area of 2,546 square feet, with openings amounting to 276 square feet. How many face brick are required for a running bond?

Solution: Find the brick area and use Table III.

$2{,}546 - 276 = 2{,}270$ sq. ft.

2,000 sq. ft. take	12,320 face brick
200 sq. ft. take	1,232 face brick
70 sq. ft. take	432 face brick
2,270 total	13,984 total

If half-inch joints are used, 14,000 face brick will be required. If a ⅝″ joint is to be used, subtract 5% from 14,000 as shown at the foot of Table III.

TABLE III. NUMBER OF STANDARD SIZE FACE BRICK AND COMMON BRICK IN MASONRY WALLS

Laid with ½" Joints in Various Bonds

Sq.Ft. of Wall	Running			Common Header Course Every 7th Course			English and English Cross* Full Headers Every 6th Course			Flemish Full Headers Every 5th Course			D'ble Headers Alternating with Stretchers Every 5th Course			Sq.Ft. of Wall
	Face Brick	Common Brick in		Face Brick	Common Brick in		Face Brick	Common Brick in		Face Brick	Common Brick in		Face Brick	Common Brick in		
		8" Wall	12" Wall		8" Wall	12" Wall		8" Wall	12" Wall		8" Wall	12" Wall		8" Wall	12" Wall	
1	6.16	6.16	12.32	7.04	5.28	11.44	7.19	5.13	11.29	6.57	5.75	11.91	6.78	5.54	11.70	1
5	31	31	62	36	27	58	36	26	57	33	29	60	34	28	59	5
10	62	62	124	71	53	115	72	52	113	66	58	120	68	56	117	10
20	124	124	248	141	106	229	144	103	226	132	115	239	136	111	234	20
30	185	185	370	212	159	344	216	154	339	198	173	358	204	167	351	30
40	247	247	494	282	212	458	288	206	452	263	230	477	272	222	468	40
50	308	308	616	352	264	572	360	257	565	329	288	596	339	277	585	50
60	370	370	740	423	317	687	432	308	675	395	345	715	407	333	702	60
70	432	432	864	493	370	801	504	360	791	460	403	834	475	388	819	70
80	493	493	986	564	423	916	576	411	904	526	460	953	543	444	936	80
90	555	555	1110	634	476	1030	648	462	1017	592	518	1072	611	499	1053	90
100	616	616	1232	704	528	1144	719	513	1129	657	575	1191	678	554	1170	100
200	1232	1232	2464	1408	1056	2288	1438	1026	2258	1314	1150	2382	1356	1108	2340	200
300	1848	1848	3696	2110	1584	3432	2157	1539	3387	1971	1725	3573	2034	1662	3510	300
400	2464	2464	4928	2816	2112	4576	2876	2052	4516	2628	2300	4764	2712	2216	4680	400
500	3080	3080	6160	3520	2640	5720	3595	2565	5645	3285	2875	5955	3390	2770	5850	500
600	3696	3696	7392	4224	3168	6864	4314	3078	6774	3942	3450	7146	4068	3324	7020	600
700	4312	4312	8624	4928	3696	8010	5033	3591	7903	4599	4025	8337	4746	3878	8190	700
800	4928	4928	9856	5632	4224	9152	5752	4104	9032	5256	4600	9528	5424	4432	9360	800
900	5544	5544	11088	6336	4752	10296	6471	4617	10161	5913	5175	10719	6102	4986	10530	900
1000	6160	6160	12320	7040	5280	11440	7190	5130	11290	6570	5750	11910	6780	5540	11700	1000
2000	12320	12320	24640	14080	10560	22880	14380	10260	22580	13140	11500	23820	13560	11080	23400	2000

*The quantities in this column also apply to common bond with headers in every sixth course.

For other than ½" joints, the following percentages must be added to or subtracted from above results:

Add: for ⅛" joint, 21%; for ¼" joint, 14%; for ⅜" joint, 7%.

Subtract: for ⅝" joint, 5%; for ¾" joint, 10%; for ⅞" joint, 15%; for 1" joint, 20%.

If the bricks used in Example 3 are not of standard size then, as previously explained, the area of the brick, plus its mortar joint, must be determined and the bricks per square foot figured and multiplied by 2,270 square feet.

Soldier courses make no difference in the count, nor do *rowlock* courses, provided half brick only are used. If full brick are used in a rowlock course, however, once again as many face brick as the course requires must be added, while the same amount is subtracted from the quantity of common brick. But for window sills, laid rowlock, no special provision need be made, as the usual allowances mentioned in the following paragraph will be sufficient.

If the workmen are careful to use *bats* (a piece of brick with one end whole, the other end broken off) for closures, instead of breaking whole bricks, no wastage need be figured. The area of small openings, not deducted in calculating quantities, and the doubling of brick at the corners, will allow a surplus of extra brick.

Since it is customary to order the brick to the quarter-thousand next above the actual number calculated, the ordinary small extras, as well as slight wastage, will be amply provided for.

For all practical purposes, common brick may be considered the same in size as the standard face brick. Any odd-sized common brick can be figured by calculating the area of the brick plus its mortar joint and dividing that area into 144 square inches. This gives the number of bricks per square foot for a 4″ wall. For an 8″ wall multiply by 2, and for a 12″ wall multiply by 3, etc.

Table III can be used to readily calculate the number of common brick, of standard size, required per foot. Because of the nature of the face brick bond used, it is only when there is a running bond that the number of backing brick in the 8″ wall equals the number of face brick. For additional thickness of the backing walls, however, this number should be added for each additional tier of backing in all bonds. This is due to the fact that the bonding face brick extending into the first tier of backing brick takes just so much more face brick and so much less backing brick; additional tiers of backing are not affected. The number for running bond applies to any common brick wall throughout, such as foundations, partitions, etc. For work of this kind, simply multiply this number by the number of brick tiers in the thickness of the wall.

While each tier of common brick takes about 20 less brick per vertical foot all around the wall, this saving may be disregarded in the case of the 8″ wall, as the extra brick will come in handy for fire stops and similar applications. If however, two or more tiers of backing brick are used, an arithmetical progression by twenties per vertical foot should be deducted for each successive tier. Thus, beginning with the first tier at 20 brick, the series runs 40, 60, 80, etc.

Many estimators use the following simple rules. These rules apply only to common brick.

½″ Joints

1 sq. ft. of 4″ wall has 6 bricks
1 sq. ft. of 8″ wall has 12 bricks
1 sq. ft. of 12″ wall has 18 bricks

¼″ Joints

1 sq. ft. of 4″ wall has 7 bricks
1 sq. ft. of 8″ wall has 13 bricks
1 sq. ft. of 12″ wall has 20 bricks

Small chimneys. Table IV is used to estimate the amount of common brick for a 4″ wall around the flue lining, and at 13 courses to 3′ in height.

Chimneys that have more than one flue lining must have 4″ of brickwork between flues.

Fireplaces. In estimating for fireplaces, figure the portions projecting

TABLE IV. BRICKS PER FOOT OF HEIGHT FOR SMALL CHIMNEYS

NO. OF FLUES	SIZE OF FLUE INCHES	NO. OF BRICKS PER FT. OF HT.
1	8 x 8	26
2	8 x 8	44
3	8 x 8	63
1	8 x 12	31
2	8 x 12	52
3	8 x 12	74
1	12 x 12	38
2	12 x 12	60
..	......	..

between the line of the wall, such as breast and ash pit, as solid areas; that is, the number of bricks for the surface multiplied by the number of tiers deep, deducting the number of bricks displaced by all flues and openings, face brick facing, and firebrick lining.

For chimneys above the roof, measure linear distance around the chimney and multiply by height above the roof. If there are 7 face brick per square foot, multiply the area by 7.

Trim. In estimating the quantity of face brick used as trim, you measure the linear feet of trim and multiply by the number of brick per linear foot.

Example 4. Find the number of standard size brick, face and common, laid in a sixth course common bond with a ⅜″ joint, for an 8″ gable wall, 25′ wide and 18′ high from grade to eaves, and then 12′ to ridge pole. The 12″ cellar wall is 7½′ high, 4½′ being below grade level. There are four windows, each requiring an opening of 3′-6″ × 5′-2″ and one window requiring an opening for 2′-6″ × 4′-2″. The cellar windows, being less than 10 square feet, are disregarded.

Solution: First find the area in square feet for face brick.

Rectangle of wall 18′ × 25′ = 450
Gable triangle 12′ × 25′ = 300
450 sq. ft.
300 ÷ 2 = 150 sq. ft.

Total area = 600 sq. ft.

Now, find and deduct the 5 window openings.

4 × 3.5′ × 5.17′ = 72.3 sq. ft.
1 × 2.5′ × 4.17′ = 10.4 sq. ft.

Total window openings 82.7 sq. ft.
(call it 83 sq. ft.)

Face brick area = 600 − 83
= 517 sq. ft.

Using Table III, you can see that 3,739 face brick are needed (for ½″ joints). Adding 7% for a ⅜″ joint,

3,739 + 262 = 4,001 face brick
(call it 4,000)

By using Table III for common brick, 2,668 brick are needed for backing.

Concrete Blocks. First find the total linear measurement of the wall in which concrete blocks are used. This includes all four sides of the basement (for most buildings), interior walls, etc. Then divide the total by the length of one block. (It is assumed here that blocks 16″ long, 8″ wide, and 8″ high are used.)

Dividing the total length by 1.33′ (16″) gives the number of blocks in one course.

Next determine the height of the concrete block wall and then divide it by 0.67′ (8″). This gives the number of courses. (Be sure to subtract openings.)

Tile. Hollow tile for backing may be obtained in the following typical sizes:

3¾″ × 5″ × 12″
(turned, 5″ × 3¾″ × 12″)
3¾″ × 12″ × 12″
6″ × 12″ × 12″
8″ × 12″ × 12″

When ordering the 5″ × 12″ shapes, which are laid on their sides, the contractor should state that the usual allowances of 6″ and 9″ length cuts be included for use in piers and other narrow places, so as to reduce the cutting of tile. Half and full closures

TABLE V. NUMBER OF TILE PER SQUARE FOOT (1/2" JOINT)

DIMENSIONS OF TILE FACE	3 3/4" x 12"	5" x 12"	12" x 12"
NUMBER OF TILE	2.7	2.15	0.94

TABLE VI. RELATIONS OF FACE BRICK AND TILE COURSES

BRICK JOINT	NUMBER OF BRICK AND TILE COURSES	TILE JOINT	BOND COURSE
† 1/8"	7 BRICK COURSES = 3 (3 3/4" x 5" x 12") TILE COURSES	1/2" +	8
* 1/4"	5 BRICK COURSES = 1 (3 3/4" x 12" x 12") TILE COURSE	1/2"	6
† 3/8"	5 BRICK COURSES = 3 (5" x 3 3/4" x 12") TILE COURSES	5/8"	6
† 1/2"	4 BRICK COURSES = 2 (3 3/4" x 5" x 12") TILE COURSES	1/2"	5
† 1/2"	6 BRICK COURSES = 3 (3 3/4" x 5" x 12") TILE COURSES	1/2"	7
† 5/8"	6 BRICK COURSES = 3 (3 3/4 x 5" x 12") TILE COURSES	3/4"	7
* 3/4"	4 BRICK COURSES = 1 (3 3/4 x 12" x 12") TILE COURSE	3/4"	5
† 1"	5 BRICK COURSES = 3 (3 3/4" x 5" x 12") TILE COURSES	3/8" +	6

With a 3/8" brick joint, 3 3/4" x 5" x 12" tile are laid on 5" side, resulting in a 5" backing.

*Not suitable for use with 3 3/4" x 5" x 12" tile.
†Not suitable for use with 3 3/4" x 12" x 12" tile.

should be ordered for use at window and door openings when 8″ × 5″ × 12″ tiles are used.

To obtain the number of required tiles, take the area to be backed in square feet and multiply by the proper number in Table V.

As the position of the brick bending course in the wall is determined by the size of the tile used for backing and, in a measure, by the width of the mortar joint, it is desirable to know how the brick courses and mortar joints are related to the height of the tiles and their mortar joints. While almost any width joint may be used with brick, it is advisable never to go below a ⅜″ nor above a ¾″ joint with tile.

Table VI gives possible combinations of brick courses and hollow tile backing, such as would be useful in average circumstances.

As the face brick is bonded with full headers to the tile backing, the amount of tile required will have to be reduced accordingly; that is, with a bond every fifth course, one-fifth less tile is needed; with a bond every sixth course, one-sixth less tile, etc.

Where the bonding courses have stretchers in them, such as Flemish bond, brick-size tile must be provided to fill the backing courses where the stretchers occur. Naturally, in bonds such as common or English, where the bonding courses are entirely of headers, no brick-size tile will be needed.

Table VII gives, in per cent, the number of brick-size tile needed for various bonds containing stretchers in the bonding courses. Remember that percentages of the total number of brick required must be taken for the face of the wall, as figured for running bond.

Note that when the tile backing is 8″ instead of 3¾″ thick, an additional course of brick-size tile should be pro-

TABLE VII. BRICK-SIZE TILE IN VARIOUS BOND COURSES

	PER CENT	
FLEMISH (full headers every 5th course)	13 1/3	(2/15)
FLEMISH (full headers every 6th course)	11 1/8	(1/9)
TWO STRETCHER GARDEN WALL (full headers every 5th course)	16	(4/25)
THREE STRETCHER GARDEN WALL (full headers every 5th course)	17 1/8	(4/35)
DOUBLE HEADERS (alternating with stretcher every 5th course)	10	(1/10)
DOUBLE HEADERS (alternating with stretcher every 6th course)	8 1/3	(1/12)

vided for behind the entire bonding course. If the brick-size tile cannot be obtained in the local market, an equal number of common brick will be sufficient. Two or three per cent should be added to all calculated tile quantities to provide for waste.

All of these calculations are based on the practice of using brick for bonding. If the builder prefers to use *metal ties,* however, these are inserted in the fifth, sixth, or seventh course as the tile heights make most convenient or as the local building codes may require. In such a case, disregard what is said about extra brick or brick-size tile for bonding courses. You simply calculate the quantities for running bond, with allowances for the different mortar joints, as the tables indicate.

Example 5. Find the number of 3¾″ × 5″ × 12″ tile for backing above the first floor in the wall, as given in Example 4.

(NOTE: A ⅜″ joint does not work out with this size tile without reducing the tile joint too much, so take a ½″ brick joint which allows 6 courses of brick to 3 courses of tile with ½″ joints, making the bond every seventh course. Using Table III, the number of face brick is reduced. Thus 520 sq. ft. take 3,661 face brick, which means that 3,750 should be ordered.)

Solution: As previously suggested, when the backing differs in different portions of the wall, it is better first to figure the backing areas separately. In our present case, the hollow tile backing begins with the first floor, which is 3′ above grade.

$$15' \times 25' = 375 \text{ sq. ft.}$$
$$6' \times 25' = 150 \text{ sq. ft.}$$
$$525 \text{ sq. ft.}$$

Deduct window openings, as before.

$$525 - 83 = 442 \text{ sq. ft.}$$

Now use Table V.

$$442 \times 2.15 = 951$$

Deduct one-seventh, or 136

815

Adding 3% for wastage,

$$815 + 25 = 840 \text{ tiles}$$

If 8″ × 5″ × 12″ tile is used, however, you deduct the area occupied by the closures from the above area. A full and a half closure are respectively 12″ and 6″ in length and about 1′ high, thereby covering an area of ¾ square feet on each side, or 1.5 square feet on both sides of the opening. As the height of openings is approximately 25 feet, you have 25′ × 1.5′ = 37.5 square feet to be deducted.

$$442 - 37.5 = 404.5 \text{ sq. ft.}$$

This amount is then divided by the unit number of tile per square foot.

Example 6. Estimate the number of closures required in Example 5 if it

TABLE VIII. ESTIMATING QUANTITY OF MATERIALS FOR VARIOUS MORTARS
Quantities Based on Laying 1,000 Brick with ½" Joints

Proportions	4" Wall				8" Wall				12" Wall			
	Cement Sacks	Lime Lump or Hydrated Bbls.	Lime Lump or Hydrated Sacks	Sand Cu. Yds.	Cement Sacks	Lime Lump or Hydrated Bbls.	Lime Lump or Hydrated Sacks	Sand Cu. Yds.	Cement Sacks	Lime Lump or Hydrated Bbls.	Lime Lump or Hydrated Sacks	Sand Cu. Yds.
Cement Mortars												
Cement 1—Sand 2	3.90	...	....	.43	5.00	...	....	.55	5.40	....	....	.60
Cement 1—Sand 2½	3.30	...	....	.43	4.20	...	....	.55	4.40	....	....	.60
Cement 1—Sand 3	2.90	...	....	.43	3.70	...	....	.55	4.00	....	....	.60
Lime Mortars												
Lime 1—Sand 2	....	.75	3.00	.43	....	.96	3.80	.55	....	1.00	4.20	.60
Lime 1—Sand 2½	....	.65	2.60	.43	....	.82	3.30	.55	....	.90	3.60	.60
Lime 1—Sand 3	....	.54	2.20	.43	....	.69	2.70	.55	....	.70	3.00	.60
Cement-Lime Mortar												
Cement 1, Lime 1—Sand 6	1.75	.43	1.75	.43	2.20	.55	2.20	.55	2.40	.60	2.40	.60

Note: The quantities given under the heading 4" wall are for all brickwork, whether backed with brick or tile, or as veneer, and for each tier of backing brick above grade and other work in which the joint between the tiers of brick is not filled. The quantities given under the heading 8" wall and 12" wall are for foundation work below grade or other places where the joints between the tiers of brick are filled with mortar.

TABLE IX. QUANTITY OF MORTAR MATERIALS FOR LAYING 1,000 TILE WITH 1/2" JOINTS

TILE DIMENSIONS	CEMENT SACKS	LIME LUMP OR HYDRATED BBLS.	LIME LUMP OR HYDRATED SACKS	SAND CU. YDS.
3 3/4" x 5" x 12"	6.50	.20	1.00	.66
8" x 5" x 12"	13.00	.40	2.00	1.33
3 3/4" x 12" x 12"	7.50	.25	1.25	.75
8" x 12" x 12"	17.00	.66	2.75	2.00

Portland cement is sold in bags of approximately 94 pounds each, or in barrels of 4 bags, running about 380 pounds. Lump line may be got by the bushel of from 75 to 85 pounds, or the barrel of about 180 pounds. The sack of hydrated lime used in the table weighs 50 pounds.

takes 4 closures, 2 full and 2 half, for each foot of height (both sides).

Solution:

25 × 4 = 100
Deduct one-seventh 14
86

Adding 3% for wastage,

86 + 3 = 89

Order 45 of each size.

Mortar. Table VIII shows the estimated quantity of materials for various mortars based on laying 1,000 brick with ½" joints.

Brickwork below grade level usually is laid with ½" joints. For brick or tile above grade, in other than ½" joints, add to or subtract from the items under "4" wall," as a unit, the following quantities:

For ⅛" joints....subtract ¾
For ¼" joints....subtract ½
For ⅜" joints....subtract ¼
For ⅝" joints....add ¼
For ¾" joints....add ½
For 1" joints....double

One cubic yard of sand, 6 sacks of cement, and 10% lime, well mixed, will lay about 1,000 concrete blocks of the 16″ × 8″ × 8″ size. The 10% lime is determined by first taking 10% of the cement volume.

Tile. Table IX is used to calculate the mortar necessary for 1,000 tiles with ½″ joints.

Estimating labor

Since there are many different kinds of bricks, blocks, and tiles, in addition to working conditions which vary greatly among regions, it is best to refer to tables and charts for your own area. The following is an estimate of the time required for the laying of 100 square feet of face brick. Remember that these figures can vary considerably in different regions.

Bricklayer	8 man-hours
Laborer	6 man-hours
Foreman	¾ man-hours
Misc. labor	¾ man-hours

Multiply each of these figures by the hourly wage rate (in your area) and then add the various labor costs. This will give you the total labor cost for 100 square feet of wall (about 7,000 brick).

For cinder blocks, one bricklayer with a laborer will lay almost 300 bricks (16″ × 8″ × 8″) in 8 hours. To this must be added about 1.5 hours of supervision and miscellaneous work.

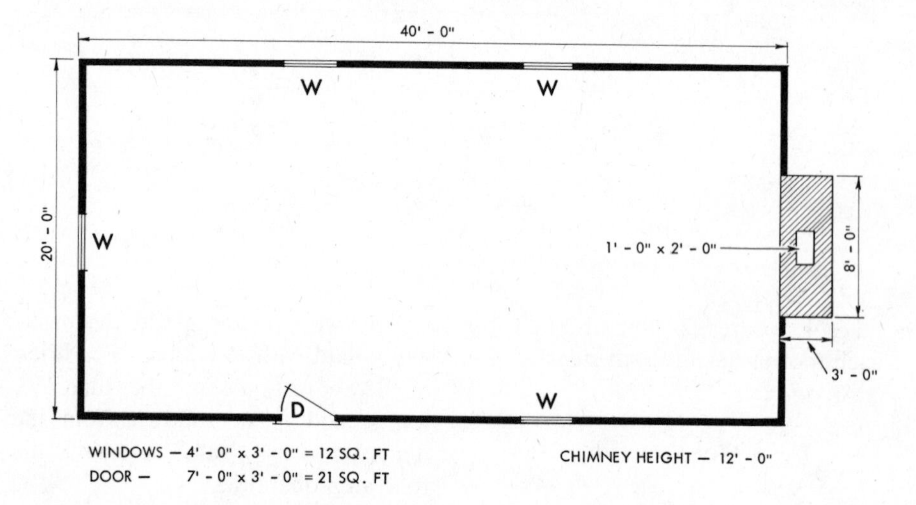

Fig. 6. House plan used for masonry calculations.

Questions and Problems

1. List the various kinds of brick used in the building trades.
2. How is scaffolding estimated?
3. How is brick trim estimated?

4. Estimate the area of the brick-work needed for the house shown in Fig. 6. (Ans.: 1,069 sq. ft.)
5. How many bricks are needed for the house in Fig. 6? (Answer: 8,018)
6. What is the volume to be bricked in the chimney in Fig. 6? (Answer: 264 cu. ft.)
7. Estimate the area of face brick needed for a building 100′-0″ × 100′-0″, whose height is 20′-0″. Window and door outs amount to 2,000 sq. ft. (Answer: 6,000 sq. ft.)
8. What is the area of backup brick needed for the building in the above problem? (The area of the spandrel beams is 1,000 sq. ft.) (Answer: 5,000 sq. ft.)

9

WET AND DRY WALLS

Wall coverings are divided into two major divisions—wet wall and dry wall. Plaster and stucco are examples of wet wall coverings; plasterboard and wood paneling are types of dry wall in common use.

It is impossible to fully explain all of the various wall coverings in a single chapter but this chapter will give you sufficient information to begin estimating this phase of construction.

Wet wall materials

Plaster. Plaster has two basic functions in a building. First, it provides a surface for finishing walls and partitions to make them acceptable from the standpoint of livability, and second, it serves as a means of sealing out dirt, cold, and heat.

Because of its smooth surface plaster gives interior areas a pleasing appearance. Moreover, plaster offers a surface upon which paint, calcimine, paper, canvas, and other decorative finishes may be applied.

The term plaster originally referred to a mixture of lime, sand, hair, and water which was widely used. Although plaster is not as widely used nowadays, improved mixtures and methods of application have kept "plastering" popular among builders. See Fig. 1.

Patented Plasters. These plasters (also called *chemical plasters*) are produced by processes patented by their manufacturers. The patent plasters are usually quick drying. The surface hardens more quickly and resists abrasure longer than the ordinary lime plastering. However, once a break occurs, the extreme stiffness of the mixture makes the crack liable to extend further and to be of a more serious nature than in the softer, more flexible lime plaster covering.

The extra stiffness of most patented plasters is caused by the cement which generally forms an important ingredient of their composition. These plasters are sold ready for use, requiring merely the addition of water. They

Fig. 1. Plaster being sprayed on a wall. United States Gypsum Company.

are, therefore, especially adapted for use by the inexperienced and are valuable for small pieces of work. Because of the quick-setting property of most patented plasters, it is inadvisable to use them on wood lath, as the subsequent drying and shrinkage of the lath will almost invariably cause cracking of the plaster.

Gypsum Plasters. These plasters are commonly called *plaster of Paris.* Glass-fibered gypsum plaster is one of the most versatile plasters ever developed. It controls the evenness of the scratch coat to produce a good bond.

Interior plastering is now applied in either two or three coatings. Three coats are required on metal or wire lath, the first serving mainly to stiffen the body of the material sufficiently for thorough working of the remaining coats. Even upon wood laths, three coats make a better job of plastering than two. Extra strength and body are obtained by the addition of the extra coat, provided time is allowed for each of the coats to thoroughly dry out before the next coat is applied. On the less expensive residential work, however, two coats are customarily applied.

The best interior plaster work is worked to a final thickness of about ⅞". Of the three coatings, the first coat is the thickest, so that, when dry, it may be strong enough to resist the pressure of working the coat or coats to follow. A large part of the advantage of three-coat plastering is obtained by *thoroughly drying out each coat before applying the next,* thus securing the added density and strength made possible by forcing the the next coating.

While it is generally the custom to add rough plaster finish on the second coat, in inexpensive work (such as summer residences), a very artistic effect is obtained by rough-working the surface of the first coat. Using but one coat, however, it is not possible to work the surface as true and as evenly as when two coats are applied.

Two-Coat Work. Most plaster work now consists of only two coats. The brown mortar used for the first coat should be made of fresh lime—used as soon as it is stiff enough to be worked. The first coat of mortar must always be put on with sufficient pressure to force the plaster through between the laths and thus ensure a good clinch. The face of this coat must be made as true and even as possible on angles and plumb on vertical surfaces. After the first coat is sufficiently set, it may be worked again, using a float that consists of a piece of hard pine about the size of the trowel. Sometimes the face of this float is covered with felt or other material to produce a rough texture on the plaster. The first coat should run a strong one-half inch in thickness, measured from the outer surface of the laths, and should be thoroughly dried out.

In general, it is inadvisable to trowel a two-coat job smooth. If the attempt is made to float the first coat when it is too thin or is insufficiently set, the tool is likely to leave marks on the wall and the plaster itself is liable to crack. If the plaster becomes too dry, it may be dampened by sprinkling water on it with the plasterer's broad calcimine brush. This is followed immediately with the float.

The finish second coat in two-coat work is the same as the final skim coat in three-coat work.

The Finish Coat. The finish, skim, or white coat should never be applied until the earler coats are thoroughly dry and hard. A simple putty coat should carry more sand than when the finish is hardened by the addition of plaster of Paris. If plaster of Paris is used, the mortar should always be *gauged* (that is, plaster of Paris should be mixed with the putty) *after* it is placed on the mortarboard. The usual method of gauging consists of making a hollow with the trowel in the lime putty lying on the mortarboard. This hollow is filled with water, the plaster of Paris is sprinkled upon it, and this mixed rapidly with the trowel and applied to the wall before the plaster has time to set. The proportion of lime and plaster of Paris, while variable, averages probably ¼ to ⅕ plaster of Paris.

The finish is skimmed in a very thin coating that is generally less than ⅛" in thickness. It is immediately troweled several times, dampened with a wet brush, and again thoroughly troweled to smooth up the surface and prevent it from chipping or cracking. The water prevents the steel trowel from staining the surface, but the plaster should not be dampened too much as it will then blister or peel. The entire surface of the finish coat, whether of putty or hard finish, finally should be brushed over once or twice with a wet brush. If a polished (or buffed) surface is desired, it is obtained by brushing—without dipping the brush into the water—until a glossy surface is obtained.

Suction. This process is absolutely necessary in order to obtain the proper bond of plaster on masonry (or concrete). It is necessary in the first and second coats so that the final coat will bond properly.

Uniform suction helps to obtain uniform color. If one part of the wall draws more moisture from the plaster than another, the finish coat may be spotty.

Obtain uniform suction by dampening, but not soaking, the wall evenly before applying the plaster. A fog spray is recommended for this work.

Curing. Since Portland cement plaster is generally exposed to severe use, it should be given every opportunity to develop its maximum strength and density through proper curing. The

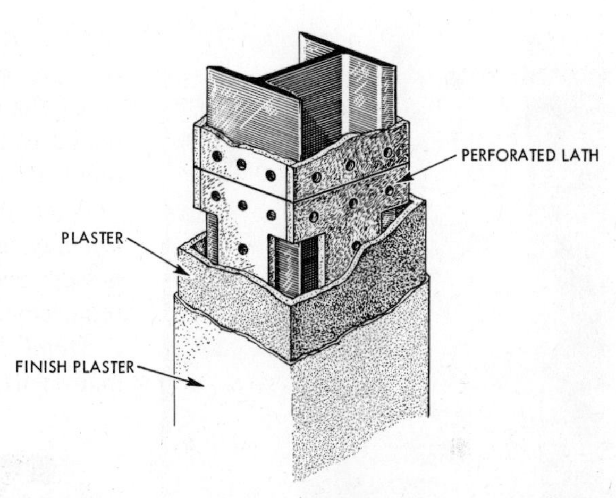

Fig. 2. Fireproofing a steel column with lath and plaster. Bestwall Gypsum Company.

method of curing Portland cement is simple: (1) keep brown and finish coats continuously damp for at least two days; begin moistening each coat as soon as the plaster has hardened sufficiently so that it won't be injured; the water should be applied in a fine fog spray and soaking should be avoided; (2) after the damp-curing period, allow the plaster coat to dry thoroughly before the next coat is applied.

Plaster on Masonry. If plaster is applied on a *stone* or *brick wall* or on tile, a scratch coat is seldom necessary as the scratch coat is used principally to insure a clinch back of the lathing and to provide a surface to receive the brown coat.

Fireproofing. Plaster is used with incombustible lathing for fireproofing walls, ceilings, beams, and columns. A 2″ solid partition of gypsum lath and sanded plaster will qualify for a one-hour fire rating. Fireproof lath and plaster ceilings are often suspended from structural members. Fig. 2 shows a fireproof steel column.

Stucco. Another wet wall material, stucco, is commonly used on the exteriors of buildings. Stucco is composed largely of cement. It is applied to a wire mesh backup which is usually nailed to furring strips. On masonry buildings, no furring is required.

Lathing. Serving as a base for plaster, lathing also may provide definite resistance to the flow of heat or cold. The most common types of laths are as follows:

1. board
2. insulation
3. metal
4. wood

Board lathing is made in large sheets, approximately ⅜″ thick. The overall dimensions are about 16″ × 32″ or 16″ × 48″. Fig. 3 shows a long-length board lath running the height of the room. Such laths can be inserted into metal floor runners and held in place by clips at the ceiling. Typical board laths are made of gypsum and are often called plasterboard,

Fig. 3. Erecting a two-inch solid plaster partition. United States Gypsum Company.

although they are used primarily as lathing.

The use of board lath saves time in lathing and can contribute some insulating value. Some contractors save one plaster coat when board laths are used, although such procedure should be indicated in the written specifications.

A *Satterly* partition is shown in Fig. 4. This is a method of erecting partitions without studs, using two layers of gypsum boards, one horizontal, the other vertical. The layers are held together by wire ties or clips.

Insulation is being used to a great extent as plaster backing. The several types are made in variously sized sheets, with the usual 16″ o.c. spacing of studs considered. Like the board types, the insulating plaster backing is nailed to the studs. As with all other types of lathing, the plaster is applied directly to the insulation. The material may be purchased in various thicknesses according to the insulating requirements.

Metal laths are manufactured in many varieties. Most laths are manu-

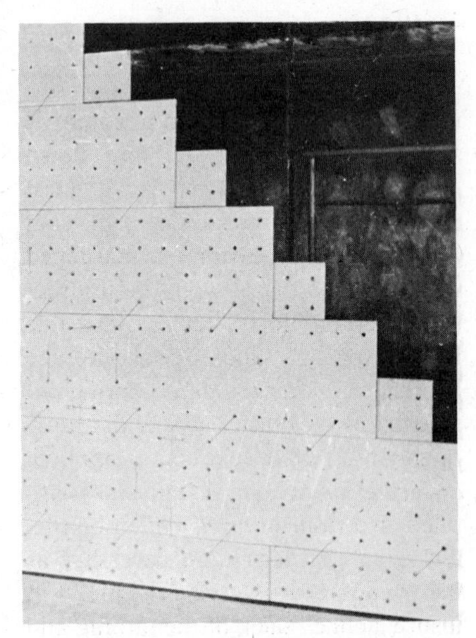

Fig. 4. A Satterly partition. Note the wire ties.

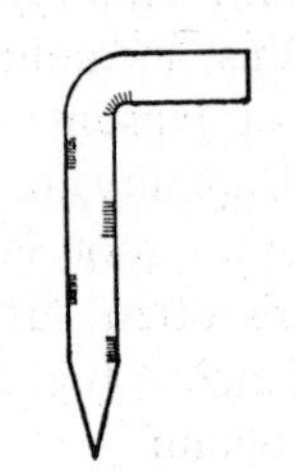

Fig. 5. A lathing nail.

factured in sheet sizes of 24″ × 96″ and 27″ × 96″.

The laths are nailed to studs or joists by a special nail of the type shown in Fig. 5. Metal lathing nails are bright, smooth, and have a long thin head especially suited for applying metal lath. There are approximately 278 such nails to the pound. Contractors often use an ordinary straight nail, drive it part way in, and then bend it over. All metal lath should be nailed at close intervals to prevent loosening or cracking of the plaster.

The lath should be well lapped at splice points and the overlap should be nailed securely.

Metal lath is especially suited for use around steel columns and beams where there is no woodwork. In such cases, it is wired on tightly. Corner beads are wired on at the same time. Metal lath is also employed in corners, on wood or steel framed arches, and where fireproof construction is required, as in garage ceilings. Where the lath is to be wired on, the estimator must include the cost of the wire and the labor involved in securing the laths.

Diamond mesh, Fig. 6, is used for all types of plastering. This metal lath weighs 2.5 or 3.4 pounds per square yard and is painted or galvanized to protect it from rust.

Flat rib metal lath is a combination of expanded metal lath and ribs, Fig. 7. Each rib does not exceed 3/16″ in depth. Its added rigidity permits wider spacing of supports.

Rib metal lath, Fig. 8, is a combination of expanded metal lath and ribs; each rib has a total depth of approxi-

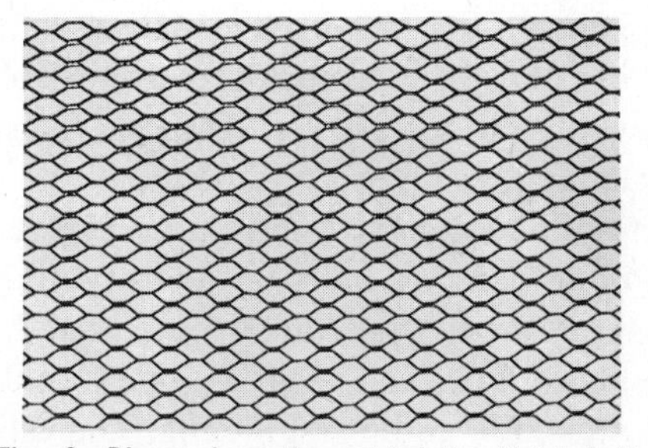

Fig. 6. Diamond mesh metal lath. Metal Lath Manufacturers Association.

Fig. 7. Flat rib metal lath. Metal Lath Manufacturers Association.

Fig. 8. Rib metal lath. Metal Lath Manufacturers Association.

mately ⅜″. It is used as plaster lathing on wider spacings than is possible with the flat rib or diamond mesh lathing.

Wood Laths. These laths are seldom used in modern construction. They have been largely replaced by board material such as sheet rock, rock lath, and metal lathing. The estimator, however, may encounter wood laths in remodeling and repair work.

Wood laths are generally made of white pine, spruce, or hemlock. Other woods containing high percentages of pitch, or woods which are too hard, are not recommended. The common size for wood laths is ¼″ × 1½″ × 4′. This size may vary somewhat in width but, in general, it is the most common size. Laths are delivered to the job in bundles of 100 but are purchased by the thousand. Laths should be new, clean, and free from knots. Old or used laths do not provide a good bond, and dirt or knots may cause a stain to bleed through the plaster.

Wood laths are usually nailed with ¼″ spacing between them, although lath spacing up to ⅜″ is sometimes allowed.

Grounds. Remember that grounds are required whenever wood or metal is nailed up close to the plaster.

Corner Beads. Plaster will not stand severe bumps without chipping or cracking. This is especially true at corners where no casings ars used. Also, it is difficult to make a perfectly uniform edge all around a plastered opening or to keep a vertical edge perfectly straight. The time involved in this work is considerable. To overcome these problems, corner beads are used, Fig. 9. Corner beads are nailed to the corners and the lathing is brought in close to the bead before plastering. After the plastering is completed, only the bead is visible and it can be painted, papered, or calcimined.

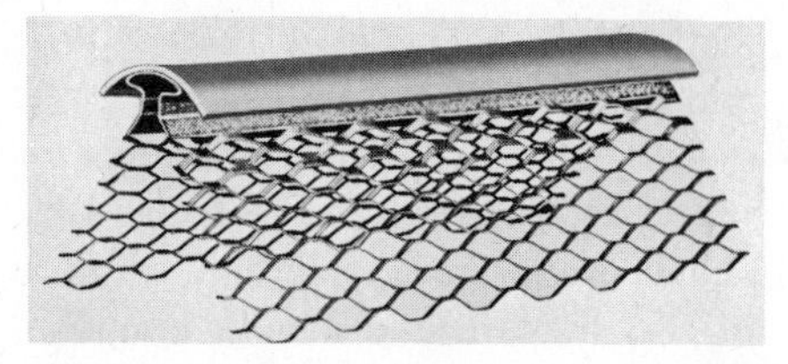

Fig. 9. Bull-nose corner bead. Metal Lath Manufacturers Association.

When corner beads are used with wired-on metal lath, binding wire is used to fasten the beads in place.

Metal Trim. There are many other metal accessories used as plastering aids and as trim. Fig. 10 shows some of the various metal accessories used in wet wall construction.

Metal Arches. As previously explained, for an arched opening 2 × 4 framing is constructed over the arch. If the arch is to be plastered, metal arches (made of heavy sheet metal containing a surface of alternate holes and solid metal) are used for plaster keying. Such arches are purchased in many shapes and widths and are complete and ready to be nailed into place. Their use saves much labor time.

Furring. As a means of preventing moisture from coming through the plaster, and also to create dead air spaces for insulating purposes, all masonry walls are furred before being plastered. Furring strips vary in size,

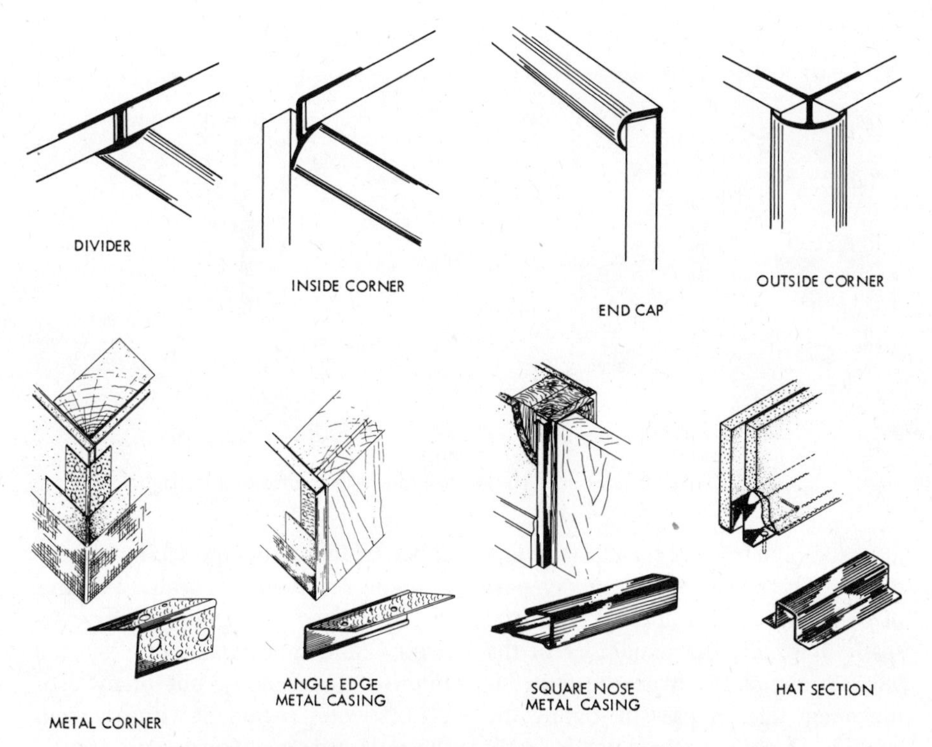

Fig. 10. Metal trim used in wet wall construction. Bestwall Gypsum Company.

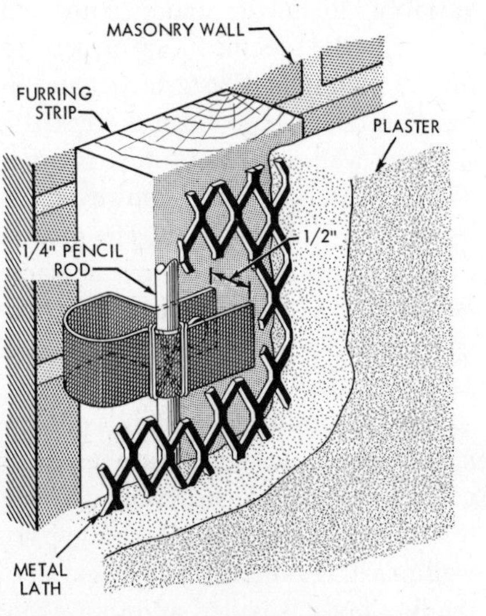

Fig. 11. Wood furring on masonry. Note the clip system.

but a 1 × 2 piece of wood is typical. These strips are nailed at intervals of 16″ o.c. directly to the masonry wall, using a cut nail. They must be plumbed to insure a uniform plaster surface. The lathing is nailed to the studs and the plaster applied to the lathing. See Fig. 11.

The wood framing at corners often is made so that it will be easier to nail the lathing. Fig. 12 shows corner framing which provides nailing surfaces for lathing and helps in furring the walls.

Scaffolding. The plasterer generally scaffolds the room with boards at a sufficient height to enable him to easily reach the ceiling without stretching his

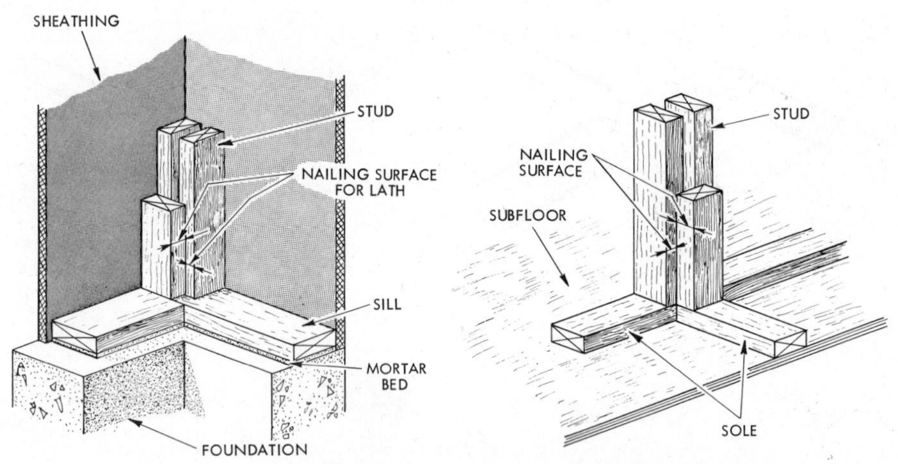

Fig. 12. Corner framing can be planned to create nailing surfaces for lathing.

arms too high to work the plaster coats evenly. The plaster for the upper part of the walls is also applied from the scaffolding, and the remainder of the work is completed from the floor. If too much time elapses in joining the coats at this point, the joint is likely to show. This is not serious unless the walls are to be left untreated. Occasionally two men working together, one on the scaffolding and one on the floor, finish their work at the same time.

Dry wall materials

Wet wall construction is being replaced by dry wall in much of today's modest-priced housing. Dry wall includes wood paneling, plasterboard, and a variety of acoustical and decorative types of paneling. (Chapter 11 discusses acoustical tile and special wall coverings.)

Wood paneling may be included either in the carpentry section of the estimate or under dry wall. We have included wood paneling under carpentry (and in Chapter 11, "Special Interior Coverings"), but many estimators would include it with dry wall. Plasterboard, wallboard, and insulation board, however, are almost invariably included under lath and plaster in the specifications or description of materials. Plasterboard and insulation board have been discussed in the wet wall section of this chapter. These materials may, however, be used as dry wall materials. The method of estimating them remains the same. When used as dry wall materials, they will require additional treatment such as taping and spackling.

Fig. 13 shows various kinds of *laminated partitions*. These are used for ceiling and wall construction. The material is glass fiber, reinforced gypsum wallboard. In the two-layer system, the first layer is nailed and the second is applied with adhesive and screws.

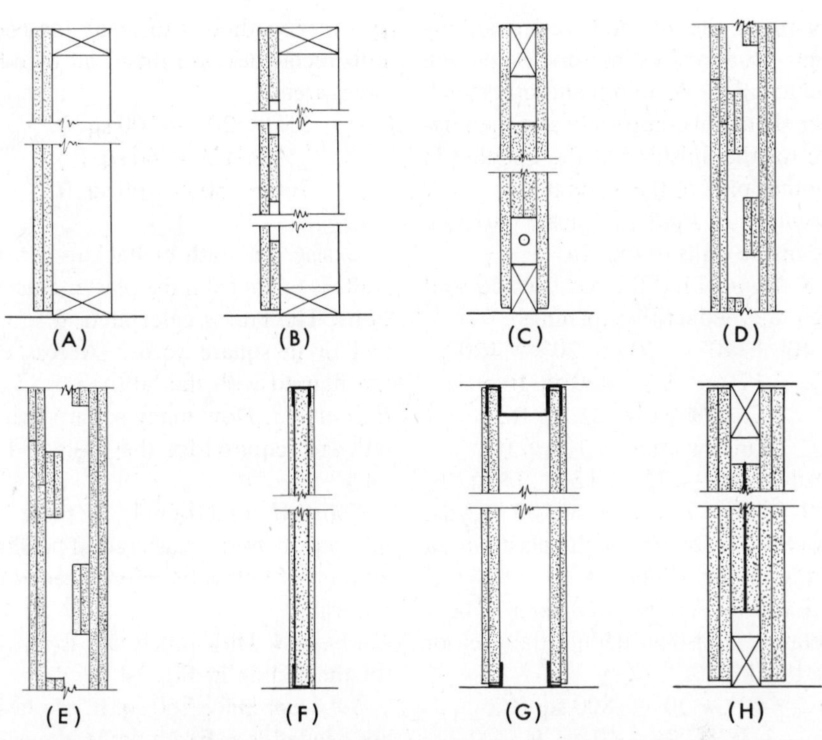

Fig. 13. Laminated partitions: (A) two-layer, (B) strip, (C) 2¼″ semi-solid, (D) 4″ semi-solid non-load bearing, (E) 5″ semi-solid non-load bearing, (F) 2″ solid, (G) 6″ double solid, (H) 2¼″ semi-solid non-load bearing (movable). Bestwall Gypsum Company.

Dry Wall Installation. This includes such costs as nailing, taping, and spackling.

Nailing. Nails used to hold plasterboard to studs or laths are driven in using a lather's hammer. This special hammer has a convex head which drives the nail beneath the surface of the plasterboard. The nail head is then filled, spackled, and painted.

Taping and Spackling. In many cases where plasterboard or some other sheet form of dry wall is used as wall covering, some method of concealing the seams is necessary. The basic material used for such concealment is gauze-perforated tape applied with a gypsum cement which acts as a filler and adhesive. This tape is available in 3″ and 4″ strips.

The tape is *spackled* after it has been applied to the seam. Spackling involves putting a very thin layer of finish plaster over the strip and blending this with the texture of the board.

Estimating material

Plaster. Plaster is calculated by the square foot or by the square yard. This is true regardless of whether it is a one, two, or three coat job. With regard to the outs or deductions for openings, the method varies among contractors and estimators. Some will deduct all the openings and others will take them out completely and then figure for the finishing of the opening in another part of the estimate.

Example 1. Find the plaster area of the inside walls in Fig. 14.

Solution: Find the total inside wall area and deduct the openings.

$$40' + 40' + 20' + 20' = 120'$$
$$120' \times 8' = 960 \text{ sq. ft.}$$
$$\text{Door area} = 21 \text{ sq. ft.}$$
$$\text{Window area} = 12 \text{ sq. ft.}$$
$$\text{Deductions} = 21 + 12 = 33 \text{ sq. ft.}$$
$$960 - 33 = 927 \text{ sq. ft. or } 103 \text{ sq. yds.}$$

Example 2. Determine the plaster area of the ceiling in Fig. 14.

Solution: Assume the room to be a rectangle and then deduct the area of the break.

$$40' \times 20' = 800 \text{ sq. ft.}$$
$$8' \times 5' = 40 \text{ sq. ft.}$$
$$800 - 40 = 760 \text{ sq. ft.}$$

Another method is to break the room into rectangles and then add together their areas.

$$35' \times 20' = 700 \text{ sq. ft.}$$
$$5' \times 12' = 60 \text{ sq. ft.}$$
$$700 + 60 = 760 \text{ sq. ft.}$$

Lathing. The lath or backup for wet walls is taken from the plaster calculations. The lath is calculated in square feet or in square yards. (Accessories are figured with the lath.)

Example 3. How many square feet of lath are required for the walls in Fig. 14?

Solution: In Example 1, there are 927 sq. ft. to be plastered. The same amount of lath is therefore needed for the walls.

Example 4. How much lath is needed for the ceiling in Fig. 14?

Solution: Since 760 sq. ft. of plaster are needed (see Example 2), the same amount of lathing is required.

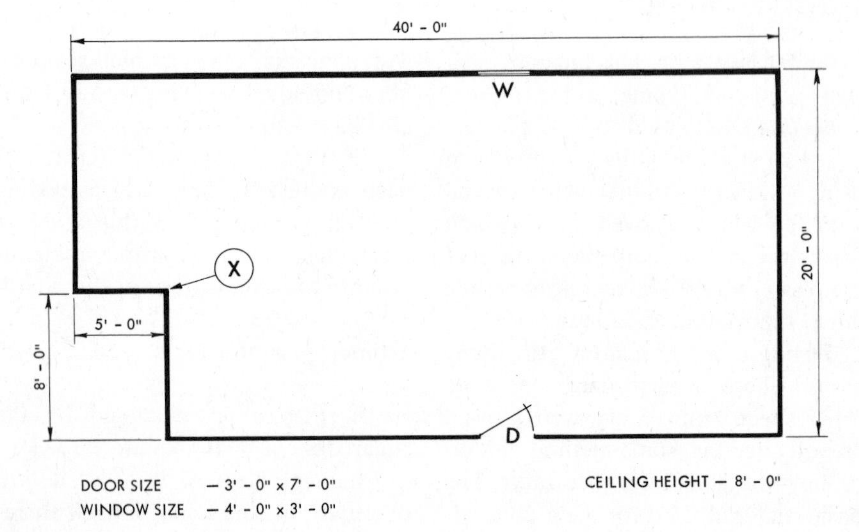

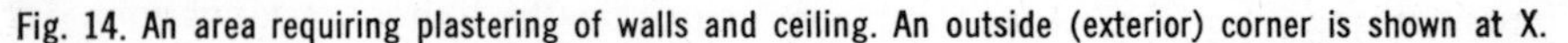

Fig. 14. An area requiring plastering of walls and ceiling. An outside (exterior) corner is shown at X.

Lathing for plasterboard and insulation board is calculated in the same manner. The sheet size used will depend on the on center distances of the studs. The cost is found by multiplying the cost per sheet by the number of sheets.

Corner Beads. These are calculated by the linear foot. The corners requiring beading are determined and the total length of bead calculated for each corner. For example, around a plastered opening rectangular in shape, the floor-to-ceiling height is multiplied by four. Curved surfaces requiring beading are estimated as to length as closely as possible.

Example 5. How many linear feet of outside corner bead will be required in Fig. 14?

Solution: In Fig. 14 there is one outside corner (at *X*). Thus 8′ of corner bead will be sufficient.

Example 6. How many linear feet of inside corner bead will be required in Fig. 14?

Solution: There are 5 inside corners.

$$5 \times 8' = 40 \text{ lin. ft.}$$

Nails. A general rule for determining the number of nails necessary for metal lathing is to allow 9 pounds of 3*d* nails per 1,000 square feet of lathing. Nailing requirements, however, vary according to the type and weight of the lath. This item is often strictly controlled by local building codes; if not specified, you will want to consult the manufacturer's specifications.

For plasterboard and insulation board (and most other board backing), allow 10 pounds of 3*d* nails per 1,000 square feet of board. The number of nails per sheet varies, of course, with the sheet size and is controlled by the local building code. In either case, you should follow the building code or manufacturer's specifications.

For corner beads, allow 1½ pounds of 3*d* nails per 100 linear feet.

Dry Wall. The square foot method is used for all dry wall materials except acoustical tile, which may be calculated by the square foot or by the piece.

Plasterboard, insulation board, and wallboard are manufactured in a number of standard-sized sheets. If screws are used (especially for plasterboard), they are spaced not over 12″ apart on ceilings and a maximum of 16″ o.c. on walls. If wall studs are spaced 24″ o.c., the screw spacing must not exceed 12″.

Tape. Seam concealing materials are estimated by the linear foot. First determine the number of sheets of dry wall material and the height of the seams. Then add the total number of feet needed for the tape to obtain the linear footage required for each room or job.

Estimating labor

Wall construction is frequently given as a unit cost which includes materials, labor, and equipment such as scaffolding. Although labor time will vary among different regions, there are some general rules which can be used.

Wet Wall. Three coats of plaster on metal lath can be calculated at the rate of 1 man-hour for every 3 square yards of plaster. Two coats on masonry require 1 man-hour for 6 square yards of plaster.

On frame construction, stucco can be applied at the rate of 2½ square yards per man-hour. On masonry walls, stucco will be applied at the rate of 2 square yards per man-hour.

A good lather can install about 1,000 square feet of metal lath in an 8-hour day.

Arches are normally installed at a unit cost per opening.

Dry Wall. For plasterboard or insulating board (which also may be used as lathing), one man can install over 1,050 square feet of sheets in 8 hours.

The estimator should check the labor times and costs for his own area. Remember that unit costs will include materials *and* labor.

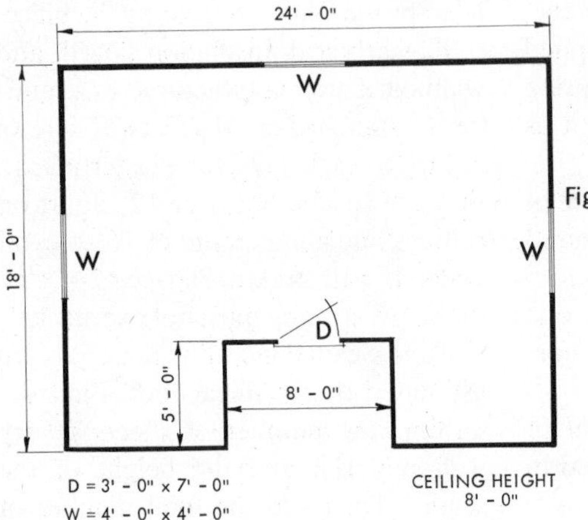

Fig. 15. A large room which is to be plastered.

Review Problems

1. How many square yards of plaster are required for the walls in Fig. 15?
   (Answer: 83.6 sq. yds.)
2. How many square feet of plaster are required for the ceiling in Fig. 15?
   (Answer: 392 sq. ft.)
3. How many linear feet of inside corner beads are needed in Fig. 15?
   (Answer: 48 ft.)
4. How many linear feet of outside corner beads are needed?
   (Answer: 16 ft.)
5. Assume that the ceiling will have tiles 18″ × 18″ installed. How many tiles are needed?
   (Answer: 175)
6. Assume that sheets of plasterboard will be used for the walls. Each sheet is 8′-0″ × 4′-0″. How many sheets are needed?
   (Answer: 24)

7. Assume the same size plasterboard sheets are to be used for the ceiling. How many sheets are needed? (Answer: 13)
8. Based on the number of sheets used in Problem 6, above, how many linear feet of taping are required for the walls? (Answer: 192 ft.)
9. How many linear feet of ceiling corner tape are needed for this room? (Answer: 94 ft.)
10. Assume that the studs are spaced 24″ o.c. How many metal studs are needed for the interior partitions of the room? (Answer: 47)

10

ELECTRICAL WIRING

The architect does not show details of the wiring on the building plans. However, he does show where the ceiling lights, wall or bracket lights, convenience outlets, wall switches, electric motors, heating appliances, etc., are to be located. Generally there is a set of typewritten specifications giving more details in regard to what is desired. The details of the particular make or kind of apparatus are usually left to the judgment of the electrical contractor who obtains the architect's approval.

Wiring methods

Estimating the cost of electrical wiring is one of the most complicated areas of the estimating field. In most cases it is left to a specialist and the building estimator is never called upon to give more than a rough idea of the cost.

Six different methods of wiring for a residence have been approved by the National Board of Fire Underwriters in the National Electrical Code and are as follows:

1. Open conductors
2. Concealed knob-and-tube work
3. Nonmetallic sheathed cable
4. Armored cable
5. Conduit work
6. Electrical metallic tubing

It is very seldom that more than one method of wiring is used in the same building. The particular method used on any job depends upon the requirements of the situation *and* local ordinances. For example, open conductor systems are, by nature, restricted to makeshift or temporary applications. Concealed knob-and-tube work is now virtually obsolete. It is specifically prohibited by the building codes of many communities and, for this reason, will not be discussed here. Nonmetallic sheathed cable and armored cable are widely used for residential wiring in smaller cities, towns, and rural areas. In large cities, and for industrial applications, however, wiring is required to be either rigid conduit or electrical tubing. (The latter is more expensive.)

The wires used for interior work are copper conductors that have been

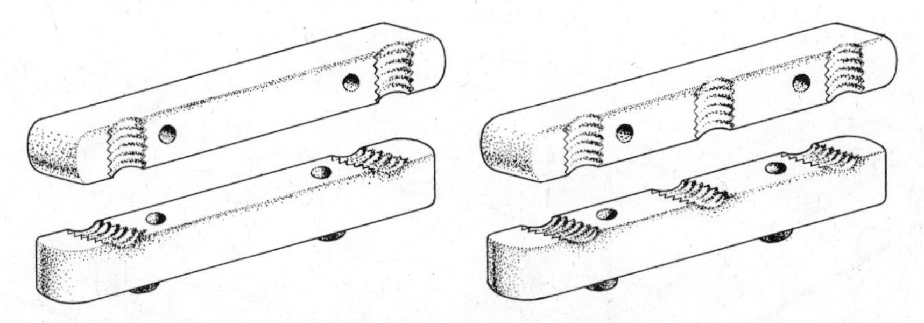

Fig. 1. Porcelain cleats are used with open conductor wiring.

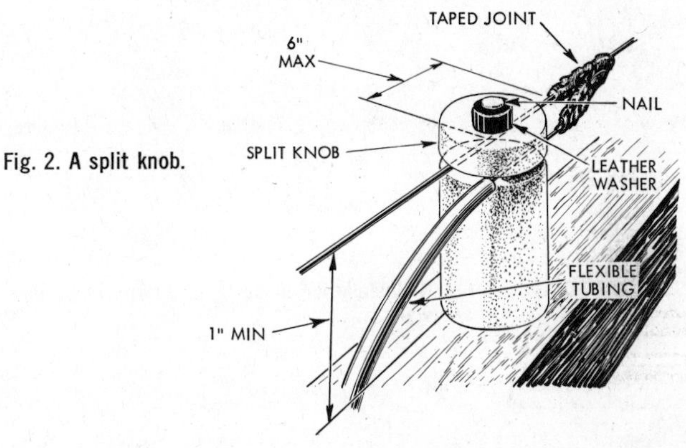

Fig. 2. A split knob.

treated with a thin coating of tin and then covered with rubber insulation. A cotton braided material is woven over the rubber insulation, and the braid is saturated with a fire-resisting compound which leaves a hard, smooth surface. The smallest size of wire that can be used in residences is No. 14 American Wire Gage.

The various methods of wiring are discussed in the following sections.

Open Conductors. In this method the rubber-covered wires are kept away from each other and from the building by the use of porcelain supports. These may be porcelain cleats, Fig. 1, split knobs, Fig. 2, or solid knobs, Fig. 3. The method of installing the wires on ceilings and walls is shown in Fig. 4. This method of wiring is used only in the smallest and lowest priced buildings, such as summer cottages or cabins which are occupied only a few months during the year, cheaply constructed sheds or buildings that are used for storage purposes, and in structures where temporary lights are needed for only a few weeks or months. When properly installed, open conductors are safe and inexpensive. This method is, of course, unsightly, as can be seen from Fig. 4.

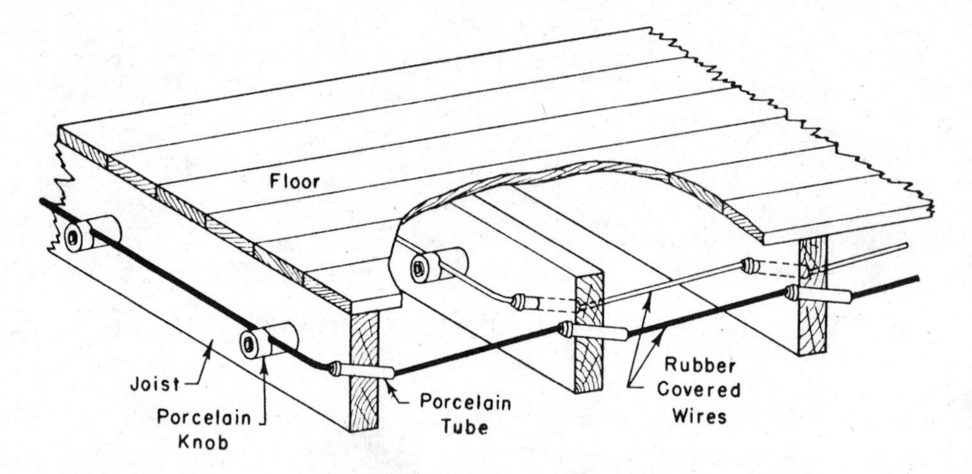

Fig. 3. Concealed knob-and-tube wiring using solid knobs. This wiring method is virtually obsolete.

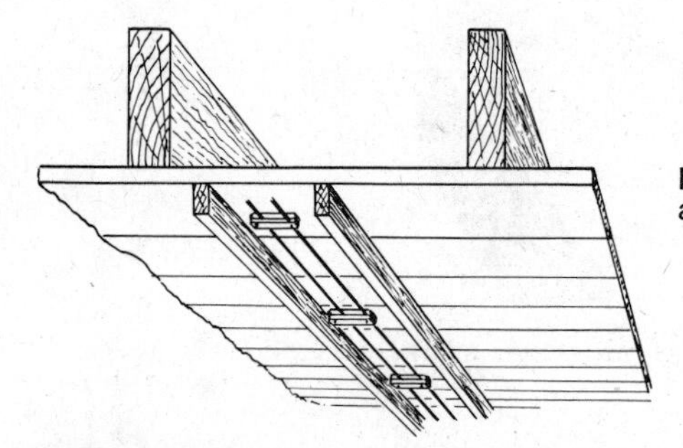

Fig. 4. Installation of open conductors in ceilings and walls.

Nonmetallic Sheathed Cable. In the nonmetallic sheathed cable wiring the rubber-covered conductors are wrapped with a covering of braided cotton or paper covering and a spiral wrapping of paper tape. Then two or three wires are enclosed in a woven cotten covering which is impregnated, Fig. 5.

Sheathed cable is installed by drilling holes in the studding or floor joists and pulling the cable through, Fig. 6. When run in the same direction as the studs or joists, it is attached to them with straps, Fig. 7, or staples driven into the wood. Either porcelain or metal outlet boxes or switch boxes, Fig. 7, may be used with this system of wiring. The porcelain outlet boxes

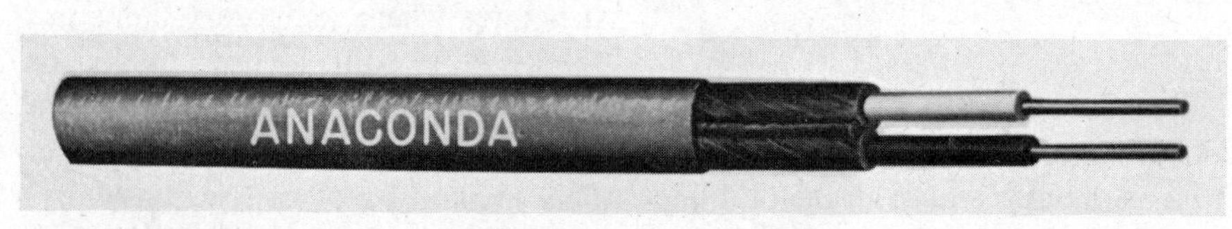

Fig. 5. Nonmetallic sheathed cable usually has 2 or 3 insulated wires. Anaconda Wire and Cable Co.

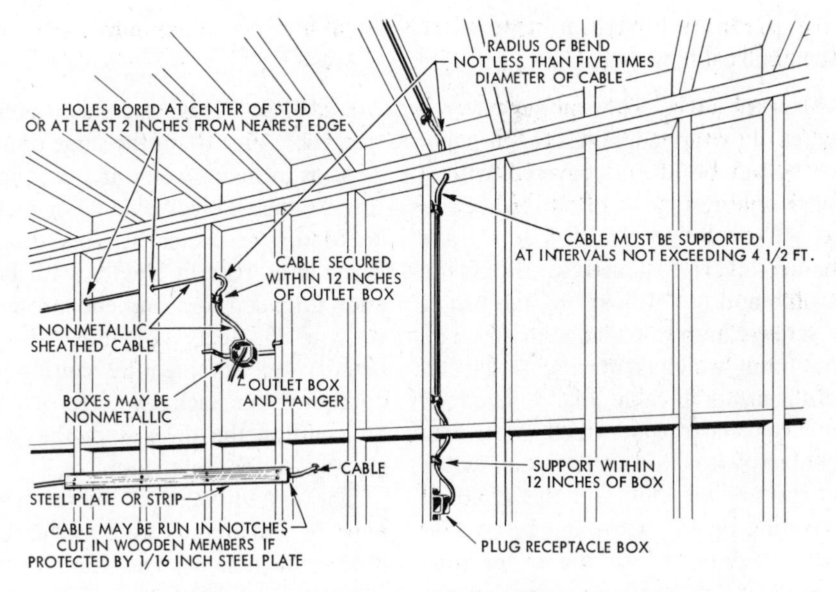

Fig. 6. Installation of sheathed cable in studding and floor joists.

Fig. 7. Method of fastening nonmetallic sheathed cable to floor joists and ceiling with straps and to switch and receptacle boxes. General Cable Corp.

Fig. 8. Armored cable should be called AC. Crescent Insulated Wire and Cable Co.

are preferred with nonmetallic sheathed cable wiring.

Armored Cable. The rubber-covered wires are wrapped with a tough paper or cotton braid and covered with a steel armor, Fig. 8, often referred to as *AC* or *BX cable.* (BX is a trade name.) It is fastened to the metal switch and outlet boxes by the use of a screw-clamp threaded stud and lock nut fitting which is attached to the end of the armored cable, Fig. 9. The lock nut has sharp teeth which make firm contact with the outlet and switch boxes so that they and the armored covering on the wires can be permanently connected to the water pipe system and the earth (in order to properly ground the wires). The armored cable is installed in the building in the same manner as nonmetallic sheathed cable, Fig. 6. This type of wiring is used in cities and towns where there is a town water system and it is desired to have the electric wires encased in metal in order to minimize the danger of fire.

Rigid Conduit. This system uses a mild-steel tube, Fig. 10, of the same dimensions as ordinary water and gas pipes. It has either a black enamel or metallic coating to protect it from rust. It is made in 10-foot lengths with both ends threaded and one end provided with a threaded coupling. Shorter lengths are obtained by cutting the conduit with a hack saw, then reaming and cutting the threads in the same manner as for water pipes.

The conduit and all outlet boxes must be installed as soon as the floor joists, rough floor, and studding are in position and before any insulation or lathing is done. The wires, switches, and receptacles are not installed until after the building is plastered.

Usually the conduit is run parallel to the floor joists and is fastened to them and to the ceiling outlet. The floor joists are notched slightly or

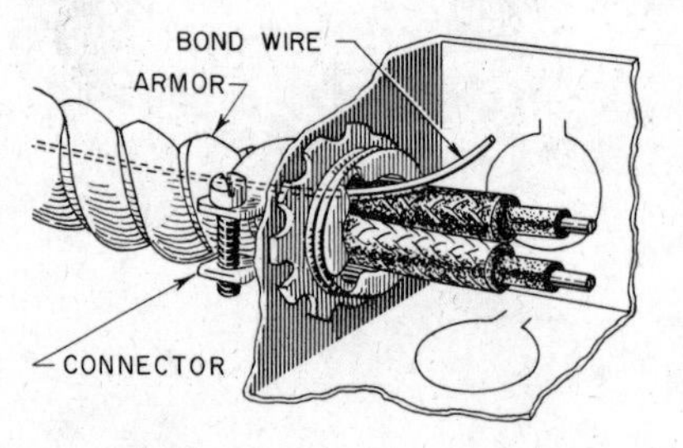

Fig. 9. A screw clamp and lock nut fitting are attached to the end of the armored cable.

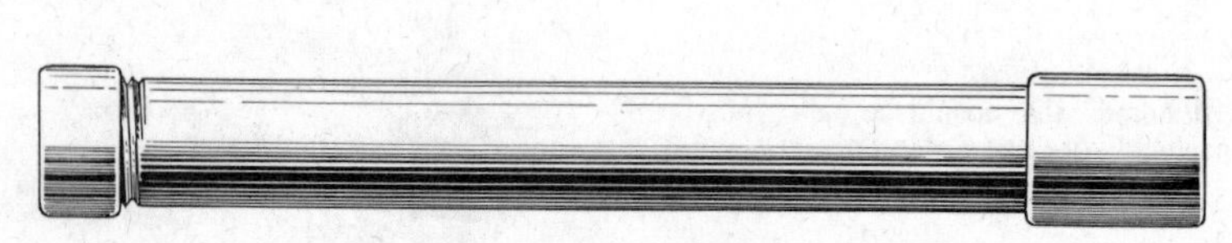

Fig. 10. Rigid conduit is threaded at both ends.

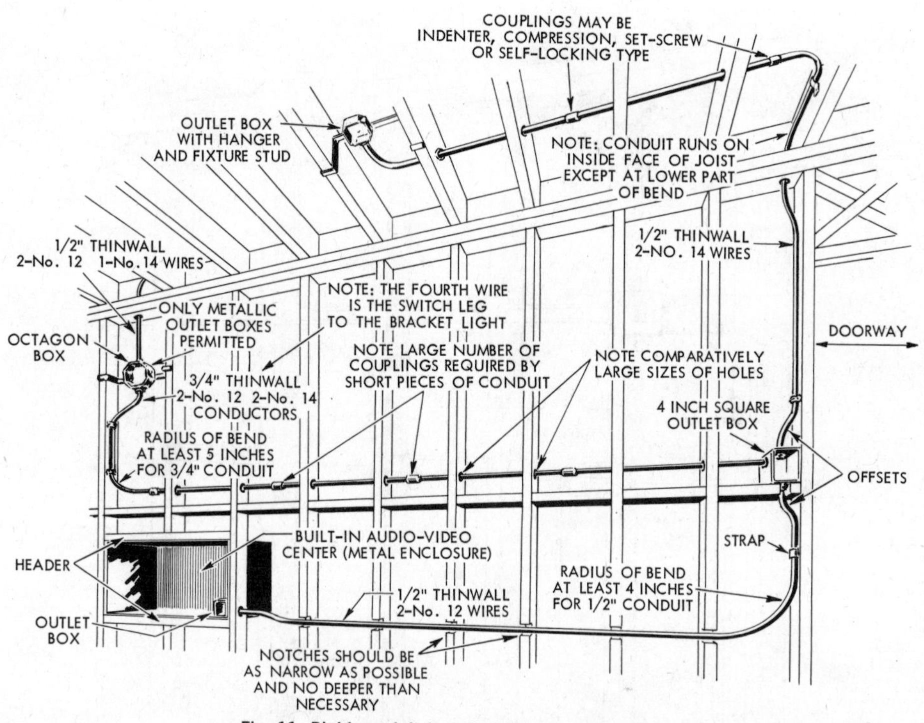

Fig. 11. Rigid conduit installed in walls and ceiling.

bored through when it is necessary to run the conduit at right angles to them, Fig. 11. When a convenience outlet is installed on a partition wall the conduit is usually run down from the floor above, or up from the floor below. For convenience outlets and wall switches a square outlet box and cover, Fig. 12, is often used instead of a switch box because it is easier to secure to the studding.

When side walls are of brick and furring strips are not used, it is necessary to channel out the brick and fasten the conduit in the channeled groove. This costs more than when furring strips are used because, with furring, the conduit is fastened to the side of the wall without channeling.

Rigid conduit wiring is the most sturdy and permanent method of wiring and it will last the life of the build-

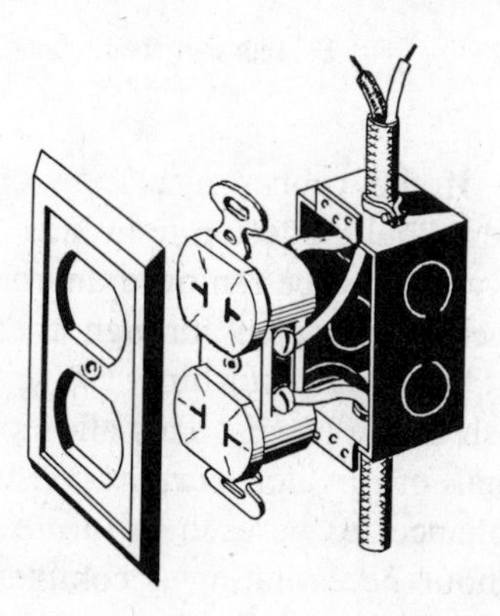
Fig. 12. A duplex convenience outlet.

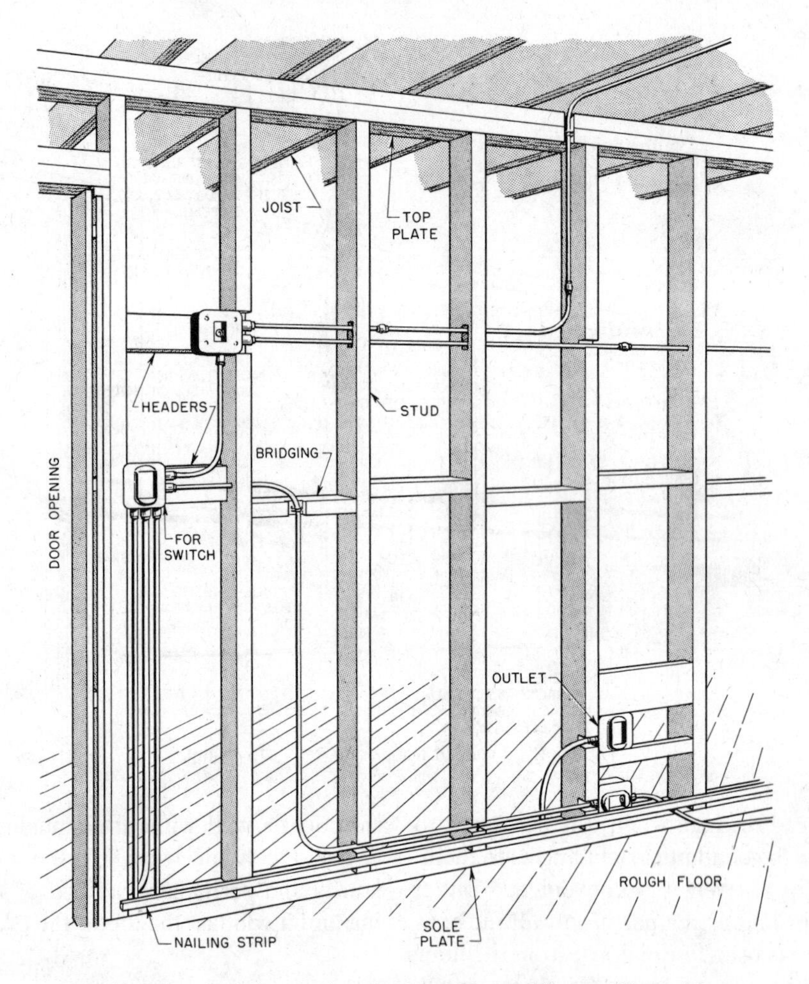

Fig. 13. Thin-wall steel tubing installed in an interior wall. Republic Steel Corp.

ing. If the rubber insulation on the wires should deteriorate in later years the wires can be removed and new or larger ones pulled through the conduit without disturbing the interior finish of the rooms. This allows additional or greater current-consuming appliances to be used in future years without necessitating a complete rewiring of the building.

Electrical Metallic Tubing. This tubing is similar to rigid conduit but it has a thin wall and is not threaded. Clamp types of fittings or connectors are used to join one piece to the next or to outlet boxes, Fig. 13. It is lighter and is easier to cut, bend, and install than rigid conduit. Thus the labor and material cost is less than with rigid conduit, yet the advantage of having all

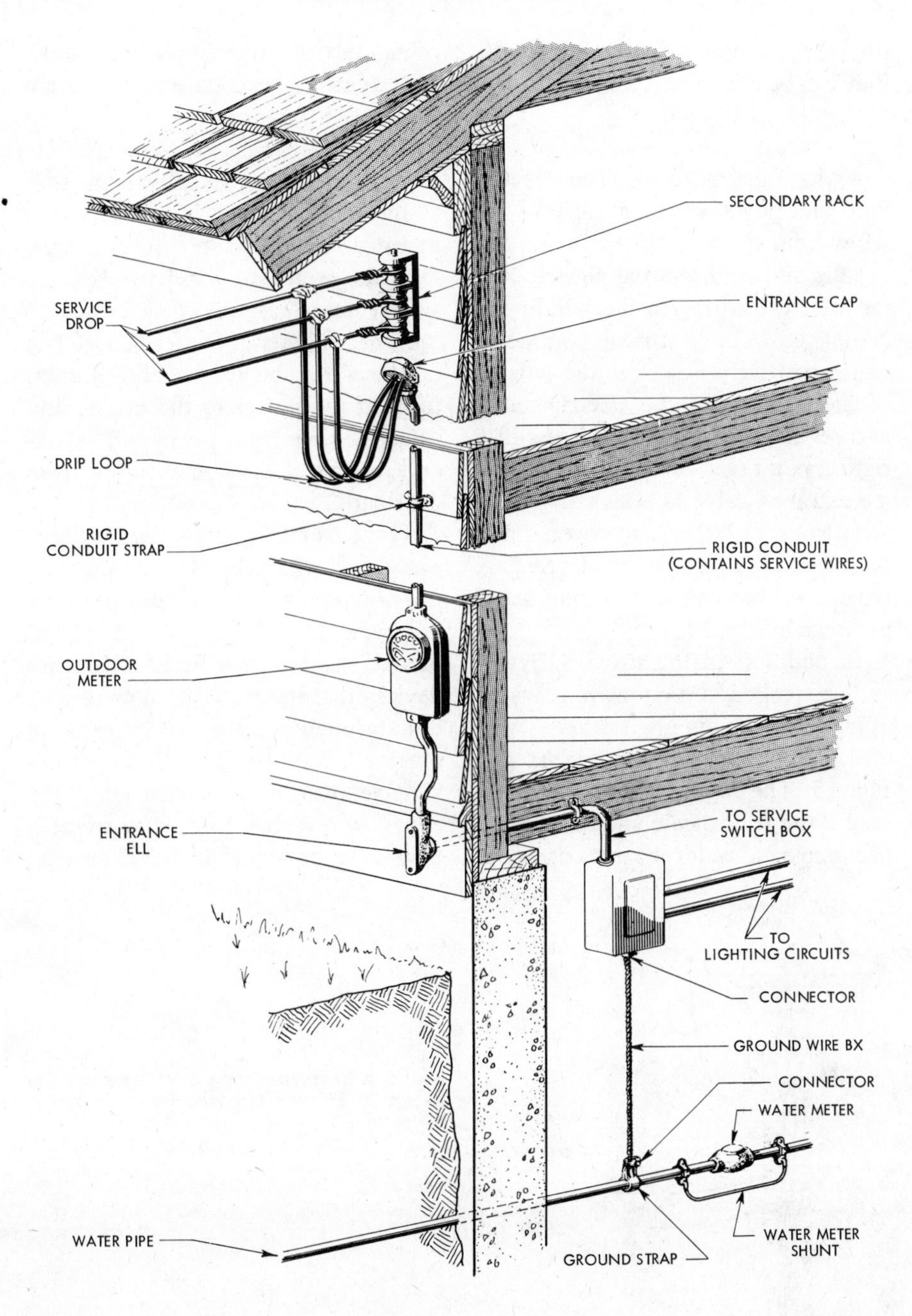

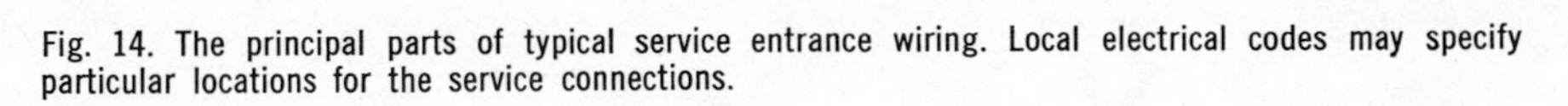
Fig. 14. The principal parts of typical service entrance wiring. Local electrical codes may specify particular locations for the service connections.

the wires inside a metal casing, where they can be removed or replaced when necessary, is retained.

Service Entrance Wiring. The electric light and power company installs the wires from the pole to the house and installs and connects the electric meter after the wiring in the building is completed. The electrical contractor will install the wires from the outside of the building to the electric meter and service switch. He will use either rigid conduit (¾″ or 1″ size) or a service entrance cable in which the wires (size No. 8 or larger) are covered and protected by a weatherproof braid. A typical service entrance wiring setup is shown in Fig. 14.

In addition to the usual 115 volt service (called a *two wire* service), homes are commonly supplied with 230 volts (a *three wire service*). See Fig. 15. The higher voltage is necessary if the residence contains an electric range, a water heater, or some other heavy duty appliance which would cause an excessive current drain on the 115 volt lines.

Service Switch Box. After passing through the meter, the electricity goes to the service switch box (also called the *distribution panel).* The branch circuits originate from the service switch box; these carry electricity to the various parts of the residence. The box also contains the switches used to turn off the power to the house, and safety devices (*fuses or circuit breakers*) to protect the house wiring from overloads.

Fuses and Circuit Breakers. Fuses may be of the *plug* type or the *cartridge* type, Fig. 16. Modern practice, however, favors the use of circuit breakers rather than fuses. These are devices that interrupt the flow of current through a circuit in case of an overload, Fig. 17. They have an advantage over fuses in that once the cause of the overload is corrected, they may be switched back on and

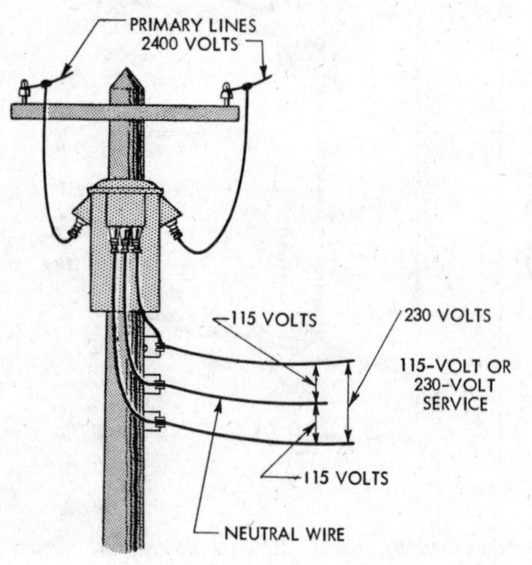

Fig. 15. A transformer steps down the power line voltage to 230 and 115 volts before it reaches the home.

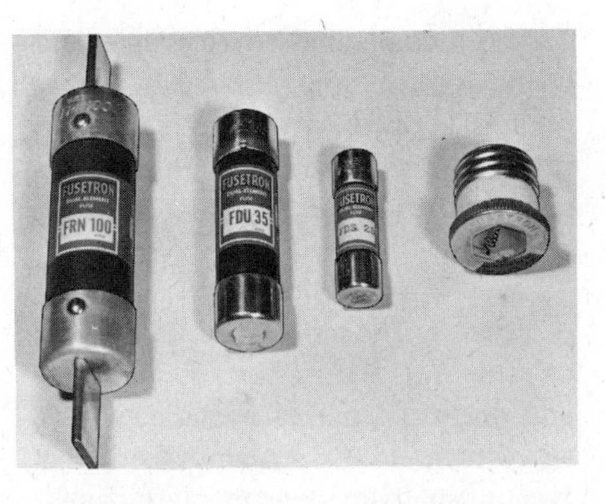

Fig. 16. Cartridge-type fuses and a plug fuse.

used over and over again. A typical service switch box using cartridge-type fuses is shown in Fig. 18.

Estimating wiring

Two methods of estimating are described below. The first is a precise method that is used for industrial work and wherever the conditions of the job differ markedly from the average. Its successful application, however, requires a great deal of experience in this sort of work. For this reason it is impractical for the average estimator and will be discussed only briefly. Though he will not use this system himself, the general estimator should know something about it. This will help him to check the estimate received from the electrical subcontractor for obvious errors due to carelessness, omissions, etc. In specialized work, involving, for example, high voltage or high-power wiring, the estimate is usually done by an experienced electrical engineer.

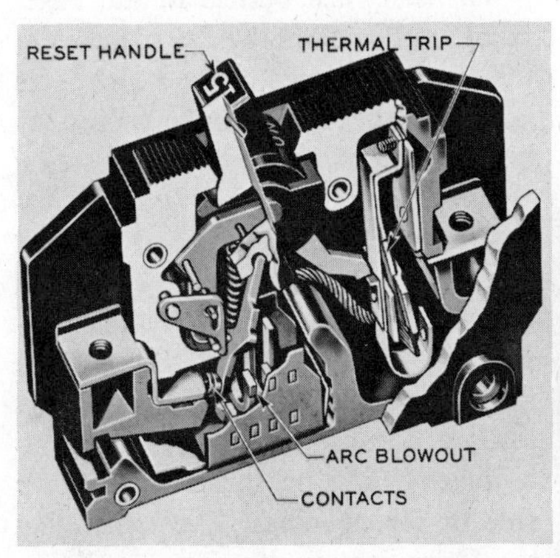

Fig. 17. A cutaway view of a circuit breaker.

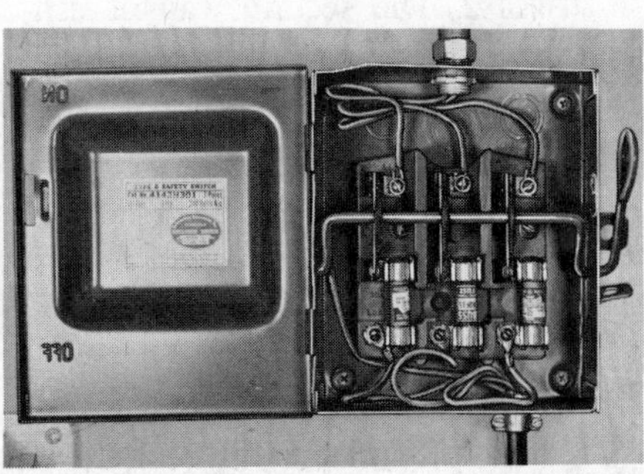

Fig. 18. A service switch box with three cartridge-type fuses.

The second method is simple and useful for the non-specialist. So long as it is applied only to average jobs involving residential dwellings, it is sufficiently accurate for most practical purposes.

Method 1. The most accurate method of estimating wiring costs is undoubtedly one which makes a detailed analysis of each circuit. An analysis of this type is usually made from plans drawn up by an electrical engineer or the architect. The materials needed for a job are determined from these plans. If such detailed specifications are not available, however, it is necessary for the electrical estimator to prepare his own wiring plans and have these layouts approved by the architect before beginning the estimate. These layouts show the number and locations of switches, receptacles, meters, etc. The estimator counts these and, from his scaled plans, measures the length of cable needed for the wiring. By consulting a manufacturer's catalog, he can determine the costs of the required materials quite accurately.

Method 2. The second way of estimating electrical wiring costs is to do the figuring on an *average cost per outlet* basis. The estimator determines the costs of the materials and labor which go into installing what he believes to be a typical outlet. He does this for each different *kind* of outlet as well as for the service switch box, meter, fuse boxes, etc. The total cost of the job is then determined from these figures. Variables which must be considered are the kinds of wiring used, local ordinances, material costs, labor regulations, and labor costs. Because of local variations, it is not possible to quote exact cost figures for these items. Outlet, receptacle, and switch costs are usually figured on the basis of a 15′ run of cable (or conduit); service to the building is figured on the basis of a 75′ overhead run.

The costs of materials can be found by consulting manufacturers' catalogs. Armored and nonmetallic cable are often purchased in 1,000′ lengths, flexible conduit comes in 100′ lengths, and rigid conduit comes in 10′ lengths. (The prices do not include fittings.) Prices of other materials, switches, outlets, fuse boxes, etc., are also obtained from such catalogs.

Quantity Take-Off—House Plan A. Fig. 19 is a *branch circuit schedule* for HOUSE PLAN A. It is a tabulation of the various outlets and fixtures shown on the plans for the house.

The electrical outlets in the basement are turned on by the switches at the top and bottom of the stairs. This is so that the lights can be turned on before descending the basement stairs.

There are no ceiling or bracket outlets in the living room. The four wall receptacles are controlled by a switch located at the foyer entrance. The passageway from the foyer through the living room and dining room into the kitchen does not have any doors, so switches could be located on either side of the openings. The switch for the foyer light is shown at the living room entrance where it is conveniently

BRANCH CIRCUIT SCHEDULE

Location	Ceiling Outlet	Wall Outlet	Heavy-Duty Outlet	Double Pole Switch	Triple Pole Switch	Exterior Outlet
Basement	6					
Living Room		4		1		
Dining Room	1	2		1		
Foyer	1			1		
Hall	1				2	
Family Room		3			2	
Kitchen	1	1	3		2	
Laundry Room	1	1			2	
Garage	2	1		1		1
Porch	1	2		1		1
Pantry	1			1		
Rear Entry				1		1
Stairway				1		
Bedroom #1		5		1		
Bedroom #2		4		1		
Bedroom #3		4		1		
Lavatory	1	1		1		
Bathroom #1	1	1		1		
Bathroom #2	1	1		1		
Closets (5)	5					
	23	30	3	14	8	3

Fig. 19. Branch circuit schedule for House Plan A.

located to switch the lights on or off when entering or leaving the house.

The kitchen light can be turned on from the family room or kitchen. A fuse box is located on the wall between the kitchen and the porch. It is located where it is easily accessible and can feed all circuits of the house. (Another fuse box is located in the basement, near the stairs.)

A ventilating fan is installed over the kitchen cabinets to the right of the sink. This fan is not installed until the wiring is finished and will probably mean a special trip to the house by the electrician. Only about one hour is required to do the actual work, but a total of two hours traveling time should be taken into consideration.

Estimating labor

Electrical work is usually divided into two parts, the *rough* work or wiring, and the *finish* work, or fixtures. An estimate of the finish work is almost never required because the selection of fixtures is done by the architect or the owner.

The rough work is done by electricians who may be assisted by apprentices or helpers. On small jobs, one electrician and one helper, or two electricians, are often sufficient. The amount of time needed for rough work depends on the size and kind of wiring as well as the special conditions of the particular job. (Labor is least with nonmetallic or armored cable, greatest with heavy, rigid conduit.) Electricians' wages vary in different parts of the country and the estimator must consider the pay scale used in his locality. The wages of helpers are usually about one-half those of electricians.

In order to compute the labor cost, the estimator once again looks at each circuit separately. He must be able to determine the time required to wire each circuit. Since this time depends on how much work (such as boring) the electrician must do for each circuit, the estimator must be able to visualize those features of each circuit (or the total job) which will tend to make the work go more slowly. This can be done with reliability only if the estimator has considerable experience in this type of work. Since it is unlikely that the average building contractor will possess this kind of experience, it is recommended that he avoid trying to make detailed estimates of this type and leave such work to the electrical subcontractor.

11

SPECIAL INTERIOR COVERINGS

In addition to plain wood or concrete floors and painted or papered wall and ceiling surfaces, one or more of a number of special materials are frequently specified. This book cannot list all such special coverings, but enough are described to acquaint the inexperienced estimator with the various materials and their application.

The use of many of these special coverings requires special treatments for floor, wall, and ceiling subsurfaces. This definitely affects material and labor costs and therefore should be given careful attention by the estimator.

Materials

Floors. Among the special floor coverings are plank floors, block floors, linoleum, terrazzo, and the various floor tiles.

Plank Floors. Fig. 1 illustrates a typical so-called colonial plank flooring. These flooring types may be obtained in such woods as oak, maple, and walnut. The planks are from 4″ to 10″ wide and come in random lengths. Sometimes nailing is done through holes in the ends of the boards, after which plugs of a contrasting color are inserted. Other types are nailed through the tongue. Some types, as shown in Fig. 1, have keys which add decoration to the floor and also prevent spreading. Plank floors usually are finished natural and then waxed. They may require sanding.

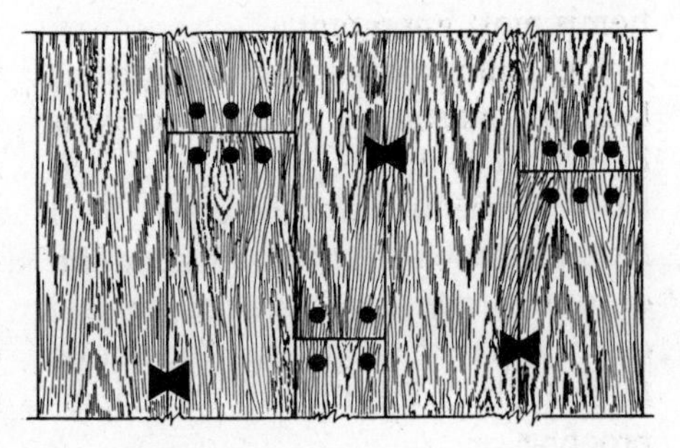

Fig. 1. Plank floors are available in various designs and dimensions.

Block Floors. Two types of block flooring are used. For installation over concrete floors, a mastic fill (hot or

Fig. 2. Block floors may be laid in a mastic fill either over concrete or over wood flooring. E. L. Bruce Company.

cold) is spread and smoothed level. The blocks are arranged carefully in courses so as not to bleed the mastic (or cement-adhesive) over the block. Sanding may be required, but in many instances the wooden blocks (or squares) are pre-finished — in which case, the need for neat, clean installation is most important.

Fig. 2 illustrates a block floor being laid on a mastic fill. If the concrete is in contact with the ground or if it is below grade level, there could be a moisture problem, so polyethylene film or waterproof paper should be laid directly over the concrete. The mastic fill is applied directly over the dampproofing.

For installation on a wood subfloor, tarred or asphalt felt is laid over the rough flooring and the blocks laid over the felt. Nailing is done through a special seat, or hole.

Linoleum. Linoleum is often specified for kitchen floors. Fig. 3 shows the procedure for applying linoleum. For wood floors, felt is laid before the linoleum is applied; for concrete floors, the linoleum can be applied directly or with a felt lining. The contractor laying linoleum follows a definite procedure which the estimator should know in order to arrive at an adequate cost estimate.

Linoleum should be exposed to a temperature of at least 70°F before laying.

A paste is necessary to cement the linoleum to the floor. Pastes should have a spreading capacity of not less

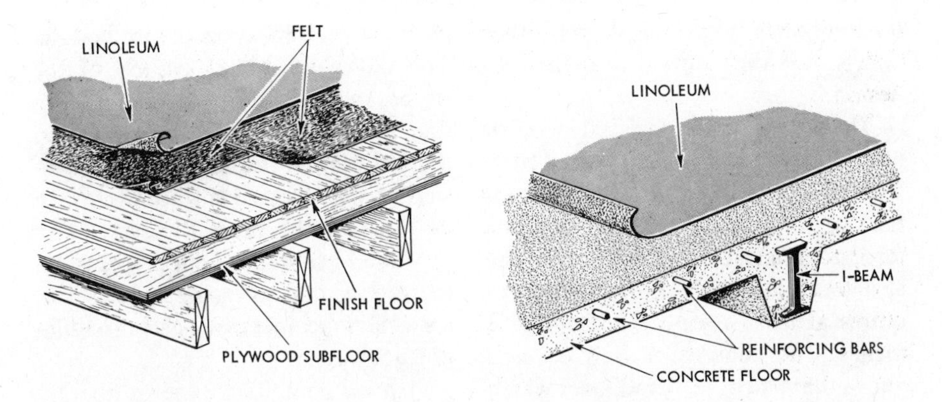

Fig. 3. A felt lining is used between the linoleum and wood. It is not always used for concrete floors.

than approximately 100 square feet per gallon.

The felt used under the linoleum should be 1½ pound saturated or 1 pound unsaturated.

Concrete floors must be smoothed and filled where necessary. Plaster of Paris is used for filling holes or scratches. Cracks or holes in wood floors must be filled with strips or plugs of wood. All flooring boards must be nailed down tightly.

Most specifications require that when the felt is cemented to the floor a roller shall be used to smooth out any bubbles. The finished linoleum surface must be protected by heavy building paper. Ordinarily wax is applied. Some waxes require polishing, whereas others dry to the desired luster. Concrete floors must be tested for dampness and waterproofed if necessary. Waterproofing cement may be required at seams or around the edges.

A base is also run around the edge of the linoleum against the wall. See Fig. 4.

Terrazzo. Two methods are used in laying terrazzo floors over concrete. One is to bond it to the concrete and the other to separate it from the structural slab.

The concrete fill can be made of cement, sand, and hard coal cinders in a mix of 1:1:6. Slag and broken stone or gravel are also good aggregates for concrete fill under terrazzo floors. Before the terrazzo contractor installs his underbed, he must see that his concrete fill is thoroughly cleaned of plas-

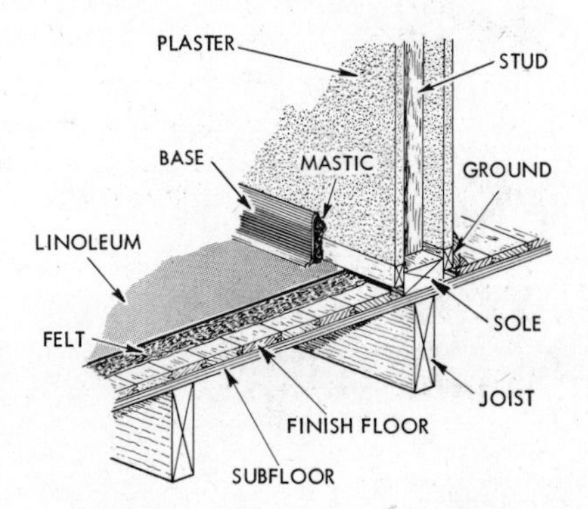

Fig. 4. A base is used around the walls, at the edge of the linoleum.

ter droppings, wood chips, and other debris; it is then wetted to assure cohesion.

The second method is used in buildings where cracking is anticipated from settlement, or from expansion and contraction, or vibration. In this case the terrazzo contractor builds upon the structural floor slab. This method requires at least a total thickness of 3 inches. The concrete slab is covered with a thin bed of dry sand, over which a sheet of tar paper is laid. Over the paper the underbed is installed, as in the first method, except that coarser aggregate can be used (such as cinders or fine gravel) where its thickness exceeds 2½ inches. See Fig. 5.

Where terrazzo is used over wood floors, a thickness of not less than 2 inches is required. The wood floor must be covered with tar paper and overlaid with galvanized wire netting. The underbed is laid on top of the paper and mesh. Brass dividers, see Fig. 5, may be required, depending on the pattern. Troweling is required to smooth the terrazzo surface even with the tops of the brass dividers. Gray or white Portland cement is spread on the surface and surfacing is done with a special machine, using carborundum grit stones.

Asphalt Tile. This type of flooring is made in patterns of various sizes, and must be laid one piece at a time. When laid on wood floors all holes, rough spots, or cracks must be filled and smoothed. A felt lining should be used if the tile is laid over a wood subfloor. When applied to concrete, the surface (if new work) should be wood trowel finished. All holes or scratches must be filled. The tiles are cemented

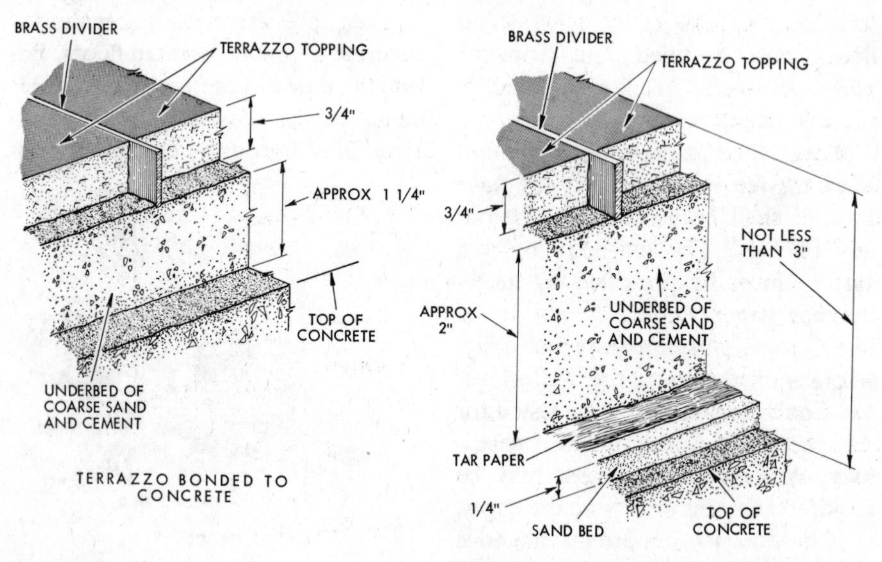

Fig. 5. Terrazzo may be either bonded to or separated from the concrete underbedding.

to the base flooring. A prepared cement is spread evenly over the subfloor (or felt); it is allowed to set and the tile is applied by hand. Nailing is not required. Exposed edges must be protected by special metal strips made for the purpose. A base may be used against the wall. Generally, the contractor is required to wax or clean the finished surface.

Vinyl, Cork, and Rubber Tiles. These tiles are made in sizes similar to asphalt tile and are applied in the same manner.

Ceramic Tile. This type of tile is used in bathrooms, powder rooms, and shower areas. A thin cement or mastic bedding is applied over the concrete base on which the tile are laid. Under very damp or otherwise severe conditions, Portland cement mortar may be used as a base. This is much more expensive. The tiles must be thoroughly soaked in water before they are used.

The final step is to fill the joints with clean Portland cement mixed with water.

If dampproofing is required, ceramic tile primer may be applied.

Mosaic Tiles. These are small tiles, of various patterns, mounted on sheets of paper of a convenient size. The sheets are laid on a bed of cement with the paper side up. The tiles are pressed into the bed and beaten down with a block until the mortar in the joints is visible through the paper. Finally, the paper is moistened and removed. White sand should be used on the surface so that as beating continues the cement will not adhere to the block. The final steps are to level the entire area and to grout the joints in the same manner as for ceramic tile. Fig. 6 shows base details for a ceramic or mosaic tile floor. These details require additional material and labor and must be included with the floor costs.

Walls. There is a great variety of in-

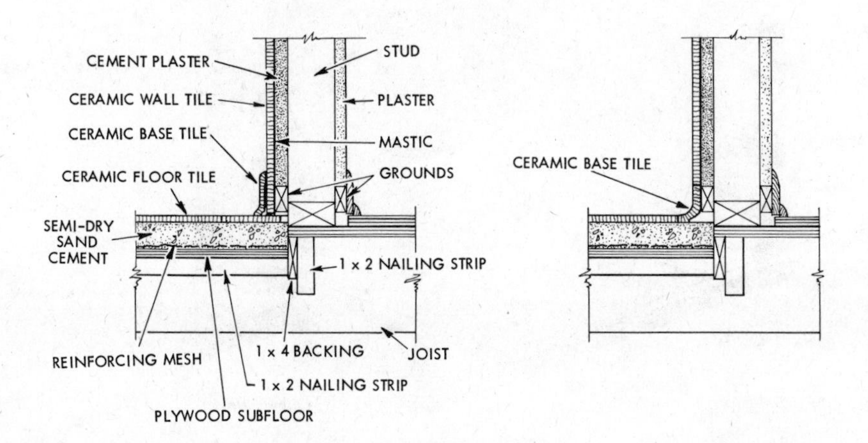

Fig. 6. Base details must also be included in the material and labor estimates.

terior wall coverings ranging from tile and wallpaper to the many varieties of wall panels. This section, as in all previous sections, aims not to teach design or application, but to give the inexperienced estimator an appreciation of the amount of material and labor necessary for the various parts of a building.

Wallpaper and Fabric. There are hundreds of wallpaper styles and many grades, from handmade papers of the finest quality to low cost conventional patterns. Wallpapers are made in narrow and wide widths in rolls of various footage. Fabric wall coverings (vinyl coated) are also applied to walls. This is almost identical to wallpaper in appearance and application.

Paste is supplied in powder form requiring only the addition of water and stirring. Plastered and dry walls are generally coated with sizing (a glue or paste) before papering or other treatment is applied.

Presdwood Tile. In bathrooms, kitchens, playrooms, etc., this style of tile may be used on the walls. One such form of tile is manufactured of treated and formed wood pulp made up in large sheets or panels embossed in various designs. These tiles may be nailed directly to studs or furring strips or may be applied over old or new plaster.

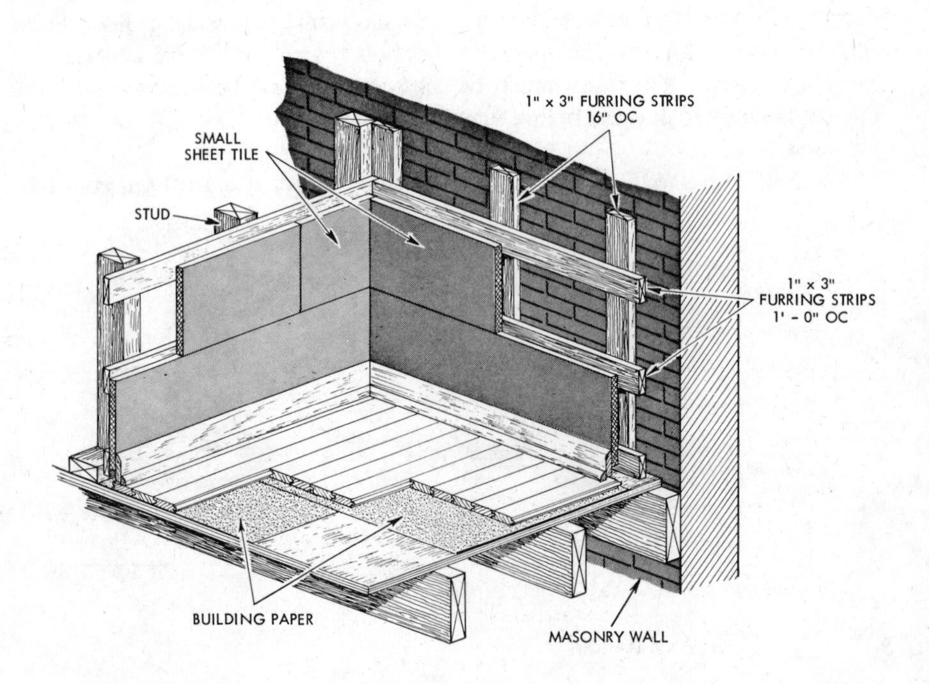

Fig. 7. Tile may be applied by using furring strips as a nailing base.

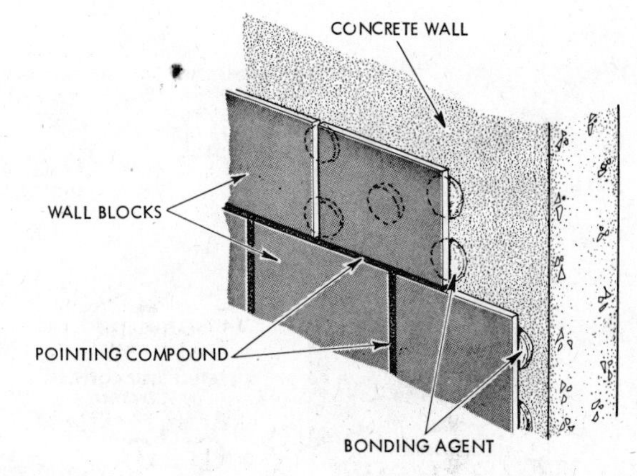

Fig. 8. Tile may also be applied by using an adhesive. This method of applying adhesive is called "buttering."

Small Sheet Tiles. These tiles are manufactured in a variety of materials characteristically rigid in form and of various shapes.

Fig. 7 shows typical applications of small sheet tile to stud and masonry walls. In both instances furring strips (approximately 1″ × 3″ in section) are used as nailing strips. Fig. 8 illustrates a different method of application to a masonry wall.

Ceramic and Mosaic Tile. The preparation of wood stud walls for tiling is simple. It is only necessary to provide a firm and continuous sheet of plasterboard upon which the tiles may be bedded.

Wall tiles are set by two methods, called *floating* and *buttering*. In floating the tiles a portion of the bed is spread on the wall and the tiles are placed in position and tamped until firmly united. (Fig. 2 illustrates floating.) Buttering consists of spreading the mortar on the back of each tile, then placing the tiles against the prepared wall and tapping gently until they are united with the bed. (Fig. 8 illustrates buttering.) When the tiles are set, by either process, the joints are carefully washed out and filled with cement. When plumbing fixtures are to be installed, a wooden strip to which the fixtures can be secured is fastened to the wall flush with the rough bed and the tiles are laid over it. The setting of mosaic tile is exactly as explained for floors. Fig. 9 shows typical uses of tile.

Insulating Boards and Tile. Insulating boards and tile are manufactured from cornstalks, wood, and other substances and have definite insulating as well as decorative qualities. They are obtainable in many shapes, sizes, and thicknesses, and can be applied directly to studs or furring strips, or over old surfaces. They may also be used on ceilings. The surfaces have varied textures and finishes.

Fig. 10 shows a typical application of insulating board in a residence. Usu-

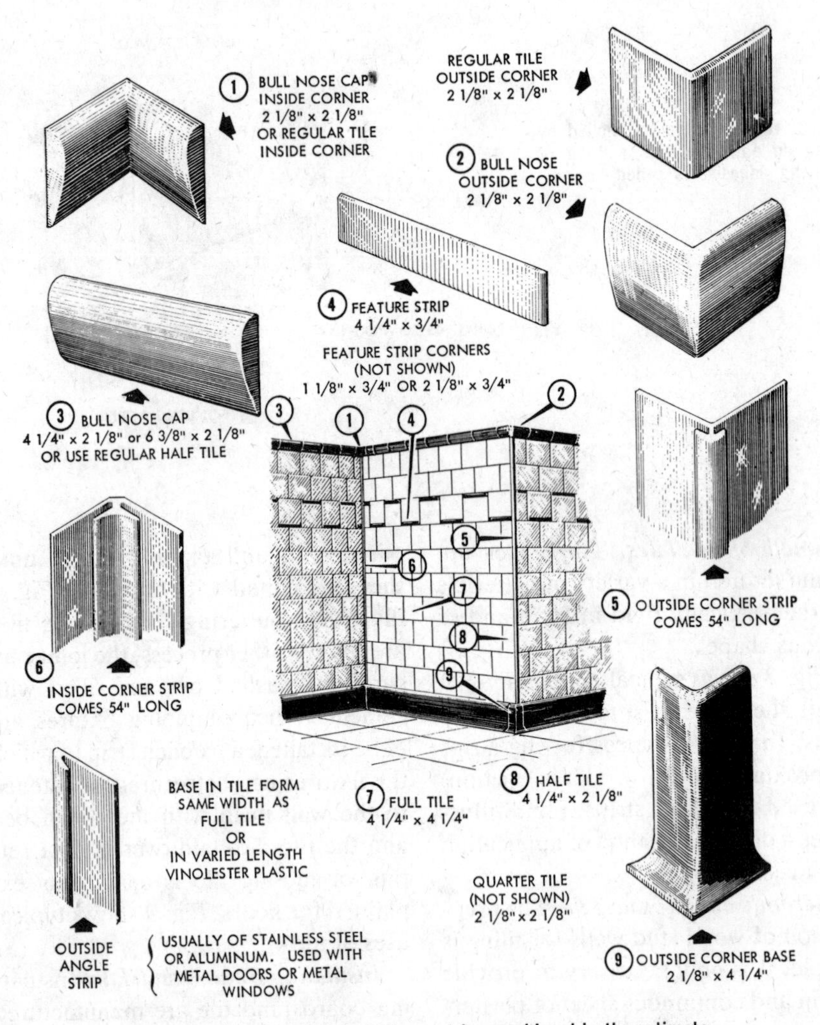

Fig. 9. Each different kind of tile must be considered in the estimate.

ally such insulation material is nailed.

In some instances large sheets are used; in others, workmen, using specially designed tools, cut or carve designs in boards, tile, or planks as they are installed.

Wood Panels. Wood paneling is used extensively for bedrooms, living rooms, and basement recreation rooms. Fig. 11 shows a typical wood paneling installation.

Furring strips are required for masonry walls and also for studs. On masonry walls, the furring prevents damage to the panels or planks from moisture; and on studs, the furring provides a nailing surface for the wood paneling.

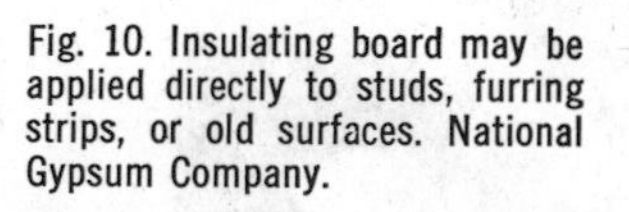

Fig. 10. Insulating board may be applied directly to studs, furring strips, or old surfaces. National Gypsum Company.

Fig. 11. Wood paneling adds an attractive appearance to this room. Note also the tile floor.

Fig. 12. Acoustical tile may be stapled to furring strips which are attached to the joists. Celotex Corporation.

Ceilings. Many of the wall coverings are also used for ceilings. This is especially true of plasterboard and tile.

Acoustical Tile. Sound absorption may be provided by installing acoustical tile on the ceiling. This commonly is sized 12″ × 12″ × ½″. Sometimes the tile is 24″ long. It may be nailed, stapled, or fixed with adhesive. Furring strips are used when it is nailed or stapled. Fig. 12 illustrates the method of stapling. Acoustical tile is also used on walls.

Estimating material

The amount of material required to finish off a given floor, wall, or ceiling is determined by the total area to be covered—in square feet. Any large openings such as windows, doors, arches, stairs, fireplaces and hearths, heating units, etc., are treated as deductible items in the take-off.

Before the estimator begins his work he must study carefully the plans and specifications of the working area. He must know of what materials the existing or proposed subsurface consists. The subsurface material (which may be concrete, plaster, or wood) determines the kind of material and the extent of the work that is to be applied to the surface. To be sure of a correct estimate, the estimator should also have a knowledge of the various methods used to apply the special coverings.

Floors. In addition to the special flooring material, you must include any base molding and paste needed for proper installation.

Plank Floors. The amount of lumber required for special plank floors is estimated in square feet. This is calculated as the total area of floor to be covered.

Example 1. If a colonial plank floor is to be laid in a room which is 8′-0″ × 12′-0″, how much flooring is needed?

Solution: Simply find the floor area.

$$8.0' \times 12.0' = 96.0 \text{ sq. ft.}$$

Block Floors. The blocking can be computed by the square foot or by the piece (for which you must know the area to be covered). This is a costly item because of the millwork needed to achieve the desired course. The kind of subflooring used will also greatly affect the total cost. If the work is to be done on a concrete bed, a mastic fill must be used. Dampproofing may also be required. These items are calculated in relation to the amount of lumber which is to be laid down, or in relation to the room area.

Linoleum. Since many kitchen plans specify a floor of this type, it will be a common item in the estimator's work. The amount of linoleum needed is calculated and ordered by the square foot or by the square yard. The dimensions (usually wall-to-wall) are taken to the deepest part of the room. Cutting is usually necessary around closets and other obstructions; however, the excess is used to line the inside of the closets.

Example 2. How many square yards of linoleum are needed for the room in Example 1?

Solution: Since the area, in square feet, is already known, convert to square yards.

$$\frac{96}{9} = 10.6 \text{ or approximately } 11 \text{ sq. yds.}$$

If felt backing is necessary, it is ordered to the exact size of the linoleum. The adhesive paste is bought by the gallon. Allow one gallon of paste for each 100 square feet of linoleum.

If special metal moldings are necessary to protect the linoleum, they are ordered by the linear foot.

When waxing is specified, allow one gallon of wax for each 200 to 300 square feet.

Terrazzo. Since terrazzo may be applied in either of two ways, the estimator must know the type of subflooring.

If it is to be a bonded job, the amount of underbed fill must be calculated. This will be taken as if it were a poured slab. The amount of terrazzo topping is again calculated as a poured slab in relation to the area of the floor and the depth of the terrazzo finish.

If the terrazzo is to be separated from the concrete slab, a thin bed of dry sand is laid and tar paper is put over it. The underbed is put over this and the finished topping is installed. All fills are figured as if they were a slab. The tar paper is calculated as the area of the floor.

When terrazzo is used over wood floors, tar paper is again used, but an overlay of wire mesh is also necessary. This is figured as the area of the floor since the mesh will run under the entire surface. If any hand troweling is necessary, it is figured into the estimate. Brass dividers are necessary to separate the floor areas and to achieve the various desired patterns on the floor surface. These are figured in linear feet. The depths of the brass strips must also be considered when ordering or figuring the costs for this item.

Floor Tiles. The tiles, whether of vinyl, asphalt, cork, or rubber, are ordered and installed by the square foot or by the piece in relation to square feet of floor area. The area of a single tile is divided into the total area of the room to determine the number of tiles needed for the room. Some cutting of tile is almost always necessary, therefore a waste factor should be used. The waste factor is determined by the size and shape of the job.

Example 3. Determine the number of 9″ × 9″ asphalt floor tiles which would be needed for the room in Example 1.

Solution: Find the area of one tile and the room area (in square inches); then divide the tile area into the room area to determine the number of pieces.

$$\text{Tile area} = 9'' \times 9'' = 81 \text{ sq. in.}$$

$$\text{Room area} = 96 \times 144 = 13{,}824 \text{ sq. in.}$$

$$\text{No. of tiles} = \frac{13{,}824}{81} = 171$$

The amount of adhesive is determined by the square feet of tile area needed to cover the floor. If a felt lining is used, it is the same size as the floor area. If any metal strippings are necessary, they are figured by the linear foot. The number of marble thresholds is taken from the floor plans and are dependent on the number of bathrooms. The size of the threshold is the same as the door width.

Ceramic Tiles. These tiles are calculated in the same manner as floor tiles. If any special molding is needed to finish the floor edges, it is calculated by the linear foot in relation to the perimeter of the base of the walls. Trim pieces such as (bullnoses, caps, base strips, etc.) are figured by the linear foot. If dampproofing is required, allow one gallon of ceramic tile primer per 125 square feet.

Walls. Since the various wall coverings may be ordered in different ways, they must be calculated in relation to the contact area of the walls. Once again, it is to the estimator's benefit to familiarize himself with the methods of installing these materials.

Wallpaper. The amount of wallpaper is taken as the number of rolls needed to finish a room. The number of rolls is determined by the wall area and the area of each roll (in square feet). If a border is required, it is ordered in linear yards. The paste is bought by the pound. The amount of paste depends on the weight of the paper. The percentage allowed for waste is affected by the pattern of the wallpaper.

Example 4. An 8′-0″ × 12′-0″ room has a ceiling height of 8′-0″. The room has one window 3′-0″ × 4′-0″ and one door 3′-0″ × 7′-0″. How much wallpaper is needed?

Solution: Find the area of the walls and subtract the window and door openings.

$$\text{Wall area} = 8.0 \times (8.0' + 12.0' + 8.0' + 12.0') = 320.0 \text{ sq. ft.}$$

$$\text{Window area} = 3.0' \times 4.0' = 12.0 \text{ sq. ft.}$$

$$\text{Door area} = 3.0' \times 7.0' = 21.0 \text{ sq. ft.}$$

$$320.0 - 12.0 - 21.0 = 287.0 \text{ sq. ft.}$$

Fabric material used as wall covering is treated in the same manner as wallpaper.

Tile. The amount of tile is determined by the size of the tile in relation to the wall area. If furring strips are used, they are estimated by the linear foot. Adhesive is figured by the pound or gallon, depending on the kind. If the walls are damp, primer is used. On the average, allow 130 square feet per gallon. Grout for filling the tile joints is figured at one pound per 15 square feet.

If the floating method is used, the take-off will be the same as for floor tile. The buttering method requires less adhesive. Trim pieces are figured the same as for floor tile.

Example 5. How many 6″ × 6″ tiles are needed for the room in Example 4? A 6″ cap and a 6″ base run all around the room.

Solution: Find the total area of the wall and then deduct the areas of the cap, base, and door; then find the area of each tile and divide into the area to be covered.

$$\text{Wall area} = 320.0 \text{ sq. ft.}$$

$$\text{Cap area} = 0.5' \times (40.0' - 3.0') = 18.5 \text{ sq. ft.}$$

$$\text{Base area} = 0.5' \times (40.0' - 3.0') = 18.5 \text{ sq. ft.}$$

$$\text{Door area} = 21.0 \text{ sq. ft.}$$

$$\text{Area covered} = 320.0 - 58.0 = 262.0 \text{ sq. ft.} = 37{,}728 \text{ sq. in.}$$

$$\text{One tile} = 36 \text{ sq. in.}$$

$$\text{No. of tiles} = \frac{37{,}728}{36} = 1{,}048$$

Insulation Boards. These are figured in the same manner as tile.

Wood Panels. These are figured by the square foot of wall area. A more accurate method is to compute the square footage, taking into consideration the height of the walls in relation to the spacing of the studs. Furring strips are figured by linear feet.

Ceilings. Ceilings are also figured in square foot of surface area. Sometimes furring strips are required, especially for acoustical tile which is nailed or stapled. The amount of adhesive depends on the area to be covered and the absorption capacity of the backup wall.

QUANTITY TAKE-OFF

SHEET 1

SPECIAL FLOOR AND WALL COVERINGS — HOUSE PLAN A

Location	Description	Unit	Amount	Amount + 10%
Kitchen				
	Plywood base	sq.ft.	137 —	151 —
	Vinyl floor	sq.ft.	137 —	151 —
	Rubber base	lin.ft.	41 —	45 —
	Paste	gal.	1.5	1.7
Laundry				
	Plywood base	sq.ft.	51 —	57 —
	Asphalt tile	sq.ft.	51 —	57 —
	Rubber base	lin.ft.	17 —	19 —
	Paste	sq.ft.	55 —	60 —
Lavatory				
	Plywood base	sq.ft.	23 —	26 —
	Ceramic tile	sq.ft	23 —	26 —
	Ceramic base	lin.ft.	16 —	18 —
	Wall tile	sq.ft.	75 —	83 —
	Bullnose caps	lin.ft.	16 —	18 —
	Angle caps	lin.ft.	27 —	30 —
	Tile adhesive	sq.ft.	80 —	88 —
	Marble threshold	lin.ft.	2 —	
Bathroom1				
	Ceramic tile floor	sq.ft.	40 —	44 —
	" " base	lin.ft.	26 —	29 —
	" " wall	sq.ft.	240 —	264 —
	Bullnose caps	lin.ft.	26 —	29 —
	Angle caps	lin.ft.	27 —	30 —
	Adhesive	sq.ft.	280 —	
	"	lin.ft.	79 —	
	Marble theshold	lin.ft.	2.33	
Bathroom2				
	Ceramic tile floor	sq.ft.	30 —	33 —
	" " base	lin.ft.	22 —	25 —
	" " wall	sq.ft	128 —	141 —
	Bullnose caps	lin.ft.	22 —	25 —
	Angle caps	lin.ft.	18 —	20 —
	Adhesive	lin.ft.	62 —	
	"	sq.ft.	128 —	
Note:	Paste and adhesive are ordered by the gallon.			

Fig. 13. House Plan A. Quantity take-off.

House Plan A—Quantity Take-Off. Fig. 13 is a quantity take-off for the special floor and wall coverings for HOUSE PLAN A. The floor and wall coverings are listed by the total area in square feet of surface to be covered. The dimensions are obtained from the plans.

The caps, bases, and special moldings are listed in linear feet. Each corner, whether inside or outside, requires an angle cap to finish off the wall.

The waste factor used here is 10%. This is added to the amounts of materials after they are listed. This figure is used for *illustrative purposes only*. Actual waste factors would vary with the various materials.

After the estimator lists the materials and quantities, he calculates the total cost of the materials, the unit cost of labor, and the total labor cost.

Estimating labor

In this type of work the cost of labor is quite high because of the special hand labor required. The estimator must visualize the type of job which is proposed and then know the various kinds of labor required to finish the job.

The following labor estimates are furnished as a rough guide. They should only be used as an outline. The amount of work done in a given period of time depends on the working conditions and the skill of the workmen. The estimator familiarizes himself with the local working laws and wages. Since workmens' wages vary regionally, the hourly pay rates will differ among regions.

Floors. Labor for installation of special flooring is costly because of the special treatment of nailing and joining, and the necessary sanding and waxing of the finished surface.

One man can nail 1″ × 6″ tongued-and-grooved *planking* at the rate of 100 sq. ft. per 1½ man-hours. If special hand tooling is necessary, this time will be tripled.

A worker can lay 100 sq. ft. of *block flooring* over a pasted floor in 1 hour. If the flooring is nailed, the carpenter may take 3 hours per 100 sq. ft.

Labor for *linoleum* is not as costly for this item. If patchwork is necessary on the floor before laying, then this must be included. A man can put down approximately 4 to 5 square yards per hour.

Labor for installation of a *terrazzo* floor is costly. High quality and skill are necessary. Hand troweling is necessary to finish the surface; special surfacing is done with a grinder. Under average conditions, 100 sq. ft. of floor can be put down in about 10 man-hours. Allow 8 hours per 100 sq. ft. for grinding.

Floor tile is also costly. Work is tedious and slow; care must be taken to keep the tiles in a straight line. If troweling is necessary before the floor is laid, this must be calculated too. A tile setter can average 100 sq. ft. of tile per 2½ hours under average conditions. If a felt lining is used, then it can be placed at the rate of 200 to 300 sq. ft. per hour. Trim pieces are included in the general setting time. It takes approximately 15 minutes to set a threshold.

Ceramic tile can be set at the rate of 100 sq. ft. per 3 man-hours.

Walls. Labor rates in some areas are dependent upon such things as the kind and size of rolls of wallpaper.

A paper hanger can paste about one roll per hour. The rate of work is dependent on the pattern and the shape of the room. The mixing time for the paste is almost negligible and need not be included in the estimate. Preparing the wall for pasting, however, should be included.

Wall tiles can be calculated at the rate of 50 sq. ft. per 3 man-hours. This is affected by the size of the tiles and the amount of cutting.

Insulating boards are figured in the same manner as wall tiles.

Wall panels average approximately 50 sq. ft. per man-hour. The time varies according to the kind of panels and the shape of the room. The time for any roughing necessary to prepare the room must also be included.

Ceiling Coverings. This is figured at approximately the same rate as comparable work on the walls. *Acoustical tile,* for example, is figured at the same rate as wall tile.

Other Work. Preparing and finishing surfaces of floors, walls, and ceilings must also be included in the labor take-off.

Pouring and leveling a *cement underbed* for a floor is done at the rate of 500 sq. ft. per man-hour.

Furring (½" wood strips) is done at the rate of 150 lin. ft. per man-hour.

Molding around the walls is installed at the rate of 100 lin. ft. per man-hour.

Hand troweling an average surface takes 1 man-hour per 100 sq. ft. Using an *electric sander* on the same surface requires 1 man-hour per 30 sq. ft.

Questions and Problems

1. How is the amount of floor tile estimated?
2. How are marble threshold sizes determined?
3. How is the labor figured for wall tiles?
4. How long does it take one man to finish 75 sq. ft. of floor with an electric sander? (Answer: 2.5 hrs.)
5. How many 9″ × 9″ tiles are needed for a floor area of 148 sq. ft.? (Answer: 264)
6. A 10′-0″ × 10′-0″ room has one 2′-6″ door. How much base molding is needed? (Ans.: 37.5 lin. ft.)
7. How much ceramic tile is needed for the floor of a bathroom, 7′-0″ × 8′-0″, with a bath tub 2′-0″ × 5′-0″? (Answer: 46.0 sq. ft.)
8. A room is 9′-0″ × 9′-0″; the height of the room is 10′-0″. There is a 3′-0″ × 4′-0″ window and a 3′-0″ × 7′-0″ door. How much wallpaper is needed? (Answer: 327.0 sq. ft.)

12

GLASS

Glass estimating is one of the easiest items for the estimator. There are, however, numerous kinds of glass, thus prices for materials and labor can vary greatly.

Materials

There are five kinds of glass of general interest to the contractor and estimator. These are cylinder glass, drawn glass, plate glass, vision-proof glass, and special glass.

Cylinder Glass. Cylinder glass obtains its name from its manufacturing process wherein molten glass is blown into a cylinder during its formulation. The United States Government has laid down certain standards for the classification of cylinder glass according to thickness and defects. Glass sheets are graded as *Single Thick* when they are approximately 1/12″ thick and *Double Thick* when they are approximately 1/8″ thick. Glass weighing 26 to 38 ounces per square foot and ranging from 1/8″ to about 3/8″ thick is called *Crystal Sheets.*

The government standards grade glass according to its defects, such as bubbles, curvatures, bows, streaks, waves, etc. Glass is graded ordinarily as *AA, A,* and *B*. Grade AA glass is the best of the blown glass while grade A is the next best and the type most generally used in residences. Grade B glass is used generally in factories and cheaper residences. The quality of glass is all the same and the classification given above refers only to waves, etc., as previously mentioned.

Drawn Glass. Another manufacturing process becoming more common is where the glass is drawn out into sheets instead of being blown. This process produces more grade AA and A glass than the blowing method.

Plate Glass. Plate glass is made by a process whereby the molten glass is poured on a hot flat surface and rolled until it is the required thickness. There are two kinds of plate glass, namely, *rough* and *polished*. The rough plate is made from plates which have been rolled and annealed and the polished plate is made by grinding and polishing the rough plate. Rough plates can be used as skylights whereas the polished plate is used for stores, residences, etc. Polished plate is divided into two classes depending on defects. These classes are *silvering* and *glazing*. The silvering class is generally used for mirrors and the glazing class for stores, etc., as mentioned above.

Plate glass varies between ⅛″ and 1½″ in thickness. Glazing thickness is generally about ¼″ thick.

Thermal (insulation) glass is commonly used in residential construction. This glass is generally made of two layers of plate glass separated by an air space.

Vision-Proof Glass. For use in bathroom windows and other places where light is required but vision obscured, there are many kinds of glass made. These are made of the same material but are cut, shaped, sandblasted, hammered, etc., to prevent or obscure vision but admit light. Glass blocks (see Chapter 8) are frequently used for this purpose.

Special Glass. This includes wire glass, tinted glass, prism glass, safety glass, and bullet-proof glass. These, however, are not used generally in residential construction.

Glazing. Fig. 1-*A* shows the generally used method of fastening glass into wood sash. The sash is given a priming coat, generally by the painters, so that putty oil will not be absorbed into the wood. A thin layer of putty is first spread on the rabbet groove to provide a bed for the glass and prevent any possibility of it rattling. The putty is spread in ample amounts so that when the glass is pushed into place some of the putty will form as back putty, as shown in

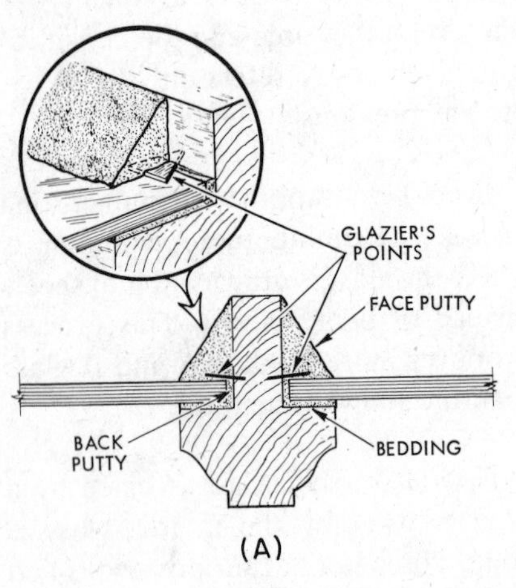

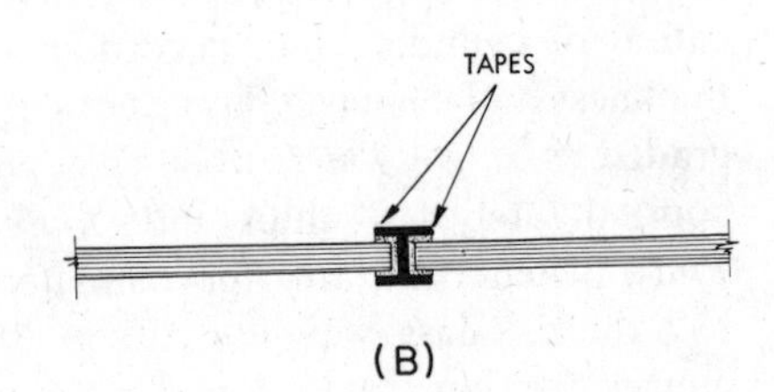

Fig. 1. Method of installing glass in wooden window frame.

Fig. 1. When the glass has been pressed into position, glazier's points are driven into the wood 8″ apart, as shown in Fig. 1, so as to hold the glass firmly against the putty bed. Finally the entire groove is filled with putty and beveled as shown. This process is called face-puttying.

On the inside of windows the glass is generally held in place by means of wood beads nailed or screwed to the wood sash.

Metal Sash. Putty or some kind of metal beading can be used for metal sash. Glass may also be placed in rubber neoprene gaskets or liners. Where heavy plate glass is used in doors and casement windows, the glass is bedded in putty and back-puttied and then held in place by wood beads. In metal settings, the glass should be somewhat smaller than the sash opening to allow for expansion.

Putty. Good putty for wood sash consists of powdered chalk, linseed oil, and about 10 per cent of white lead. More lead would cause trouble if reglazing became necessary, because it causes the putty to harden too much, and its removal from the wood is almost impossible. When putty is used with steel sash, 5 per cent of litharge is added.

Fig. 1-*B* shows the common method of setting leaded glass. The pieces of glass are held in place by pieces of H-shaped lead which has flexible tapes. The glass is puttied into the tapes.

Estimating material

In some localities, and especially for small jobs, windows, doors, etc., are glazed at the mill and shipped to the job ready to install. In such cases it is not necessary to estimate glass material and labor, because the carpenters install the windows and doors, and the cost of glass is included in the cost of the windows and doors. For large buildings such as high-rises, however, the windows are not glazed until the building is nearly finished. This is to avoid breakage.

Although most windows for high-rises can be installed from the inside, sometimes they cannot, and scaffolding is necessary. This item must then be included in the estimator's costs.

The first thing a contractor or estimator does is to study the working drawings and specifications. From these he obtains information about the grade and thickness of the glass to be used, kind of windows, what doors require glass, where plate glass is to be used, sizes of all windows and doors, number and size of panes, etc.

Lights. With a general idea of the job and specifications, the sizes and numbers of lights (individual panes of glass) is next determined. In glass take-offs (counting the amount of each size and kind), full inches are used to describe the sizes of the various lights. For example, if one light scaled 7½″ × 10¾″, the estimator or contractor calls it 8″ × 11″. In other words, all lights are listed in the take-off to the next full inch in excess of the actual size required. When the glazier is putting in the lights, he cuts them to the exact size.

Glass Sizes. There are many standard light sizes, all of which are noted in standard or official published lists as for example, that published by the National Glass Distributors Association. Standard sizes start with a minimum of 6″ × 8″.

Ordinary double-strength grade A *(DSA)* glass as used for window lights is generally priced per piece of glass. That is, there is a list price for the variously-sized lights. There is generally some reduction allowed from this list price. Such reductions vary widely, but whatever the reduction is it can be subtracted from the given list price. Any contractor or estimator can easily determine what the list price and reduction percentage is by consulting his local glass dealer.

Plate glass and mirrors are estimated by the *square foot*. The allowance for breakage is usually between 3 and 6 per cent.

Putty. The amount of putty used per light or per window is difficult to determine. It is not profitable to do such

QUANTITY TAKE-OFF — GLASS — SHEET 1 — HOUSE PLAN A

Location & Description	No.	Type	Window Size (in.)	No. of Lights	Kind of Glass	Light Size (in.)
Dining Room window	2	DH	32X24	24	DSA	10X12
Living Room window	1	BOW	10′-10″X6′-8″	15	DSA	
Bedroom #1 window	2	DH	28X24	24	DSA	8X12
window	1	DH	32X24	12	DSA	10X12
Baths window	2	DH	36X18	12	DSA	12X9
Bedroom #2 window	2	DH	32X24	24	DSA	10X12
Bedroom #3 window	1	DH	36X24	12	DSA	12X12
Family Room window	2	DH	36X30	24	DSA	12X15
Kitchen window	1	SLIDING	24X36		DSA	
Laundry Room door	1	REAR		9	PG	8X12
Lavatory window	1	DH	24X36	8	DSA	12X8
Garage window	2	DH	36X20	24	DSA	12X10

Note: All window are to be ordered from the window schedule complete in gangs to include all frames, sash, trim, mullions, etc.

Fig. 2. House Plan A—Quantity take-off.

a take-off so the estimator uses an average figure based on his experience. Some estimators allow one pound of putty for every two square feet of glass.

House Plan A—Quantity Take-Off. Using HOUSE PLAN A as an example, typical take-offs for glass quantities are shown in Fig. 2. This list by rooms is a more detailed list than most estimators or contractors would make, but it is given so that you will better understand the procedure used in estimating glass. The next step would be to determine the actual prices for the glass.

Labor estimates

Most contractors estimate their labor by assigning certain costs to each of the various-sized lights of glass. This unit cost system is fast and accurate.

In some localities, the glass contractors have an association which has fixed the costs for all of the various-sized lights. These costs are based on actual cost analysis data, obtained over a long period, and upon the prevailing wage scale for glaziers.

In localities where no such cost studies have been made, the contractor or estimator should keep his own cost records and after a time be able to make up such standard costs, based, of course, on the wage scale prevailing in his locality.

A glass installation in a wooden window would differ in labor cost from a glass installation in a steel or aluminum window. These costs also vary with the size of the job and climatic conditions. Outside work in the winter may add as much as 25% to the labor cost.

13

HARDWARE

A hardware schedule for a building must be compiled from both specifications and drawings. To do this requires the ability to read plans easily and thoroughly.

Sometimes a quantity take-off is given to the estimator; the hardware is listed room by room and it is a simple matter to price the items and then estimate the labor time. If such a take-off is not available, the estimator should make one for himself.

Materials

The hardware estimate includes *finish* hardware and not the rough hardware such as nails, which are usually estimated under the carpentry and millwork section.

Hardware schedules do not always contain all the hardware necessary for a residence. They will contain, however, the hardware necessary for doors and windows. (Sometimes this hardware is installed at the mill and the doors and windows are bought as complete units.)

Items not found on many hardware schedules, but which are costs, include the following: mailboxes, street numbers, foot scrapers, door knockers, etc. The bathroom might include one or more towel racks, hooks, etc. The pantry might include hooks for utensils and dishes, knife sharpeners, special knobs or fasteners for cabinets, brackets to support shelving, etc. First-floor windows may require special locks which allow windows to be raised only a few inches. The front and rear doors may have special bolts or chains. Closets may include special devices for hanging up ties, suits, dresses, etc. Storm windows may require special locks and devices allowing them to be only partially opened. These and similar items are usually listed under hardware.

Estimating hardware

The first step in estimating hardware is to study the hardware specifications along with the carpentry and millwork

QUANTITY TAKE-OFF HARDWARE		SHEET 1 HOUSE PLAN A	
Hardware	Location: Description	Location: Size	Amount
EXTERIOR DOORS			
BUTTS 4½"x4½" STL DULL BRASS	DUTCH DOOR	3'-0"x7'-0"x1¾"	2 pair
BUTTS 4½"x4½" STL DULL BRASS	DUTCH DOOR	2'-6"x6'-8"x1¾"	2 pair
HINGES, 1 SET OH DOOR (RAIL sets)	OH DOOR (4 PANEL)	8'-0"x6'-8"x1¾"	4 pair
BUTTS 4½"x4½" STL	9 PANE, 1 PANEL	2'-8"x6'-8"x1¾"	1½ pair
BUTTS 4½"x4½" STL	SCREEN DOOR	2'-6"x6'-8"	1 pair
INTERIOR DOORS			
BUTTS 4½"x4½" STL	FLUSH HC	2'-6"x6'-8"x1⅜"	4 pair
AUTOMATIC DOOR CLOSER	FPSC	2'-6"x6'-8"x1⅜"	1 set
BUTTS 4½"x4½" STL	FLUSH HC	2'-4"x6'-8"x1⅜"	1 pair
BUTTS 4½"x4½" STL	FLUSH HC	2'-0"x6'-8"x1⅜"	5 pair
SLIDING HARDWARE	SLIDING	2'-0"x6'-8"x1⅜"	4 sets
BUTTS 4½"x4½" STL	FLUSH HC	1'-6"x6'-8"x1⅜"	1 pair
BUTTS 4½"x4½" STL	LOUVERED	1'-6"x6'-8"x1⅜"	2 pair
NOTE: 1 door bumper for each door—total 23. Armored pick-proof cylinder door locks for front and rear door—total 2.			
WINDOWS			
SASH LOCKS	GANG OF 2 DH	2(2'-8"x4'-2")	2
" LIFTS	" " " "	"	4
(FIXED)	BOW WINDOW	10'-0"x6'-8"	1
SASH LOCKS	GANG OF 2 DH	2(2'-8"x4'-2")	2
" LIFTS	" " " "	"	4
SASH LOCKS.	DH	2'-8"x4'-2"	3
" LIFTS	"	"	6
SASH LOCKS	DH	3'-0"x3'-2"	2
" LIFTS	"	"	4
SASH LOCKS	DH	3'-0"x4'-2"	1
" LIFTS	"	"	2
SASH LOCKS	GANG OF 2 DH	2(3'-0"x5'-2")	2
" LIFTS	" " " "	"	4
(ALREADY EQUIPPED)	SLIDING	4'-0"x3'-0"	1
SASH LOCKS	DH	2'-0"x3'-2"	1
" LIFTS	"	"	2
SASH LOCKS	GANG OF 2 DH	2(3'-0"x3'-6")	2
" LIFTS	" " " "	"	4
LATCH	TH	2'-8"x1'-8"	5
TOP HINGES	"	"	5 pair

Fig. 1. House Plan A. Hardware take-off.

specifications to see if the latter contains anything which should properly be included under hardware. The window and door schedules should also be used. A detailed hardware schedule is needed for a complete and accurate estimate.

In many construction projects, there is an allowance for hardware in the specifications. This permits the builder or owner to choose his materials after the construction contract has been awarded. For example, if there is a $10,000 hardware allowance for a construction project and the builder chooses materials for $12,000, the contractor is entitled to an extra $2,000 for the materials plus a percentage for installation, overhead, and profit.

House Plan A—Quantity Take-Off. A hardware schedule for HOUSE PLAN A is shown in Fig. 1. This schedule contains the finish hardware needed for the doors and windows. It is based on the door and window schedules for the house plans.

As previously explained, a complete hardware schedule includes more than door and window hardware. This schedule is sufficient, however, to give the inexperienced estimator an idea of how an actual hardware take-off is compiled.

Estimating labor

The installation of hardware is usually done by carpenters and the cost is therefore included under the carpentry contract. The cost of the hardware materials, however, is established under a separate hardware contract.

Labor rates vary in different regions, but the estimator can determine the prevailing rates by contacting the local union.

Installation time for the various pieces of hardware can vary from one minute to several hours. You should therefore keep accurate cost records as you gain experience in this kind of estimating.

14

PAINT

Painting is another one of the finishing touches in building construction. The estimator must study the specifications to determine just what is to be painted and how many coats of paint are needed for the job.

Materials

Specifications. The specifications may designate all areas to the painted, varnished, enameled, or stained—or they may merely state that all exterior wood, for example, shall be painted. This means that the estimator must study the plans to determine the exact areas which require painting. A typical list of outside areas might be as follows:

siding or shingles
trim
cornice
sash
window frames
porches
fences
steps
floors
garage

If one or more outside items are omitted, it can easily reduce the contractor's profit.

In a similar manner, you must consider all interior painting, varnishing, enameling, waxing, calcimining, etc.

Paint can be purchased in many grades. For this reason a specification should give definite instructions regarding the paint, either naming the brand or listing the various materials used in making the paint.

Often, due to climatic or other conditions, the specifications give exact materials and formulas for all areas to be painted.

Typical material specifications are as follows:

(a) *White Lead.* All white lead shall be pure white lead of *X* brand, which conforms to the latest specifications of the American Society for Testing Materials.

(b) *Colors in Oil.* All colors shall be of *X* brand.

(c) *Turpentine.* All turpentine shall be pure gum spirits of turpentine, of pure steam-distilled quality, and shall conform to the specifications of the above-mentioned society.

(d) *Putty*. Putty shall be composed of pure white lead and whiting, mixed with pure linseed oil to proper consistency, and shall conform to white-lead putty, Class B, of the latest specifications of the above-mentioned society.

Miscellaneous Items. Some items belonging in a painting estimate, while seemingly unimportant, can reduce the contractor's profit should they be omitted from the estimate.

Cleaning. The painting contractor must remove all paint spots from finished work and leave the premises free of rubbish. Sometimes the specifications require the painters to clean the windows.

Protection of Work. The painting contractor has the responsibility of protecting his work and the work of all other contractors during the time that his work is in progress. He is responsible for any and all damage to the work or property of others caused by his employees or himself.

Sanding. On window sash, for example, sanding may be necessary to remove rough places left by machine sanders at the mill. Other items of trim, splices, joints, etc., often require sanding to give them the smoothness of recommended work.

Major sanding operations are frequently required on floors before finishing can proceed.

Puttying. All nail holes, cracks, etc., should be filled with putty to make a first-class finished job.

Scraping. Scraping is a time-consuming item which should not be hurried. It is hard work and the workmen usually require frequent rests.

Preparation for Painting. The estimator must know of the preparations necessary for painting. First, there is the work of cleaning and preparing surfaces; second, the time spent in mixing paints; and third, the erecting of scaffolds, covering with canvas, etc.

Preparing Surfaces. New wood surfaces must be smooth. All rough places must be filled with putty. Dirt or other spots must be removed. Previously painted surfaces may be lightly sanded and dusted. Old blisters should be removed. Sometimes old paint must be removed with a blow torch. Old surfaces from which paint has been removed require a primer coat in addition to the usually specified second and finish coats.

Plaster surfaces, if new or not aged, require treating by a solution made of two pounds of zinc sulphate dissolved in one gallon of water. This is applied with a rag or a large brush. Aged or previously painted plaster does not require a priming coat.

Cracks in plaster should be filled; if this is to be done correctly, then considerable time must be allowed on the labor estimate.

Stucco, concrete, bricks, and stone also require special preparations. If stucco or concrete is not aged, a solution of zinc sulphate and water should be applied before any painting is done. New brick must be allowed to dry several days. On old brick surfaces, the joints frequently require tuckpointing before painting can be done. Small defects in the brick surfaces are sometimes filled with putty.

Scaffolds. In many cases, ladders and planks serve this purpose, but in others special scaffolds must be erected and taken down. The estimator must not forget to consider this item in his painting estimate.

Estimating material

Areas to be painted are calculated in *square feet* but the paint is bought by the *gallon*. The estimator should know how much area can be covered by a gallon of paint, varnish, stain, etc.

Painting Materials. In addition to paint, there are a number of products which are included under the paint estimate. These include calcimine, glusize, putty, stain, varnish, and shellac.

Calcimine is bought by the pound. One and one-half pounds of calcimine is enough for 100 square feet.

Glusize is made by mixing glue with water. One pound of glue in ½ gallon of water makes ½ gallon of glusize, which will cover 350 square feet.

Putty can be estimated at 5 pounds for a 6 room residence.

Floor varnish. Most contractors assume that 1 gallon of varnish will cover 700 square feet.

Stain covers about 500 square feet per gallon.

Shellac can be estimated at 1 gallon for every 700 square feet of surface.

Trim varnish will cover about 700 square feet per gallon.

Doors. If exterior door frames are to be the same color as the outside walls, then no extra paint need be calculated. If the door (or window) frame is to be a contrasting color, the exterior paint can probably be tinted in sufficient amount for the frame color. In such a case, about 3 quarts are figured for each 20 openings.

For interior door frames (and window openings) about 1 quart of paint is required per coat for 4 openings. The same allowance is made for bases, picture moldings, etc.

Stain and filler for doors cover about 500 square feet per gallon.

On the average, allow 1 quart each of stain, shellac, and varnish for every 5 doors. Although doors can be estimated individually, this method is sufficient for most jobs.

Interior paint covers about 700 square feet for priming coat and 800 square feet for second and third coats. The areas of doors are figured by multiplying length by width and then multiplying by two because of two sides. Sometimes an additional two or three square feet are added to take care of the bevels on panels, etc.

For *enamel* paint on doors, most estimators figure the coverage as the same as regular paint or up to 10% more.

For screen doors, one gallon of screen paint cut with ½ gallon of benzine generally will be ample to paint both sides of 25 screen doors with one coat.

Windows. The sash and frames are painted at the same time as the wall surfaces and the amount of paint figured for these surfaces is ample for the window sash and frames. No additional paint need be considered.

For brick walls window frames and sash are considered as separate jobs because the exterior walls usually are not painted. About 3 quarts of paint per coat is required for 20 windows, or 20 openings. Ten openings require about ½ gallon of paint per coat.

About ⅓ the amount of paint required for wood sash usually is sufficient for steel sash. One quart of paint should be enough for one coat on 20 steel sash.

One half gallon of oil will prime the sash for about 25 windows.

When enamel is to be used for windows and other openings, allow 1 quart of undercoating for every four openings. The enamel provides about the same coverage as ordinary paint; sometimes it covers slightly less.

If window sills require spar varnish, allow 1 pint per 6 room residence. One pint stain and one pint shellac will also be needed.

For painting storm windows, allow 1 quart of paint for every 10 windows.

If the building has shutters, then the total area of the shutters, including both sides, must be considered when they are to be painted. The paint is assumed to cover 500 square feet per gallon.

Exterior Painting. Most painting contractors figure that 1 gallon of paint will cover about 700 square feet of *wood siding*. For *shingled walls,* allow about 600 square feet per gallon.

Example 1. A residence 30′-0″ × 50′-0″, 20′-0″ high, has shingles on all walls. How much paint is needed for each coat?

Solution: First find the total wall area.

$$30.0' + 50.0' + 30.0' + 50.0' = 160.0'$$

$$160.0' \times 20.0' = 3{,}200 \text{ sq. ft.}$$

$$\frac{3{,}200}{600} = \text{approximately } 5.3 \text{ gallons}$$

5½ gallons should be sufficient

If the *roof shingles* are to be stained, allow 1 gallon per 150 square feet (per coat).

Cornice paint is included in the paint for exterior wall surfaces on frame buildings. For brick buildings the cornice area is calculated and the paint bought on the basis of 700 square feet per gallon.

Two quarts of paint will usually cover 250 linear feet of *gutters* and *downspouts* when new if the metal has been primed with a red lead coating. Old gutters and downspouts require about 3 quarts per 250 linear feet.

Exterior *porch* floors are covered with at least three coats of paint in most instances. The priming coat covers about 600 square feet per gallon, and second and third coats cover about 700 square feet per gallon. The areas are calculated by multiplying the length by the width. If the floors are irregular in shape they can be broken into squares, rectangles, triangles, etc., for easier calculation.

For porch railings, columns, etc., the estimator must learn to judge by sight or else go through the tedious work of calculating the various items. For a picket railing, the pickets are assumed to be a solid surface and the areas of both sides are totaled. A hand railing is assumed to be cylindrical and the area is figured as the lateral area of a cylinder. Most paint used for such purposes will cover about 700 square feet per gallon.

For *brick, stucco,* and *concrete* walls, allow about 600 square feet per gallon of paint. Do not deduct the areas of openings, unless there are several, each as large as ordinary windows.

Example 2. How many gallons of paint are needed for each coat of a brick building which has 2,400 square feet of surface area?

Solution: As in Example 1, divide the total area by the coverage of one gallon of paint.

$$\frac{2,400}{600} = 4 \text{ gallons}$$

Allow about 400 square feet per gallon for concrete floors.

Interior Painting. When calculating the areas of interior walls (and ceilings), deduct the areas of windows and doors if they are not to be painted. One gallon of primer covers 800 square feet of plaster area. One gallon of second or third coat paint also covers 800 square feet.

The type of paint generally used for wood *basement partitions* usually covers about 700 square feet per gallon. If both sides of the partition are to be painted, multiply the area by two.

Paint for *interior stairs* will cover 700 square feet per gallon of primer, and 800 square feet per gallon for second and third coats. The areas of treads, risers, and stringers are calculated by multiplying the length by the width.

Thin enamel can cover 800 square feet per gallon, but if it is thick it may cover as low as 450 square feet per gallon, depending on the type of wood used for the stairway.

For an average flight of around 14 risers 3′-0″ wide, some contractors allow the following:

½ gallon of paint.
1 quart of enamel
1 pint of stain
1 pint of varnish
1 pint of filler (if open grain wood)

Undercoating for enamel walls is assumed to cover 500 square feet per gallon. The enamel for plaster walls will cover about 700 square feet per gallon.

Kitchen cabinets are usually estimated roughly in regard to area. The paint or enamel is then calculated, depending on the covering capacity. When cabinets are to be stained and varnished, some estimators will figure the cost at a fixed price per square foot.

If *pipes* are to be painted, allow one gallon of bronze for every 3,000 square feet of 3-inch pipe—other sizes in proportion.

House Plan A — Quantity Take-Off. The quantity take-off for the interior of HOUSE PLAN A is shown in Fig. 1. Each room has been estimated separately, but the areas and quantities of paint are rough figures. This estimate

QUANTITY TAKE-OFF

INTERIOR PAINT

SHEET 1

HOUSE PLAN A

Location	sq. ft.	lin. ft.	gallons	remarks
NOTE: 3 coats for all surfaces listed				
Dining-Living Room				
walls	500		2	
ceiling	325		1 1/3	
floor	325		1 1/3	shellac
base molding		64	1/4	
window trim		14	1/10	
bow window trim				not specified
Foyer				
walls	200		1	
ceiling	80		1/3	
floor treatment	80		1/3	
base molding		30	1/8	
door trim		35	1/8	
Family Room				
walls	200		1	
ceiling	170		1	
exposed beams, 6"x4"	320		1 1/3	shellac
floor treatment	170		1	
base molding		50	1/8	
window molding		30	1/8	
door molding		17	1/10	
Kitchen				
walls	280		1 1/3	
ceiling	135		1/2	
exposed beams	150		3/4	shellac
window trim		15	1/10	
door trim		35	1/8	
Laundry				
walls	170		1	
ceiling	50		1/4	
door trim		70	1/4	
Hall				
walls	325		1 1/3	
ceiling	63		1/4	
floor treatment	63		1/4	
base molding		20	1/10	
door trim		85	1/4	

Fig. 1. Quantity take-off. House Plan A.

QUANTITY TAKE-OFF
INTERIOR PAINT

SHEET 2
HOUSE PLAN A

Location	sq. ft.	lin. ft.	gallons	remarks
Lavatory				
walls	150		3/4	
ceiling	23		1/8	
door molding		16	1/10	
window trim		11	1/10	
Stairwell				
walls	160		1	
ceiling	40		1/5	
stair treatment	52		1/4	
Bedroom #1				
walls	500		2	
ceiling	180		1	
floor treatment	180		1	
base molding		70	1/4	
window trim		27	1/8	
door trim		42	1/8	
Bedroom #2				
walls	380		1 1/2	
ceiling	121		1/2	
floor treatment	121		1/2	
base molding		50	1/8	
window trim		28	1/8	
door trim		33	1/8	
Bedroom #3				
walls	360		1 1/2	
ceiling	100		1/2	
floor treatment	100		1/2	
base molding		60	1/4	
door trim		33	1/8	
window trim		15	1/10	
Bath #1				
walls	152		3/4	
ceiling	57		1/4	
door trim		17	1/10	
window trim		12	1/10	
Bath #2				
walls	220		1	
ceiling	30		1/8	
door trim		16	1/10	
window trim		12	1/10	

TABLE I. PAINTING TIME SCHEDULE

Item	Amount	Time (man-hours)	Remarks
Exterior siding	100 sq. ft.	1	
" windows	20	6	Included with siding unless brick walls
" doors	—	—	Included with windows
Exterior trim—contrasting color	—	—	no additional labor
Gutters—on ground	500 ft.	1	
" —hung	"	2	
Basement partitions	200 sq. ft.	1	
Stairs—interior	1 flight	5	1 coat
Interior walls and ceilings	200 sq. ft.	1	Includes trim
Interior windows and doors	4	1	1 coat
Steel sash	6	1	allow 25% more for old sash
Calcimining	200 sq. ft.	1	
Glusizing	200 sq. ft.	1/2	
Varnishing	1,500 sq. ft.	6	
" window sills	6 rooms	1	
Floors	200 sq. ft.	1	1 coat
Doors	4	1	same for stain, shellac, varnish
Pipes	2,500 ft.	6	3-inch diam.
Shingles—walls	150 sq. ft.	1	longer for priming coat
" —roof	200 sq. ft.	1	staining
Shutters	each	1/2	1 coat

is accurate enough, however, for most contractors.

Estimating labor

Table I gives average times for various painting jobs. These times will vary with the working conditions, the skill of the workmen, and the set-up times, that is, the time necessary for erecting scaffolds, mixing paints, cleaning brushes, etc.

Example 3. How long will it take to paint 250 feet of gutters if the gutters have not been installed before painting?

Solution: Table I shows the average time as 500 linear feet per hour.

$$\frac{250}{500} = 0.5 \text{ hours}$$

15

WINDOW COVERINGS

Window coverings include shades, venetian blinds, and curtains. These are among the last items installed in a new building. Although these items are a relatively minor portion of the total construction cost, they are included in the estimate when they must be installed by the contractor.

Materials

Window Shades. There are many kinds and qualities of shade cloth. All of them cannot be explained here but a few of the more common types are given in the following:

(1) *Machine-painted*—oil paint finish applied by painting machines.

(2) *Hand-painted opaque*—base of good quality woven cloth, with no filling; linseed oil paint brushed on by hand.

(3) *Hand-painted tinted cambric*—base of finely woven cloth with no filling; tinted with linseed oil paint brushed on by hand.

(4) *Pyroxylin* — a finely woven cloth base to which is applied Pyroxylin (a cellulose product) which coats the cloth and combines with it.

(5) *Holland*—after the color is applied it is calendered (ironed between hot rollers) giving it a smooth, glossy finish; in light colors it is especially translucent.

Shade cloth is manufactured in various standard widths as shown in Table I on the following page.

Rollers. Rollers for shades are furnished in two types as follows: *Type A* — Wooden Rollers and *Type B* — Metal Rollers. The construction of roller ferrules and ends are of a grade known as standard in any of the following: malleable iron, coppered finish, stamped steel, nickeled finish, or aluminum.

The wooden roller, Type A, is made of a seasoned, straight-grained wood, and finished to a smooth, straight roller.

The metal roller, Type B, is made of tinned metal of sufficient thickness and strength to minimize bending.

Rollers should be selected and used

TABLE I. STANDARD WIDTHS OF SHADE CLOTHS

SHADE CLOTH	WIDTH—INCHES													
MACHINE-PAINTED	..	..	..	36	38	42	..	45	48	54	..	..	..	..
HAND-PAINTED OPAQUE	..	..	..	36	38	42	..	45	48	54	..	63	72	82
HAND-PAINTED CAMBRIC	..	..	..	36	38	42	..	45	48	54	58	63	72	82
PYROXYLIN	..	..	..	36	38	42	..	45	48	54	..	63	72	..
HOLLAND	28	30	32	36	38	40	44	..	48	54	60	..	72	82

according to the various widths of shade cloth required. Typical recommendations are shown in Table II.

Measuring for Shades. When measuring for window shades, make sure where the brackets are to be placed and measure accordingly—from jamb-to-jamb, stop-bead to stop-bead, or on the face of the casing. A steel or wood measuring tool is usually used for this measurement. Measurements should be made with care so that the shades will fit correctly when they are installed. The length of a shade is the distance between the ends of the round and rectangular shaped rods which enter the brackets. This length should be just enough less than the distance between stop-beads, for example, so that there is little play in the roller after it has been put in the brackets. This allows the roller to revolve freely. The brackets do not affect the length of the roller or end rods as the rods protrude through the brackets.

Fig. 1 shows the top portion of a

TABLE II. SELECTING ROLLERS FOR SHADES

LENGTH OF SHADE—FEET	WIDTH OF SHADE—INCHES				
	28 TO 41	41 TO 46	46 TO 61	61 TO 73	73 AND OVER
4' TO 7' (INCL.)	1" W	1 1/8" W	1 1/4" W	1 1/2" M	1 3/4" M
8' TO 10' (INCL.)	1 1/4" W	1 1/4" W	1 1/4" M	1 1/2" M	1 3/4" M
11' TO 12' (INCL.)	1 1/2" M	1 1/2" M	1 1/2" M	1 1/2" M	1 3/4" M

W—indicates mounted on wood roller.
M—indicates mounted on metal roller.

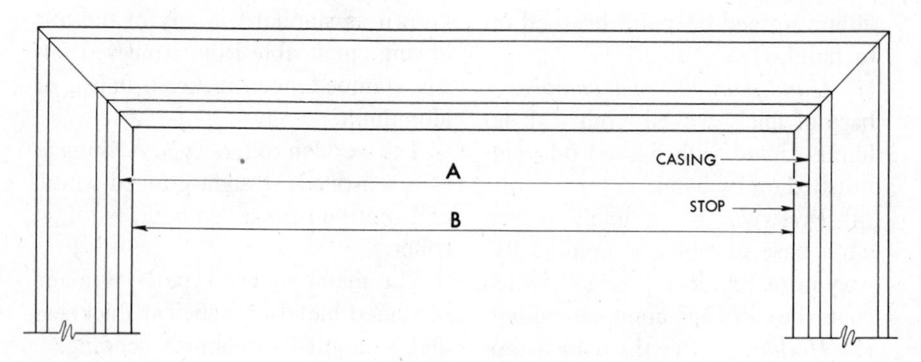

Fig. 1. How to measure lengths of shade rollers.

typical window. Line *A* shows the distance to be measured if the roller brackets are to be put on the casing. Line *B* shows the distance to be measured if the brackets are to be put on the stops. The allowance for free rolling need not be over $1/16''$.

To determine the shade length, measure the distance from the top of the sash (allow room for roller to operate when it contains the full amount of shade) to the window sill. Then add at least 10″ so that the shade will not be strained and possibly jerked off the roller when the shade is pulled all the way down.

For casement windows, French doors, etc., it is usual practice to measure 1″ beyond the glass area on each side so that the shade width will be sufficient to cover the entire glass area without any uncovered spaces on either side.

Venetian Blinds. Venetian blinds can be made to operate horizontally or vertically. They are made of either wood, metal (baked enamel), or vinyl plastic. The strips of the blinds are usually connected by cloth and operated by cords. The blinds may be made to fit either inside or outside the window casing.

Curtain Rods. Curtain rods require brackets for support somewhat the same as shades do. The rods are generally of brass about $5/16''$ diameter. Sometimes two rods are supplied at each window (one for curtains and one for drapes). Where two or more windows occur in a group, a single length rod is used.

Estimating material

Window shades are available in standard sizes. The cost of the shades is determined by the quality of the roller and the shade material. (The cost of making a shade, except material, is the same for all sizes. This includes cutting, accessories, brackets, rollers, slats, and cords.)

Venetian Blinds. These are usually estimated by the square foot of window area which they will cover. They are usually ordered in standard sizes but custom-made blinds may be ordered at extra cost.

Estimating labor

Once they are assembled, window shades, venetian blinds, and curtain rods can be quickly installed. Window shades can be installed at the rate of 12 to 15 per hour depending on the type of window or door. On ordinary windows and doors, the rate of 15 per hour is usually assumed.

One man can generally install 10 complete curtain rods in one hour.

(Sample specifications are on the next page.)

Sample Specifications

A set of window shade specifications can be as detailed as these sample specifications. Note that these specifications also give details of materials and assembly for the shades.

GENERAL. (A) Each window is to be equipped with a window shade hung at the top of the opening and must operate smoothly and efficiently. Each shade is to hang perfectly level and have the spring tension of roller properly adjusted.

(B) All necessary measurements are to be taken after casings and jamb linings are in place.

(C) Shades are to be made in a thoroughly workmanlike manner, cut perfectly square and true and mounted on rollers in the same manner, using suitable substantial fasteners. All materials used in the manufacture of the shades are to be new.

CLOTH. All shades are to be made of *X* shade cloth.

COLOR. As selected by the owner, architect, or superintendent.

LENGTH. Finished length of each shade is to be 10 inches longer than the actual height of the window opening.

ROLLERS. All shades are to be mounted on guaranteed wood or metal rollers to carry shades of the respective widths and lengths properly.

HEMS. Hems are to be of ample width to include free inclosure of the slat; shall be turned, sewed, and the end backstitched to reinforce hems.

SLATS. Smooth white pine slats of the proper size are to be inserted in hems of all shades.

EYELETS. Aluminum rust proof eyelets for shade pulls are to be affixed to each shade, being inserted through the center of the slat.

RING PULLS. Crocheted ring pulls for operating shades are to be attached to each shade.

16

PLUMBING

The purpose of this chapter is to explain the principal items to be considered in estimating a typical plumbing job. Before a beginner, or even an experienced man, can successfully do a plumbing estimate he must be able to visualize all of the work and material likely to be needed on a construction job. Some specifications carry clauses stating that one or more items may possibly have been omitted despite a careful check of plans and specifications. This is one reason why an estimator must be able to completely visualize the plumbing requirements for a new building.

Materials

Septic Tanks. In small towns, at summer resorts, in summer cottages, and throughout rural communities the septic tank is used extensively as a means of treating and safely disposing of sewage. There is considerable labor involved in the installation of a septic tank as will be seen once the complete system is understood.

Septic tanks are manufactured in various sizes and styles. They are made of metal, of reinforced concrete, and of glazed tile. Sometimes tanks are built on the job. The smaller tanks generally are made of metal and are simple to install. For private residences, usually the small manufactured tank is installed. Where facilities are needed for a large number of people, tanks made to specifications or any number of ready-built units are installed.

Pipes vary in size depending on the size of the installation. Specifications or directions sent with septic tanks should indicate pipe sizes. The pipe leading from the residence to the septic tank is generally of salt glazed tile having cemented joints. Pipes leading from the septic tank to the diverting box, Fig. 1, and from the diverting box to the sewage field are also salt glazed.

Fig. 1 illustrates a typical septic system. The tile is of a porous material laid with open joints. The amount of tile laid in the field naturally depends on the size of the septic tank

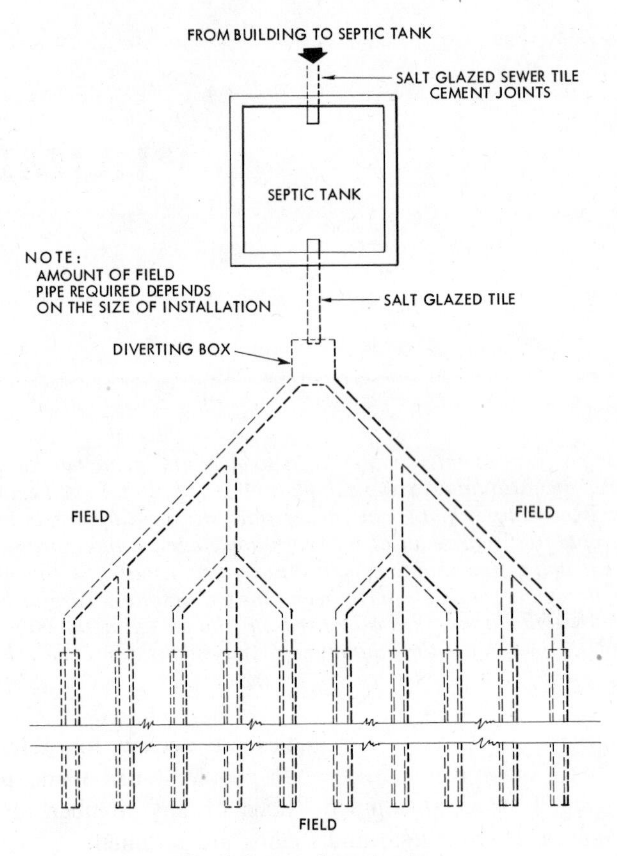

Fig. 1. Sewage disposal system.

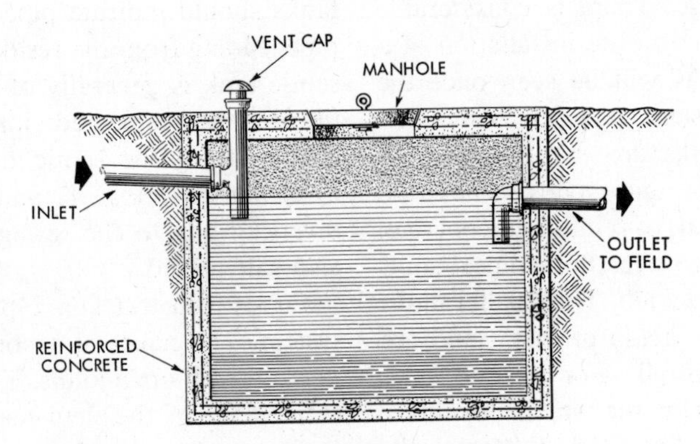

Fig. 2. Typical concrete septic tank.

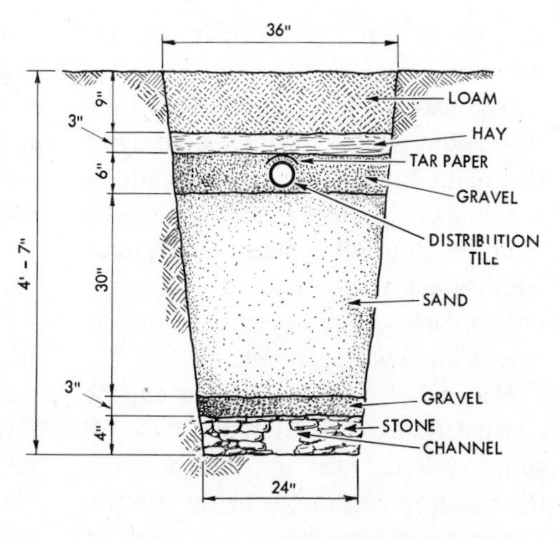

Fig. 3. Leaching trench employed with tight soils.

and the amount of sewage to be disposed of. The tile laid in the field is between 3″ and 6″, as required. The grade must be at least 1 inch for every 20 feet of run and must be made uniform. Tile pipes should be spaced at least 3 to 4 times their diameters. The lengths of drains should be given by specifications or directions sent with the tanks.

Fig. 2 shows a typical septic tank of the type suited to ordinary residences. There are many styles and kinds so the estimator must acquaint himself with the style indicated for the job.

Figs. 3 and 4 show cross sections of trenches which are used for sewage pipes. In some instances the specifications or directions recommend that clover, because of its affinity for water, be planted over the field.

Drainage. Every residence requires some means of disposing of rain water from the roofs. There are a great many ways of doing this, among which are (1) draining to sewer, (2) draining to cistern, and (3) draining to open ground.

The first method is used extensively in city residences where no use is made of soft (rain) water. In smaller towns and throughout rural districts, rain water is drained into a storage cistern for future use. Such cisterns may be a short distance away from the house, in the basement, or under the garage or other ordinarily unexcavated areas. The third method sometimes uses tile to drain the water a short distance from the house, where

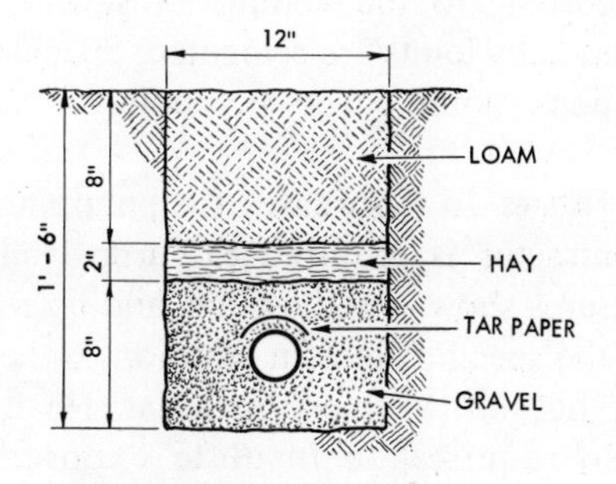

Fig. 4. Typical trench for distribution tile.

it empties on the surface of the ground.

The estimator obtains his information about such systems by studying the plans and written specifications.

Cisterns. Cisterns are sometimes built of poured concrete and lined with cement to make them watertight. Others are constructed of concrete blocks lined with cement.

Because of the materials used, the construction of tile systems, cisterns, sump systems, etc., may be done by the masonry contractor or the plumbing contractor, or both, according to the local regulations.

The estimator must determine of what a cistern is to be made, its shape, depth, diameter, openings, and other details. If such information does not appear in the plans or specifications he must consult the architect or owner.

Sump Pumps. The specifications should give definite instructions about sump pumps. The sump can usually drain directly into the main sewer.

Sometimes the areaways around basement windows are drained by tile and empty into the main sewer.

Sewers. All underground sewers are of some form of tile and vary in size according to the volume of sewage. Generally joints are cemented. Specifications should cover such points.

Fixtures. In most cases the plumbing contractor is required to furnish all fixtures shown on the plans and listed in the specifications in addition to the piping and accessory material. He is also required to insulate exposed pipes.

The plans should indicate all main fixtures and their exact locations. The specifications should give the quality or brand for fixtures, piping, faucets, valves, etc. The specifications should also describe the system and give directions as to hot and cold water outlets, soft water outlets, etc.

Some fixtures require special consideration. For example, specifications may direct that bath tubs shall be supported by a hanger rather than rest on the joists. Lavatories and closet tanks require planking or other woodwork between the joists to which they can be fastened or held securely in position. The same thing is true for kitchen sinks. Laundry tubs often require a means of support on side walls.

Piping Systems. The estimator must determine the amount and sizes of pipes required for water supply, sewage, etc. Occasionally an architect may supply a completely designed plumbing system as part of the plans. In such cases the estimator has but to refer to the drawings and calculate the lengths of the various pipes. Where no plumbing design is supplied, the estimator must, after noting all the fixtures, determine and plan the piping system himself. This will require a careful check of the plans and specifications to determine the size and and length of each and every pipe indicated. Most architects, when designing a building, will make it a point to specify the use of 2 × 6 studs in one partition near both the bathrooms and the kitchen. The reason for this is to provide a wall space ample for 4″ soil pipes having bell connections almost 6″ in diameter. If a plumbing design is not supplied, this one feature indicates the position, at least, of the soil and vent stacks.

Estimating materials

Plumbing estimates are usually based upon cost analysis records kept by the estimator over a period of time. Some general explanations concerning plumbing estimating were given in the materials section.

Experienced plumbers study the plans, locate all plumbing requirements such as water closets, tubs, etc., and then visualize where the various piping would be required. The inexperienced estimator, however, will profit by making a plumbing layout. The layout is made by first tracing the architect's basement plan. Fig. 5 is a plumbing diagram made from the foundation plan of HOUSE PLAN A.

Sewer Pipe. The sewer pipe which runs beneath the finished floor is shown in Fig. 5 by a heavy dashed line. In this instance it is composed of 4″ cast iron. If no building code requirements need be followed, a 4″ pipe is recommended for most residences.

When calculating the lengths of sewer pipes, first estimate to the outside wall of the building, keeping in mind that additional pipe is needed to reach the existing street sewer.

Most building codes allow branch sewers to be 4″ in diameter, but require a 4″ deep seal trap at each drain. The lengths of branch sewers can be scaled and the traps, one for each floor drain, can be counted on the drawing.

Miscellaneous Sewer Equipment. Sewer pipes, L's, 45's, curves, traps, etc., can be counted directly from a plumbing plan if the plan has been carefully and completely laid out.

Rain Water Drains. These drains can be indicated by long dashed lines on the *exterior* of a plan. These are drain tiles which carry the rain water from the downspouts to a cistern.

If the plumbing contractor is to install the tiling, he scales the lengths of the long dashed lines and then adds a small percentage for breakage. The number of Y's, T's, etc., can be counted by studying the plan. Ordinarily these tile lines would also be shown in the plumber's plan. For 4″ tile just about the same amount of mortar is assumed as for 6″ tile.

Catch Basins. Catch basins are generally constructed of concrete blocks and have concrete bottoms ranging from 6″ to 8″ in depth. The lid is generally of cast iron set in a prepared stone cap. The total depth of the catch basins range from 6′ to 8′ and the inside diameter is generally 2′-6″. At this diameter, about 5 blocks are required per layer. If the blocks are 6″ high, 12 layers (60 blocks) are required for a 6′-0″ depth.

Catch Basin Pipe. Most large cities require that the sink drain be carried directly to a catch basin. The catch basin then empties into the sewer. If the architect has not specified the catch basin location, then the plumber

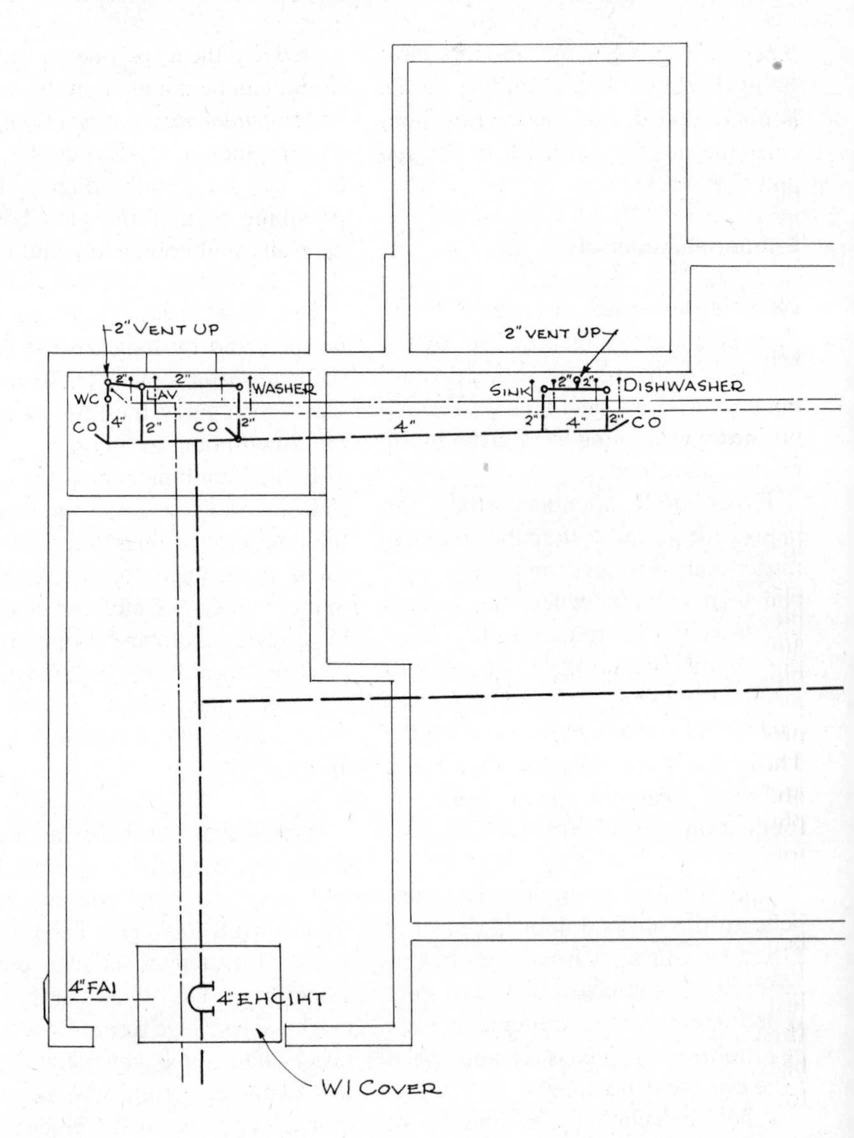

Fig. 5. Plumbing diagram for House Plan A. The following abbreviations are commonly used: FAI—fresh air intake, CO—cleanout, WI—wrought iron, and EHCIHT—extra heavy cast iron house trap.

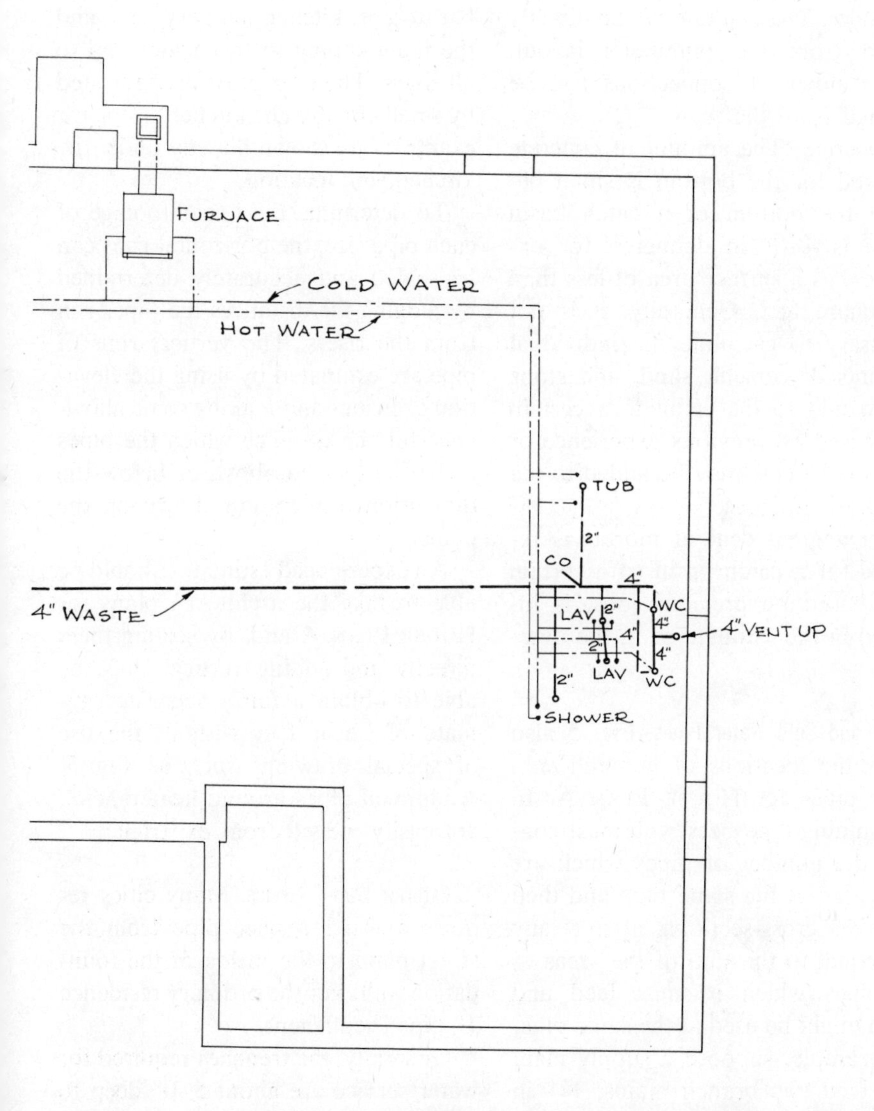
FURNACE
COLD WATER
HOT WATER
TUB
2"
CO
4"
4" WASTE
LAV
2"
WC
4"
4"
4" VENT UP
2"
4"
LAV
WC
2"
SHOWER

(or estimator) must plan its location. Usually a 6″ pipe is required by building codes. The amount can be directly scaled from the plumber's layout. The number of connections can be counted from the plan.

Concrete. The amount of concrete required for the bottom is small because the bottom of a catch basin which is 4′-0″ in diameter, for example, has a surface area of less than 13 square feet. Generally, it is not necessary to calculate the individual amounts of cement, sand, and stone but to add to the estimate a certain cost based on previous experience or jobs, or the cost may be added to the overhead expenses.

Not a great deal of mortar is required for a catch basin so a certain cost, based on previous jobs, is included in the estimate or in the overhead.

Hot and Cold Water Pipes. Fig. 5 also shows the locations of hot and cold water pipes for HOUSE PLAN A. In determining pipe sizes, you must consider the number of pipes which are to be used at the same time and then make the cross-sectional area of any pipe equal to the sum of the areas of the pipes which it must feed and which might be used at the same time. For example, suppose a supply main must feed two branch mains, ¾″ in diameter and ½″ in diameter. The supply main must have a cross-sectional area equal to the sum of the cross-sectional areas of the two branch mains. The diameter of the required pipe is 9/10″ or nearly 1″.

Fig. 5 shows the arrangement of the piping for hot and cold water to the bathrooms, kitchen, laundry, etc., and the main supply to the heater and to all risers. The riser pipes are indicated by small circles. The kitchen risers, for example, are shown directly under the kitchen sink location.

To determine the linear footage of each pipe size, the horizontal runs can be scaled and accurately determined by judging the distances the pipes run from the risers. The vertical runs of pipe are estimated by using the elevation drawings and making some allowance for the distance which the pipes are to be located above or below the dimension termination shown on the plans.

An experienced estimator should be able to take the architect's plans for HOUSE PLAN A and, by scaling them directly and adding vertical runs, be able to obtain a fairly accurate estimate of linear feet without the use of special drawings such as Fig 5. Additional pipes around heaters, etc., are easily judged from experience.

Exterior Water Service. Many cities require a water service pipe from the street main to the inside of the foundation wall. For the ordinary residence 1″ pipe is sufficient.

Generally, the trenches required for water service are about 5′-0″ deep to keep the pipe below the frost line (which is above 5′-0″ in most areas).

Permits. If parkways (areas between sidewalk and curb) are city controlled, a deposit for digging a trench is generally required. This also

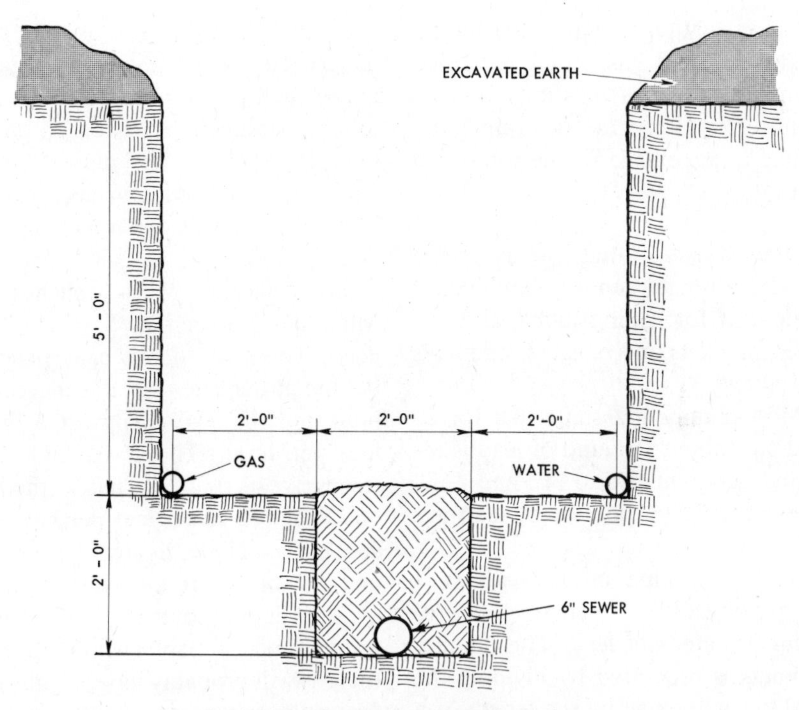

Fig. 6. Trench for sewer, water, and gas pipes.

applies to sewer trenches. The street opening permits for water service are the same as for sewers. For connecting the pipes to the water main a charge of about one-third of this amount is generally made by the city. Charges for this item, however, vary among cities. In some cities the water main is so situated, or ready-to-use connections provided, that the street does not have to be torn up. Thus the street permit would apply only for the sewer. Such information can be obtained from the sewer department of the city prior to making your estimate.

Trenches. Sometimes the same trench can be used for gas, water, and sewage pipes. Such a trench is usually wider than normal. Fig. 6 shows one method for placement of the pipes. Here a trench 6 feet wide is dug down to a depth of about 5 feet. From this depth a secondary trench 18″ to 24″ wide and 2′-0″ is extended down for the sewer pipe. The exact depths depend on the street depths of sewer, water, and gas mains. This trench plan may not save any time (depending on the required depths, widths, etc.) but in some instances it could result in a saving in permits and fees.

In most cases, however, separate water and sewer trenches are required.

Gas Pipes. The gas companies generally install the gas supply pipe between the street main and the house free of charge. The plumber has only to run the necessary pipes within the house.

The required gas piping is scaled from the drawings or the plumber's sketch. The fittings are determined by adding a percentage of the total pipe cost.

Other Material. Other items which must be estimated under plumbing include lead for wiping joints, and soil pipes, brackets and hangers, and nails and screws.

Most plumbers assume that for 1″ lead pipe about 1 pound of lead is required per joint. For 1¼″ and 1½″ pipes, allow 1½ pounds of lead per joint.

A 4″ soil pipe requires about 4 pounds of lead. A 2″ soil pipe requires about 2 pounds of lead. The number of joints is estimated by dividing the total linear footage by the length of a standard piece of pipe and adding for special connections.

Brackets and pipe hangers are estimated by studying the plans and noting in the basement, for example, where there are long runs of pipe to be hung from the ceiling or between joists.

Nails and screws cannot be estimated exactly and are usually considered as an addition to the overhead expenses.

Estimating labor

Estimating plumbing labor can be confusing because of the numerous ways in which it can be done. Contractors may either estimate their labor by a certain percentage of material costs plus detailed calculations for trenching and such other items as have little material, or they may estimate labor at so much per portion of an average job. Sometimes labor is based on the number of pipe joints plus a certain percentage of fixture values and detailed calculations for trenching.

Trench Digging. Most trenches for water pipes are either 5′-0″ or 7′-0″ deep. The 5′-0″ depth can generally be dug at the rate of 4 linear feet per hour and the 7′-0″ depth at 3 linear feet per hour. The 5′-0″ and 7′-0″ trenches can be backfilled at the rate of 15 and 12 linear feet per hour, respectively. These figures for digging and backfilling are the average figures of a particular contractor. Each contractor (and estimator) who keeps such records will probably have slightly different times recorded for this type of work.

When figuring the cost of digging and backfilling, a certain price is allowed per linear foot.

Roughing Labor. Soil and waste stacks are considered part of the roughing-in job along with hot and cold water pipes and downspout connections. For residences such as HOUSE PLAN A, the labor can be based at a certain percentage of the material cost, say 80%. On better quality homes where more work is involved and where greater accuracy is required, the percentage rises to 125%. The percentages vary, of course, among various areas.

Copper Pipe. Ordinarily contractors would raise their labor figure about 20% when pipes are made of copper because of the additional installation time.

Sewer Pipe. There is not much difference between the time required for laying 4″ and 6″ sewer or drain tile because one can be handled as easily as the other. The actual laying time can be taken into consideration along with trench digging.

Catch Basins. On the ordinary catch basins, the time for digging and backfilling as required is about 6 hours. This work can be done by a laborer. The basin itself requires about 6 hours for installation by a sewer builder.

Gas Pipes. The labor for gas pipe installation is usually figured as about 150% of the material costs.

Fixture Labor. The installation of fixtures for nominal cost residences is usually taken as 50% of total fixture cost. For higher quality residences where the fixtures are more elaborate only about 40% is assumed for labor.

Labor is figured at about 30% of the material cost for hot water heaters and pumps for cistern water.

17

SHEET METAL

Sheet metal has a variety of uses in building construction. It is utilized in all stages of construction, from foundation and framing work to roofing and interior work. This chapter discusses the most common sheet metal applications.

Uses of sheet metal

Because of the variety of sheet metal applications in building construction, a careful check of both specifications and drawings should be made. Some applications, such as termite shields, Fig. 1, and water table flashing, are near the foundation. Other applications, such as chimney flashing and ridging, are on the roof. Sheet metal is also used on and around doors and within the house.

The most common applications of sheet metal are as barriers and shields, flashing, stops, drops and chutes, ridg-

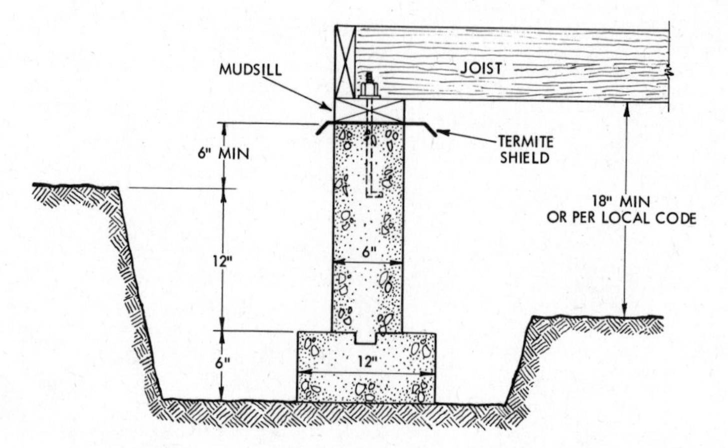

Fig. 1. A termite shield used between the concrete foundation and wood sill.

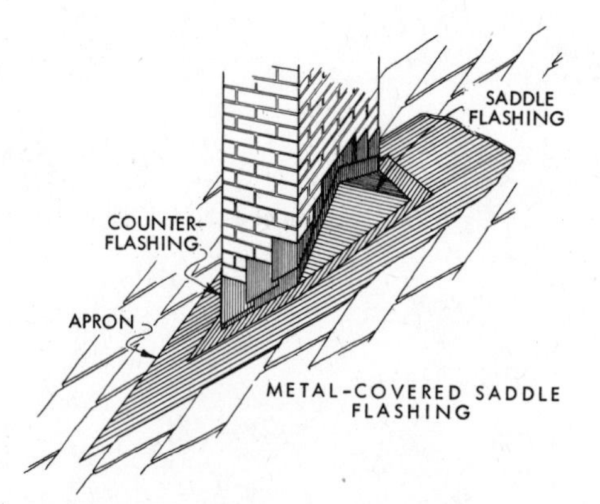

Fig. 2. Saddle flashing and sheet metal around a chimney.

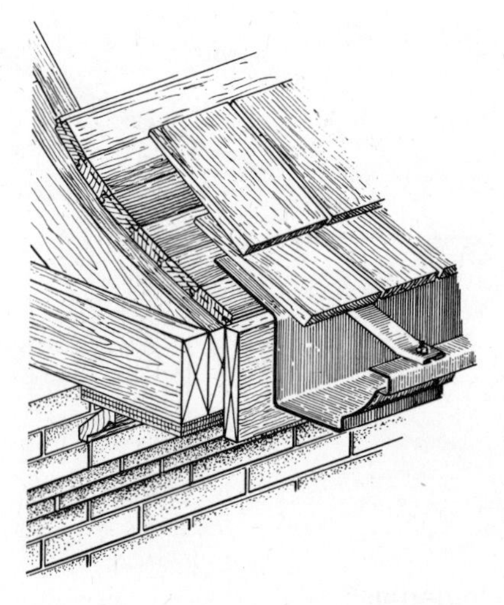

Fig. 5. A sheet metal gutter attached to the edge of a roof.

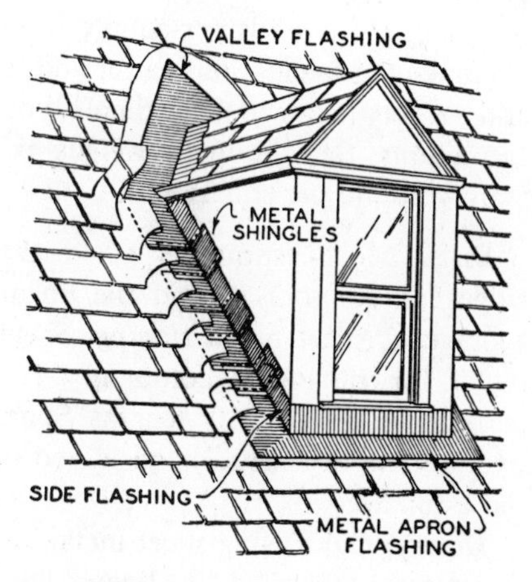

Fig. 3. Valley flashing and sheet metal around a dormer.

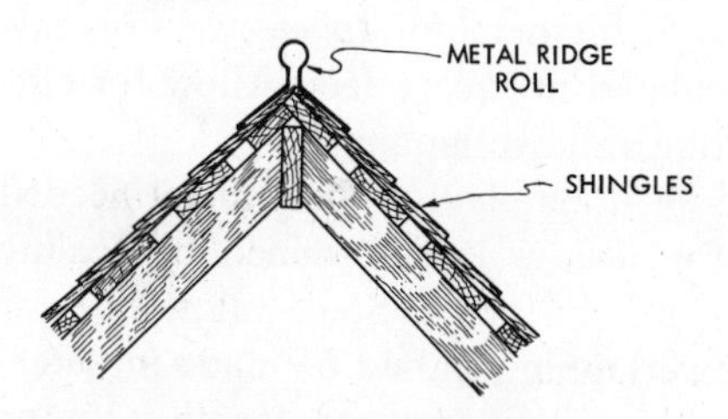

Fig. 4. Sheet metal is used along ridge boards to waterproof the top of the roof.

ing, gutters, leaders (downspouts), and roofing.

Fig. 2 shows sheet metal being used as saddle flashing to divert water around a chimney. Fig. 3 shows valley flashing at the intersection of a roof and dormer. It serves the same purpose as saddle flashing.

Fig. 4 illustrates how sheet metal is used along the lengths of the ridge boards.

Fig. 5 shows how a sheet metal gutter is attached to a wood roof. Note the metal strap bolted to the gutter. Fig. 6 is a cross section of a sheet metal gutter which hangs free of the roof. The metal straps are the only support for the gutter.

Some sheet metal work such as heating and air-conditioning ducts and flashing for plumbing vents are usually figured as part of the heating and plumbing estimates.

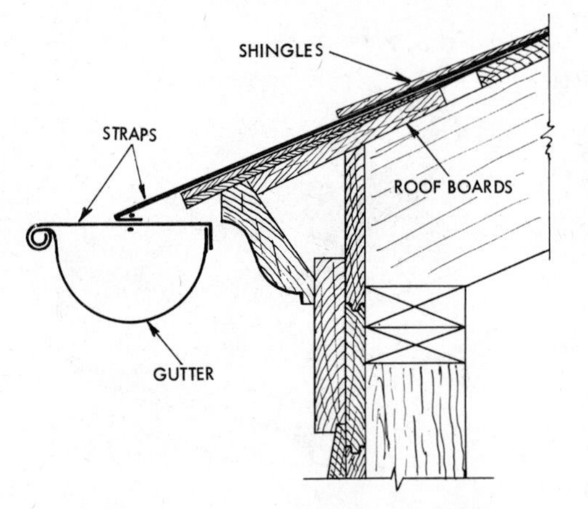

Fig. 6. Cross section of a sheet metal gutter.

Materials

A wide variety of metals are used for the various sheet metal applications. Although the majority of jobs will specify galvanized sheet steel for most work, any of the following metals could be specified for a portion of a sheet metal job: aluminum, copper, lead, monel, tin, or zinc.

The various metals can differ greatly in cost and labor required for installation. It is therefore important to know not only where sheet metal is to be used, but the kind of metal which is to be used.

Along with sheet metal are various screws, bolts, nails, paints, and solder necessary for proper installation and finishing.

Estimating materials

Table I shows the various thicknesses and weights of iron and steel sheets and plates. Contractors and estimators assume round numbers for weights in order to simplify calculations. For example, 26-gage sheet metal is shown as weighing .75 pounds per square foot. To make the calculations easier and to add a safety factor to the estimate, you can assume that the metal weighs exactly 1 pound per square foot.

Roofs. Most estimators scale the ridges of the roof to find the linear footage of sheet metal *ridging*. Sheet metal for ridging is commonly 13″ wide and comes in 8′-0″ lengths. Some overlapping of sheets is considered in the estimate.

Dormer pans using sheet metal are measured or scaled in linear feet. The material generally comes in 8′-0″ lengths and in various standard widths.

Sheet metal for *dormer decks* is calculated in square feet. Allow for cutting and overlapping.

The amount of sheet metal needed for *valleys* is determined by scaling from the plans. Some allowance for overlapping should be made in determining the required length; 2″ for overlaps should be sufficient.

TABLE 1
UNITED STATES STANDARD GAGE
FOR SHEET AND PLATE IRON AND STEEL

Number of Gage	THICKNESS		WEIGHT		Number of Gage
	Approximate thickness in fractions of an inch	Approximate thickness in decimal parts of an inch	Weight per square foot in OUNCES avoirdupois	Weight per square foot in POUNDS avoirdupois	
0000000	1-2	.5	320	20.	0000000
000000	15-32	.46875	300	18.75	000000
00000	7-16	.4375	280	17.5	00000
0000	13-32	.40625	260	16.25	0000
000	3-8	.375	240	15.	000
00	11-32	.34375	220	13.75	00
0	5-16	.3125	200	12.5	0
1	9-32	.28125	180	11.25	1
2	17-64	.265625	170	10.625	2
3	1-4	.25	160	10.	3
4	15-64	.234375	150	9.375	4
5	7-32	.21875	140	8.75	5
6	13-64	.203125	130	8.125	6
7	3-16	.1875	120	7.5	7
8	11-64	.171875	110	6.875	8
9	5-32	.15625	100	6.25	9
10	9-64	.140625	90	5.625	10
11	1-8	.125	80	5.	11
12	7-64	.109375	70	4.375	12
13	3-32	.09375	60	3.75	13
14	5-64	.078125	50	3.125	14
15	9-128	.0703125	45	2.8125	15
16	1-16	.0625	40	2.5	16
17	9-160	.05625	36	2.25	17
18	1-20	.05	32	2.	18
19	7-160	.04375	28	1.75	19
20	3-80	.0375	24	1.5	20
21	11-320	.034375	22	1.375	21
22	1-32	.03125	20	1.25	22
23	9-320	.028125	18	1.125	23
24	1-40	.025	16	1.	24
25	7-320	.021875	14	.875	25
26	3-160	.01875	12	.75	26
27	11-640	.0171875	11	.6875	27
28	1-64	.015625	10	.625	28
29	9-640	.0140625	9	.5625	29
30	1-80	.0125	8	.5	30
31	7-640	.0109375	7	.4375	31
32	13-1280	.01015625	6½	.40625	32
33	3-320	.009375	6	.375	33
34	11-1280	.00859375	5½	.34375	34
35	5-640	.0078125	5	.3125	35
36	9-1280	.00703125	4½	.28125	36
37	17-2560	.0066406	4¼	.265625	37
38	1-160	.00625	4	.25	38

Flashing. For vertical members, flashing lengths are scaled from the plans. The flashing is generally in 8′-0″ lengths and at least 12″ wide. Allow 2″ for overlap at each joint of the material. Little or no waste need be added to the quantity take-off.

Right-angle flashing is also measured in linear feet, allowing 2″ for overlap. Usually the sheets are purchased in 8′-0″ lengths and then bent into the shape of a right angle. No waste need be considered.

Chimney flashing, on a large chimney, generally is applied using pieces 10″ × 10″, or using stock 10″ wide and 8′-0″ long. Where the 10″ × 10″ pieces are used, allow 1″ for overlap.

Roof openings, such as those made for access to flat roofs, usually have the appearance of an inverted but shallow rectangular or square pan.

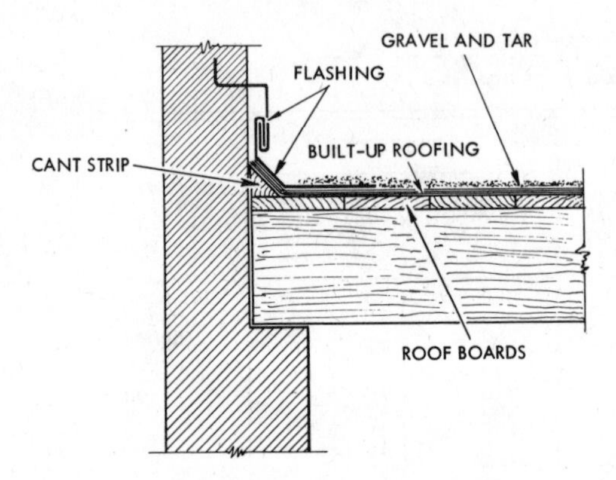

Fig. 7. Flashing used at the intersection of a brick wall and a flat roof.

The combined areas of sides and top are calculated, allowing a little for overlap, and the sheet metal estimated in square feet.

Flashing for flat roofs and side wall intersections is measured in linear feet. Fig. 7 shows how flashing is constructed at the intersection of a flat roof and brick wall.

Gutters and Leaders. The working drawings should show where the gutter is to be placed — or the specifications should explain in detail where and how the gutter is to be installed. The drawings can be scaled to determine the linear footage. Porch and dormer gutters are measured in the same manner.

Leaders should be indicated on the drawings. If they are not indicated, then the contractor or estimator must decide for himself where such piping is required. On repair jobs it is simply a matter of locating the original pipes and following the layout. The vertical dimensions on the elevation drawings can be used, or actual distances can be scaled.

Drip caps over windows, doors, etc., are scaled in linear feet. Then add about 25% more for wastage, because the material comes in 8′-0″ lengths and is not usually spliced.

Doors. Sheet metal is used to frame doors and to cover wood doors. For door frames, the jamb and casings must be covered all the way around. Sheet metal 10″ wide would accomplish this in most cases. The linear footage is the length of the 2 sides plus the top. The ordinary 6′-8″ × 2′-8″ door will require 16′-0″ of sheet metal around the frame.

For covering a wood door, the amount is first calculated in square feet and then in pounds. The door area should include both sides and enough extra to lap the sheet metal of one side around the edge of the door and over the metal on the other side; an inch or two overlap is sufficient.

Accessories. Gutters often hang free of the roof, so they must be connected by *brackets* or *hangers*. Gutter brackets are generally spaced 3′-0″ apart. Other gutter hangers have the same spacing, or less when the gutter is exceptionally large. To find the required number, divide the total length of gutter by the hanger spacing; add 4 to 6 to cover wastage and loss.

Elbows are often needed for leaders and should be specified and/or shown on the working drawings. These are counted by studying all four elevation views of the building. If the drawings do not show the leaders and elbows, the estimator must consult the architect or make his own layout.

Leader brackets and *clamps* are ordinarily spaced about every 10 feet. On a one-story building, the leader is fastened at the top and bottom. On a two-story building, perhaps 3 brackets or clamps are used. All short lengths, such as for low dormers, should have the brackets at the top and bottom of the leader.

Waterproof felt is sometimes installed on roofs, under the sheet metal for dormer decks. For one layer, it is only necessary to measure or scale the dimensions and find the area. No waste allowance is necessary as the felt can be cut and laid as required.

Sheet metal is often covered with *red lead paint*. Two quarts of paint will generally cover about 250 linear feet of gutter, ridge, valley, leaders, etc. This is for new work. For old work an extra quart is allowed for the same amount of work. Paint goes a long way on metal. (Chapter 14, "Paint", discusses paint materials and labor in detail.)

Screws used to fasten sheet metal to a wood door or door frame are difficult to estimate because the spacing between screws can vary with different doors and frames. The spacing is generally about 3″ for most doors. Allow about 75 screws for a door of 21 square feet, 150 screws for a 42 square foot door, etc. The cost is usually so small that it is often considered a part of the general overhead expenses. For the metal frame allow about 75 screws for a 21 square foot door.

Nails for applying sheet metal are not usually calculated in detail because all exposed metal is usually nailed every 2″. Allow 6 or 8 pounds of nails for an average job. This can also be added to the overhead.

Estimating *solder* is also difficult and not worth the time which it would take to estimate. For most jobs you can assume 1 pound of solder for the average residence, 2 pounds for large residences, etc.

House Plan A. The sheet metal used in HOUSE PLAN A is of several different materials. An aluminum shield is used for termite protection under all wood construction. Copper flashing is used on the roof. The gutters and leaders are also of aluminum, as is the window (and entrance door) head flashing. Zinc weatherstripping is used for the entrance door.

Estimating labor

Labor is estimated by multiplying the workman's hourly salary by the number of hours it takes him to complete his work. Typical labor times are given in this section. Use these times only under average conditions.

Generally one man can put on about 20 linear feet of *ridging* per hour. When there is a great deal of cutting and fitting to be done, reduce the work rate to about 15 linear feet per hour.

For *dormer pans* averaging 10″ wide, where 6″ of metal is under the frame and 4½″ on the roof, the labor time is about 1 hour for every 10 linear feet.

One man can lay about 25 square feet of metal *dormer deck* per hour.

Valley flashing is installed at the approximate rate of 27 linear feet per hour.

Flashing for vertical members is done at the rate of about 12 linear feet per hour. The flashing for right angles is installed at the rate of about 25 linear feet per hour. For a chimney the size of that in HOUSE PLAN A, the flashing time is approximately 2 hours. For a small, one-flue chimney, the labor time can be taken as 1 hour.

The labor time for *gutter* work can vary widely depending on the lengths of each gutter run. For example, on a roof where a great many short gutter runs are installed, the time might be as much as 1 hour per 15 linear feet of gutter. This is because the ends of the gutters must be closed, and on short runs this requires considerably more time than when the runs are long. On a roof where there are few dormers, and where the gutter runs are long, one man can install about 40 linear feet per hour. On the average residence you can assume at least 17 linear feet per hour.

Gutter hangers are included in the gutter labor and require no special labor consideration.

Labor for *leaders* varies according to the lengths. A long run naturally requires less labor per linear foot than a short run. A leader pipe must be soldered to the gutter and sometimes fitted into a sewer pipe connection. The labor for fastening the leader to the wall or other place amounts to very little per linear foot. Thus a number of long leaders can be installed cheaper than a like number of short ones. When many short runs are used, a rate of 10 linear feet per hour is a good average; for long runs, up to 30 linear feet per hour can be used as an average figure.

Drip caps are installed at the rate of 17 linear feet per hour.

For an average-sized *door* of 2′-8″ or 3′-0″ width, it takes approximately 2 hours to cover the door with sheet metal. The sheet metal angles used for a metal door frame require about 1½ hours to install for an average-sized door opening up to 3′-0″ wide. Larger doors take a proportionally longer time.

Waterproof felt can be installed over dormer roofs at the rate of about 75 square feet per hour.

Where *clothes chutes* are used, the labor is figured at ½ hour per foot for a chute which is 12″ in diameter.

18

HEATING AND AIR CONDITIONING

No matter where a residence is located, some consideration should be given to heating and/or air-conditioning systems. This chapter is primarily concerned with estimating heating systems but it also includes basic information on central air-conditioning systems. Modern air-conditioning and heating systems are frequently combined in a year-round system; they use the same ducts, fans, control equipment, etc.

Heating and air-conditioning systems

In southern areas of the United States, heating units for occasional heating are popular, and complete central heating plants are rare. Electric heaters are used extensively for occasional heating. If complete summer cooling is required, however, then a central plant for both cooling and heating is more practicable and economical.

In northern areas where severe winter conditions exist, central heating plants are absolutely necessary where the full comfort of modern heating is desired.

Central plants are located in basements or utility rooms of homes and the heating medium is distributed through ducts or pipes to the various rooms. Types of central plants include mechanical (forced air), steam, hot water, and vapor systems. (The warm-air and hot-water systems are replacing steam systems in modern construction.)

Mechanical Furnaces. The mechanical or forced-air furnace has a blower which creates positive air circulation. The furnace is purchased complete but is usually assembled on the job. Besides being heated, the air may be filtered and humidified. The filtering and humidity control equipment can be used with refrigeration equipment or cold water to supply cleaned and cooled air during uncomfortably warm weather.

Mechanical furnaces are fired automatically by stokers or oil or gas burners.

A mechanical heating or air-conditioning system should be designed and

Fig. 1. A cut-away view of the duct systems in a typical residence. The supply ducts are in the interior partitions and the return ducts are in the outside walls.

selected by a reputable heating engineer who understands the calculations of heat losses and gains, etc. This service may be an additional expense but is most likely an architectural service charge rather than a builder's expense.

Duct Systems. In mechanical heating and air-conditioning systems the hot and cold air pipes are known as ducts. They are rectangular or circular in cross section.

Fig. 1 is a cut-away drawing of a residence, showing the supply and return ducts, grilles, various fittings, etc. Note that the supply ducts are carried to the upper parts of the various rooms; that the return ducts are at the baseboard level; that all supply ducts are in interior partitions; and that all return ducts are in exterior walls.

Duct Lengths. The layout drawings generally are traced from the regular working drawings. If they are not traced, the layout drawings should be

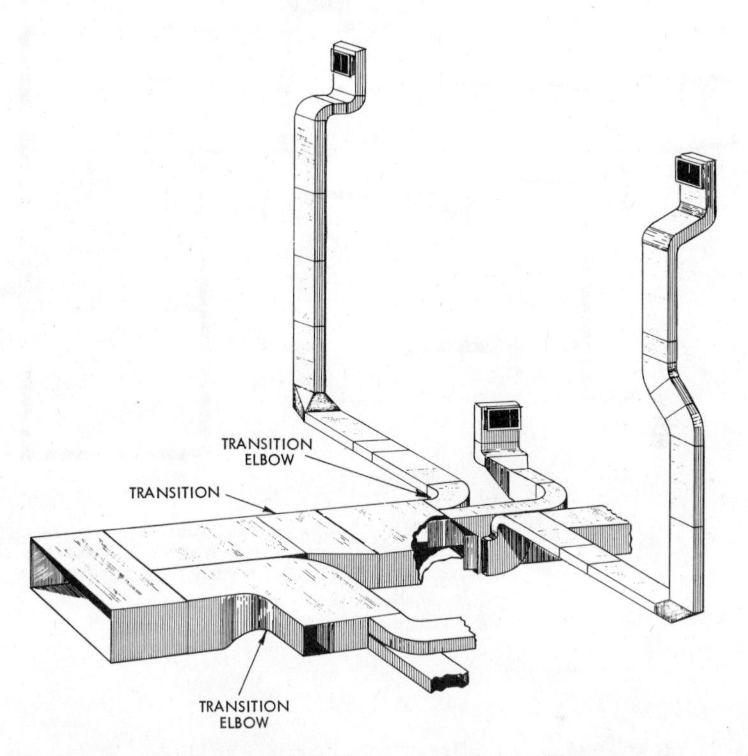

Fig. 2. A diagram of a typical duct system used for a mechanical heating system. Learn to visualize duct systems in this manner. Henry Furnace Co.

drawn to scale. In either case the layouts are scaled to determine the lengths of various ducts in the basement or wherever the central heating plant is located. The stack lengths of both warm-air and cold-air ducts can be determined easily from vertical dimensions shown on the drawings plus the known distance above the floor.

The number and kinds of various fittings required are counted by tracing the various ducts throughout their lengths. If you learn to visualize a duct system in the manner shown in Fig. 2, you will not be likely to make omissions.

Fig. 3 shows the floor plan for House Plan A, and section views of typical duct work which can be installed in this house. Try to visualize all the ducts in the walls of the house and compute their lengths.

You must be careful to note anything irregular in any of the duct runs. Sometimes one or more ducts may require insulation or, because of structural conditions, other expensive operations may be necessary. You should know whether prefabricated or a combination of prefabricated and job-made ducts are to be installed. The use of prefabricated ducts has brought about a practically standard method of estimating.

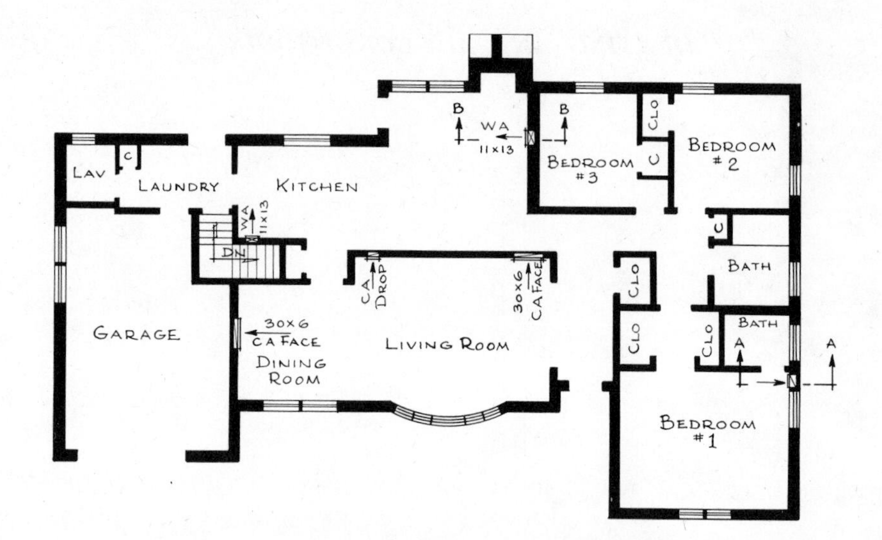

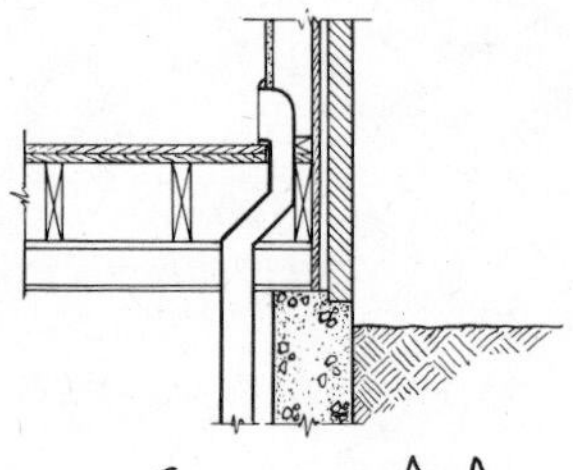

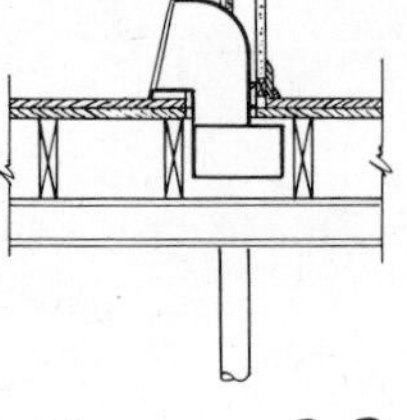

Fig. 3. This is a plan view of House Plan A with section views of air ducts in interior and exterior walls.

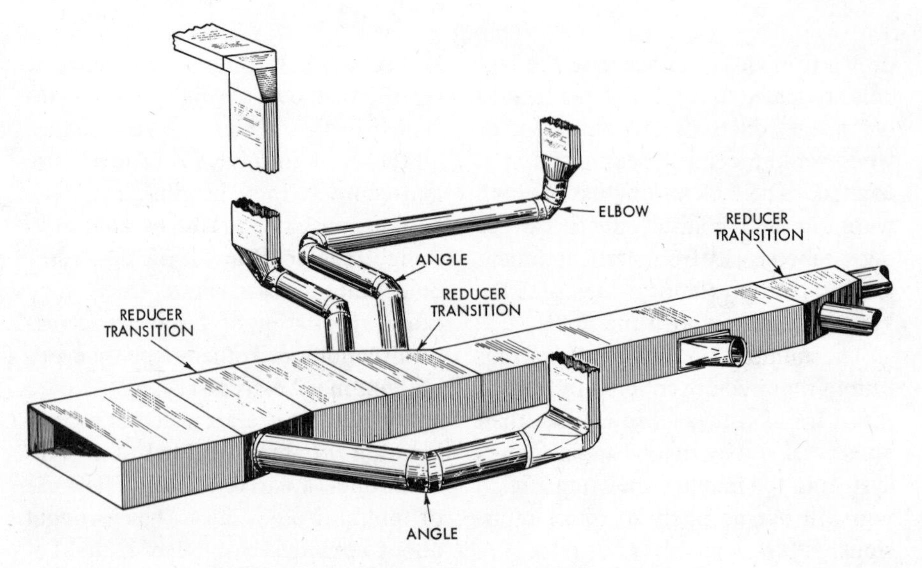

Fig. 4. Typical fittings used in a duct system. Henry Furnace Co.

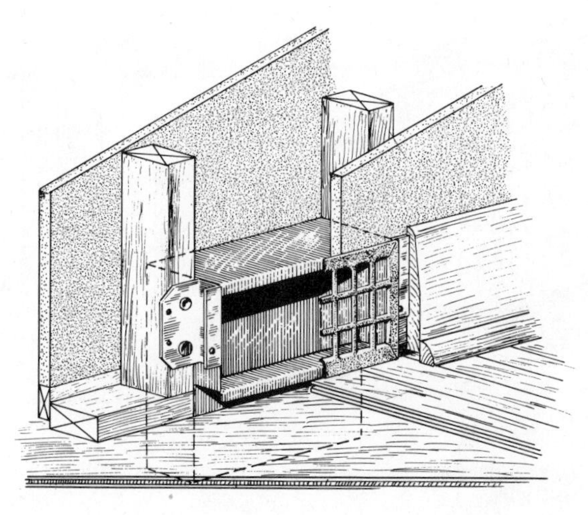

Fig. 5. A box for a baseboard grille is mounted in an interior wall as shown in this illustration.

Fittings. Figs. 2 and 4 give some idea of the various fittings used in mechanical heating and air-conditioning systems. You should obtain catalogues showing all types of fittings and their separate parts. In the system illustrated, the basement trunk duct is suspended from the ceiling. Where individual duct systems are specified the ducts are often run between joists, where they are out of sight. Various types of heating and air-conditioning equipment require widely differing connections for the ducts to the equipment. Cold-air ducts are generally laid out so as to enter the equipment in one or two trunk ducts. However, these items will be shown on the layouts. Study Figs. 1, 2, and 4 until you are able to visualize the entire system.

Grilles. Fig. 5 shows a box for a baseboard grille. A sheet metal cover may be used during construction to keep out dirt. This cover is later replaced by a grille. Fig. 6 illustrates the fitting for a wall grille. The details of

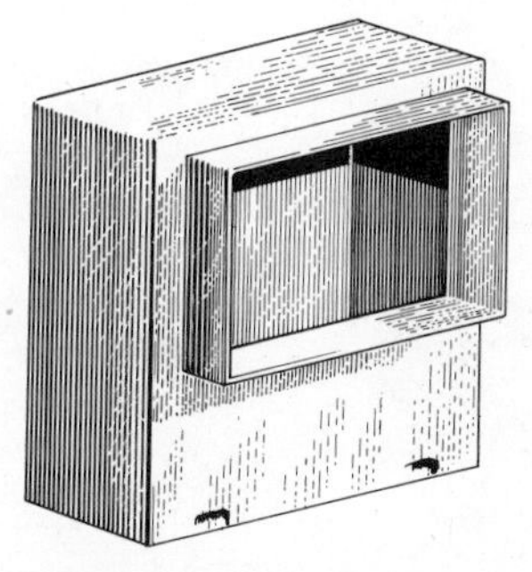

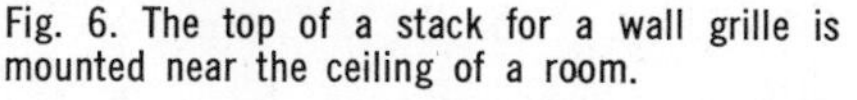

Fig. 6. The top of a stack for a wall grille is mounted near the ceiling of a room.

these boxes vary according to stack and grille sizes.

Baseboard and wall grilles are similar to those shown in Fig. 7. They are delivered ready to install.

Dampers. The typical damper in Fig. 8 can be used in any rectangular duct. Dampers are built to fit only one size of duct, but many damper sizes are available. Dampers are sometimes used behind grilles as a means of controlling the air flow in a room. In mechanical furnace systems, dampers (in

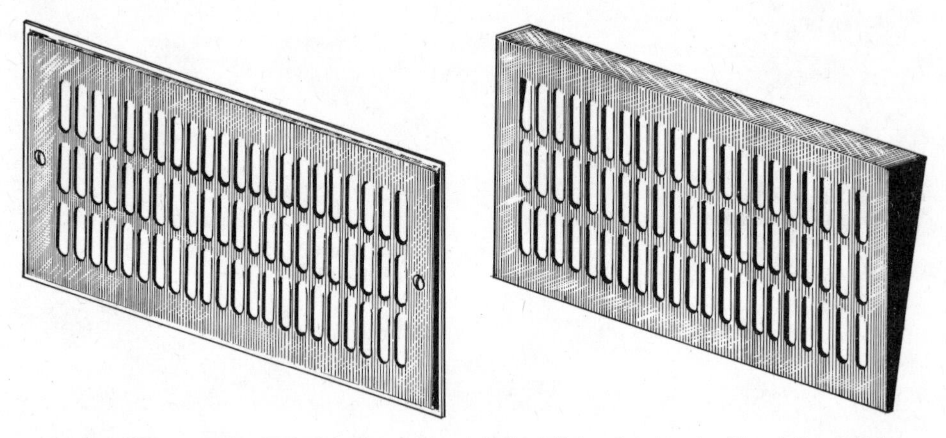

Fig. 7. Grilles are easily attached to the openings. At the left is a baseboard grille; at the right is a wall grille.

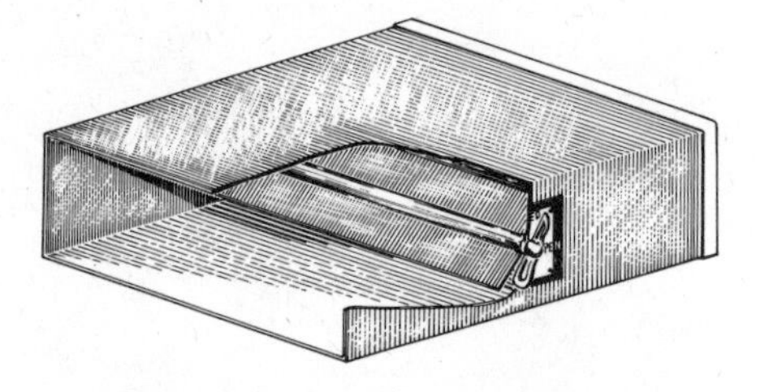

Fig. 8. This is a typical damper which is mounted in a rectangular duct.

the ducts) balance the flow of air through the system.

Controls. Where heating, cooling, and air conditioning are specified, the matter of automatic controls becomes complicated and requires a great deal of additional costly equipment and labor. Fig. 9 illustrates a typical control system for a year-round air-conditioning plant having a mechanical furnace as the primary heat source.

Steam Heating. Steam systems require a boiler, piping, radiators, valves, vents, and miscellaneous boiler and control equipment. The boilers can be automatically fired, using gas or oil burners. In many cases the pipes are insulated with a specially shaped and more or less rigid insulation.

Pipe Lengths. Layouts for steam heating are seldom made. Generally the positions and sizes of radiators or convectors are indicated on the working drawings and the estimator makes his own calculations for the amounts of piping and various fittings, such as elbows, nipples. tees, etc. If full architectural service is engaged for a new residence, it is probable that a heating layout will be furnished. If a layout is not furnished, the contractor must de-

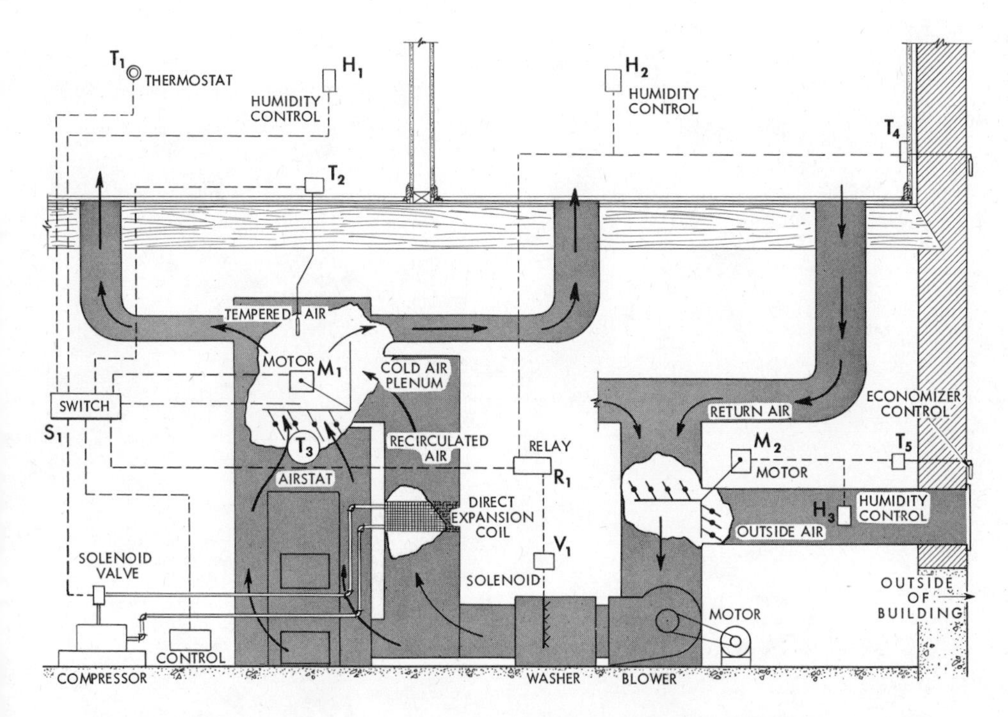

Fig. 9. A typical control system for a year-round air-conditioning system. A mechanical furnace is the basic heat source.

sign the pipe sizes before the estimator can do a quantity take-off. The pipe lengths are determined from dimensions on the drawings and by scaling.

Hot-Water Systems. A hot-water system is essentially the same as a steam system except that hot water instead of steam is the heating medium. A heating layout for a hot-water system will usually indicate the following: correct radiator and convector sizes and styles, pipe lengths and sizes, slopes of pipes, fittings, radiators, valves, expansion tank, etc. With such a layout and the working drawings, all dimensions can be found.

Fig. 10 shows various types of piping systems, some of which require two connections to each radiator. The two-pipe system requires considerably more piping and consequently a greater amount of labor.

Radiant Heating. The radiant heating systems of the Romans and Koreans were of the floor panel, hot-air variety, and similar systems are in use today. However, by far the greatest majority of installations in use and under construction at the present time are of the hot-water variety.

Panels. Panels for a hot-water radiant system may be located in the walls,

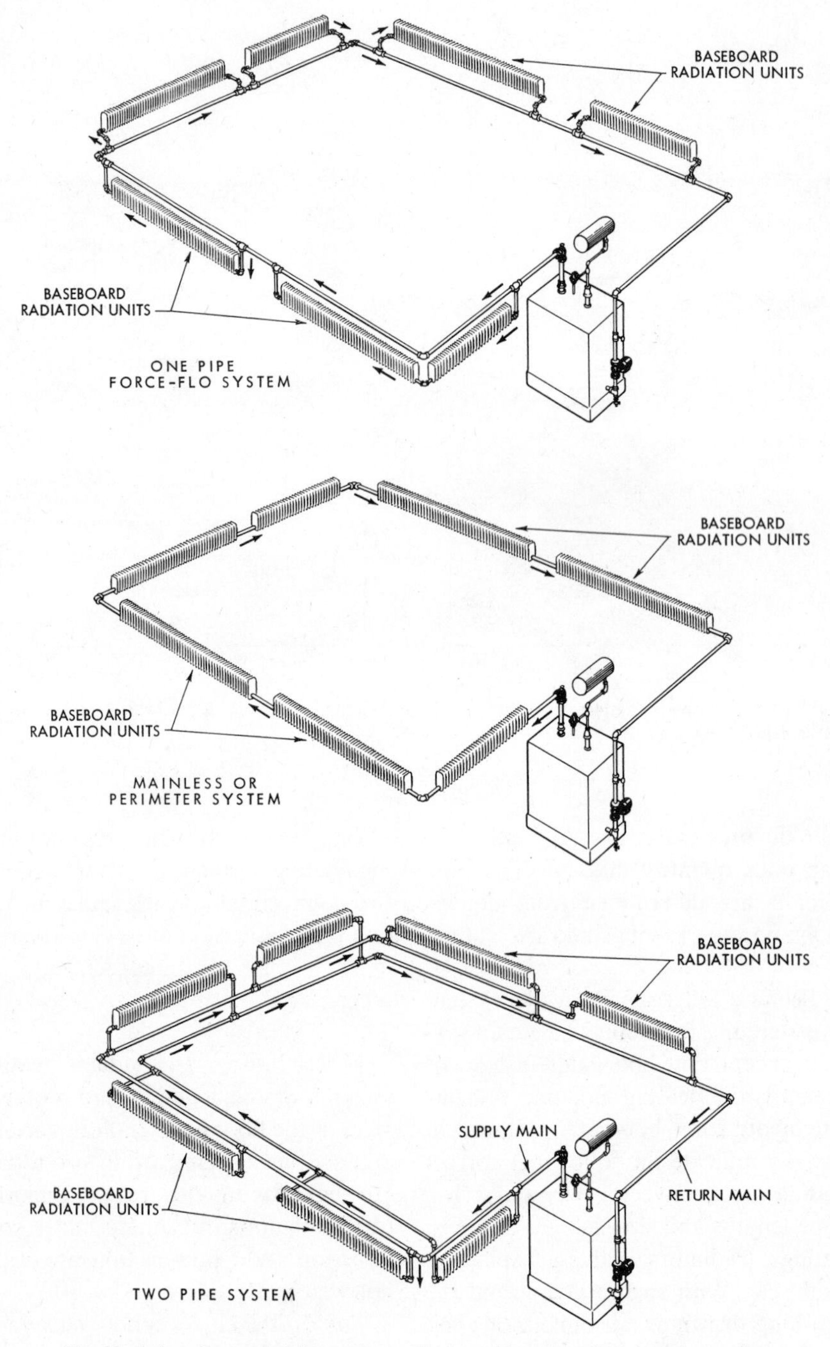

Fig. 10. Various types of piping systems. The two-pipe system is the most complex.

TABLE I
PERCENTAGE OF TYPE OF HEAT TRANSFER BY PANEL LOCATION

Panel Location	Convection Transfer	Radiant Transfer
Wall	48%	52%
Ceiling	35%	65%
Floor	52%	48%

the ceiling, or the floor. While the actual location of the panels depends to some extent upon the architectural design of the building, the differences in the percentages of heat transferred by convection and radiation for each panel, as shown in Table I, also should be a factor in the selection of the panel location.

Baseboard Radiation. This is a system of heating in which the heating units occupy space ordinarily taken by the wooden baseboard, projecting into the room only a little more than the conventional baseboard and blending into the wall. Baseboard radiation units consist either of hollow, cast-iron panels, with or without fins on the back, or of ferrous or nonferrous finned tubing behind metal enclosures.

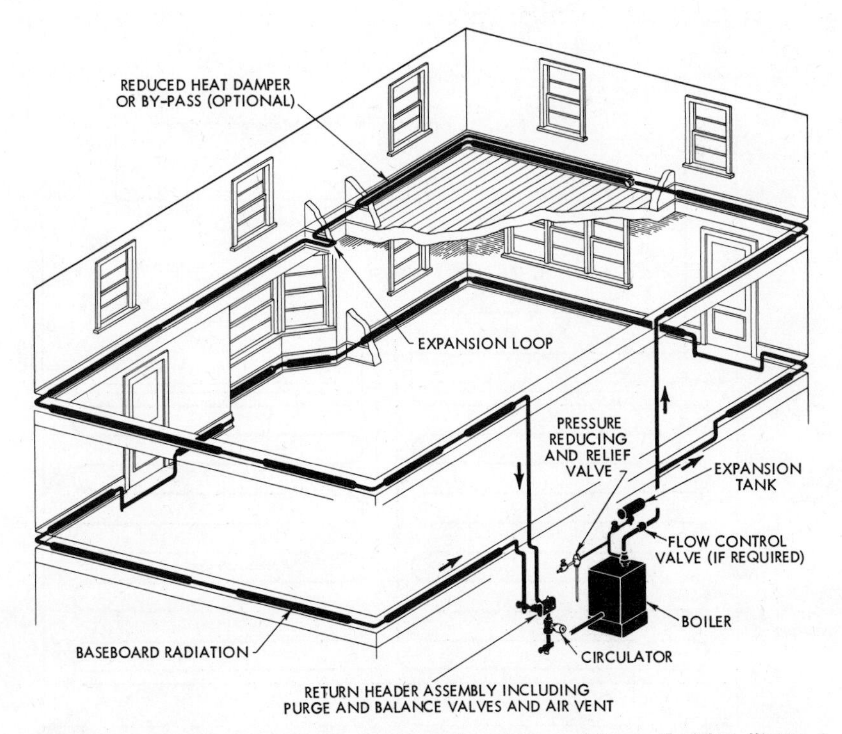

Fig. 11. The perimeter system uses no mains and requires a minimum of piping. Warren Webster and Company.

Heat is radiated from the baseboard units to the walls and furniture from which it is reradiated. Due to its location at floor level, drafts, cold floors, and hard-to-heat areas are positively eliminated. Furniture can be placed wherever desired. Since heat is distributed near the floor, the floor-to-ceiling variation in temperature is reduced to about 2° F. Thus, there is a tendency for more uniform temperatures throughout the room. Finally, baseboard radiation units are clean in operation.

Baseboard radiators can be installed as part of a two-pipe system, a conventional one-pipe or monoflow system with basement mains, or as a perimeter (mainless or series) system in which the baseboards themselves are used as mains. These systems are illustrated in Fig. 10. An example of the perimeter or series system is shown in Fig. 11.

Snow Melting. In localities where the annual snowfall is such as to present a problem with regard to its removal, an adaptation of the radiant panel coil or grid has been used successfully and is becoming more popular.

A snow melting system consists simply of a coil or grid embedded in the concrete of the walk or driveway which is to be kept free of ice and snow. Through these coils or grids is circulated water to which an antifreeze has been added. The water may be

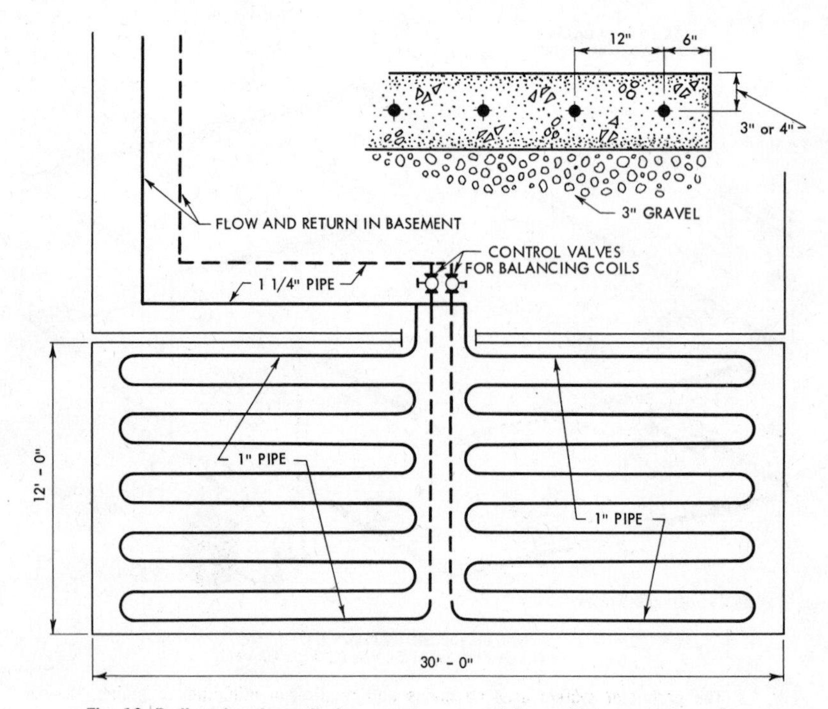

Fig. 12. Radiant heating coils in concrete are used for melting snow. Taco, Inc.

heated in any of several ways, but the most common and convenient method is by means of a heat exchanger or heater which is attached to the boiler.

A typical snow removal installation is shown in Fig. 12. The similarity between a snow removal system and a radiant panel is such that a more detailed description is unnecessary.

Summary. Other heating systems are so similar to those explained herein that special explanations are unnecessary, because the estimator is interested only in the amounts of materials and labor involved.

Unless detailed layouts are a part of the regular working drawings, the inexperienced estimator should either have a complete knowledge of all heating principles or seek aid from a more experienced person in estimating materials and labor.

This chapter does not include every material or labor item encountered in heating installation. Each job presents its own special features which the experienced estimator knows how to anticipate and estimate. Some estimators specialize in heating and air-conditioning estimating just as others specialize in electrical work.

In heating work, as in any other trade, a checklist is of great value if fully detailed. The estimator can make such a list easily, and he may add to it as much detail as he deems necessary or helpful.

Estimating materials

The new types of hot-air furnaces being installed are mostly of the air-conditioning type although older style gravity furnaces will undoubtedly continue in use for some time.* This section discusses the air-conditioning types of furnaces, but the installation methods and costs involved could be figured for gravity furnaces in almost the same manner.

The previous section explained common types of air-conditioning/heating systems in addition to ducts and fittings. Therefore no further explanation is necessary other than to explain that duct work is being installed in an increasing number of cases, using prefabricated ducts and fittings. This shortens the installation time and naturally lessens the cost by a considerable amount. With prefabricated ducts and fittings there is little cutting and fitting of sheet metal other than around joists when the joists are used as ducts for cold-air returns. See Fig. 13.

The use of prefabricated ducts makes estimating much more simpli-

*With the cooperation of Henry Furnace Company.

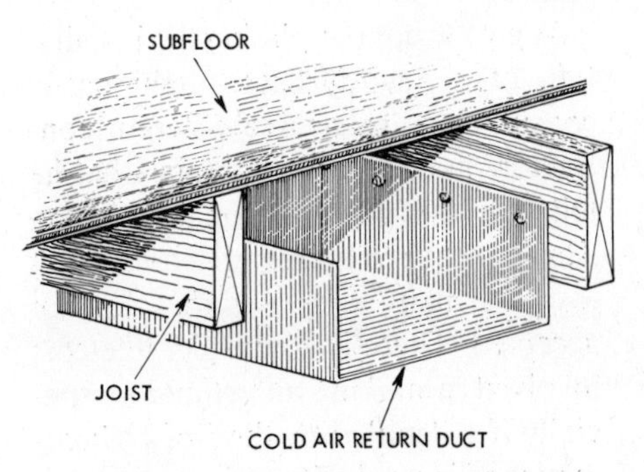

Fig. 13. Joists can be used as part of the duct for cold-air returns.

fied and the keeping of cost records easier. If a contractor has kept a good time record for installing ducts on previous jobs, he can easily determine the cost of installing ducts per foot; he can make up an average cost for ducts based on the number of hot-air registers, or make up an average cost of installing complete duct systems, depending on the number of cold and hot-air ducts or registers.

Some sheet metal and furnace contractors and estimators break up their estimates into small sections and besides considering the material cost in a detailed manner go even further and base their labor cost on a per foot of duct, per special fitting, per control instrument, etc. Other contractors and estimators think it is just as accurate to base their estimates, so far as labor is concerned, on the number of runs plus certain other charges which will be explained later. If a labor estimating method could be worked out whereby the labor could be accurately determined without the necessity of scaling all runs, etc., not only could much time be saved, but the resulting estimates would be lower.

A person contemplating the installation of a new furnace might easily consult as many as five different contractors in an effort to obtain what he felt was the most reasonable figure. This practice requires all contractors to make many estimates that are not accepted. There is a considerable cost involved in making an estimate, especially if a detailed method of estimating labor is used. Thus if a contractor makes five or six estimates in one week and is fortunate enough to get one job out of the five or six estimates, the cost for making the other estimates is lost. This has caused many contractors to do less actual work on estimates by making the figure high enough to insure its being more than sufficient to cover the cost of the estimate. This is poor practice, however, and it penalizes the home owner.

The specifications made by the architect must be read and understood so that every direction in them will be followed exactly. In the specifications there often is a section directing that the heating contractor is to make a full analysis of heat losses and prepare a complete layout. Sometimes the contractor can obtain this service from his supplier or manufacturer. The manufacturers usually maintain an engineering department which is at the service of contractors. The engineering department will use the drawings and specifications of the job to provide the contractor with a complete design of the job including layouts, heat loss calculations, size of furnace, and in some cases even a complete list of materials. Then the contractor has only to estimate his labor and add this to the quantity take-off.

In other cases where the contractor has no opportunity of obtaining such information from a manufacturer, he must do the work himself or have it done by a heating engineer. This will increase the total costs for the job and must be considered in the estimate.

Whether the work is done by a manufacturer or a contractor the process is the same and the following description, schedule, and layouts are applicable. Not many contractors have the facilities for doing all the engineering work themselves. They can figure a great deal of the work easily enough but the design of air-conditioning systems requires complete engineering preparation, which some of the smaller contractors are not in a position to have.

Sometimes the architects maintain a mechanical engineering department as part of their service. In such a case they can design the complete system and call for bids from contractors to supply material and labor. For our purpose, it is assumed that the manufacturer makes the layout.

Duct Work. Ducts can be either fabricated as required on the job, using the gages designated in the manufacturer's specifications, or prefabricated using standard parts and fittings. The use of prefabricated ducts and fittings is naturally considerably cheaper because they are made on a production basis, whereas job-fabricated ducts and fittings require considerably more labor.

Job-Fabricated Ducts. If the duct work is to be made on the job, the total weight of sheet metal required is figured as explained in Chapter 17, "Sheet Metal." The total weight is usually multiplied by a figure which includes both material and labor costs. This figure varies in different regions. The estimator's cost analysis records, if they have been carefully compiled and kept up to date, can be readily used to create such a figure.

Prefabricated Ducts. The manufacturers of prefabricated ducts have catalogues in which all parts and fittings are shown together with figure or part numbers. As each new size or part is found on the layout it is written down in the quantity take-off. Each time this part is repeated on the layout it is noted on the take-off (usually by a check mark after the first entry).

Insulation. Sometimes one or more ducts run through outside walls or other places of excessive heat or cold. For such ducts some form of insulation is necessary as illustrated in Fig. 14. Prefabricated ducts can be purchased already insulated. If the contractor is to do the insulating, he must calculate the amount of insulation required and add this to the material take-off.

Additional Parts. These include such items as controls, electrical work, and plumbing equipment.

Controls. These often come as part of the regular equipment. Where a more elaborate control system is required, all additional instruments must be calculated as extras and their costs added to the estimate.

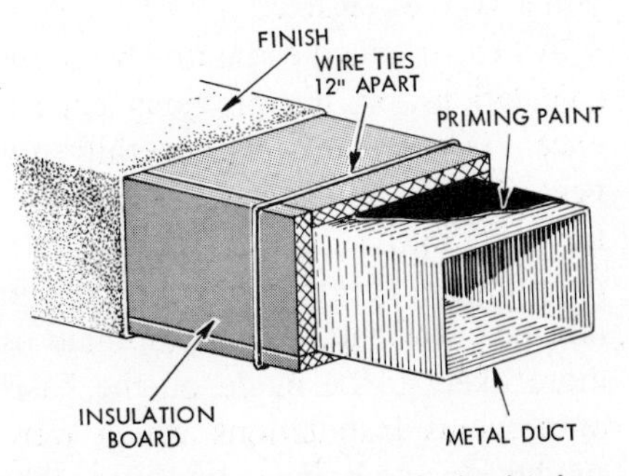

Fig. 14. Sometimes ducts must be insulated, especially when they are in outside walls.

Electrical Work. The controls used on furnaces and especially air-conditioning equipment are electrically operated. In most instances an electrical contractor is asked to bid on this part of the work (installing all electrical equipment) and his bid should be added to the cost of the job.

Plumbing Work. Where furnaces have automatic humidifiers or air washers as part of the general equipment, a plumbing contractor may be asked to bid on that part of the work. His bid is added to the cost of the job.

Permits. In most cities it is necessary to purchase a permit before installing a furnace. Permit costs vary among cities so each estimator must determine the costs in his locality.

Radiant Heating. A typical radiant heating job for an average residence is either a floor panel or a ceiling panel installation. While the costs of the completed systems is approximately the same regardless of panel location, some difference exists because the nature of the installations is different. For this reason, the estimator must be familiar with every step involved in either type of system.

An experienced estimator bases his calculations on his previous experience. As a result, it is generally unnecessary for him to itemize each factor going into a job as a means of attempting to approximate the cost in advance. Instead, a job estimate is more likely to be made on the basis of previous installations which were similar. It is obvious, however, that there must be a starting point somewhere for the individual who has no experience in estimating. The following material, therefore, may differ somewhat from the methods used by experienced estimators, the main difference being the detail into which the estimating is carried.

Gravel Fill. In order to achieve the maximum efficiency from a radiant floor panel installation, it is necessary to combine several building materials and construction techniques which are peculiar to radiant heating alone. While a gravel fill, moisture barrier, asphalt topping, and insulation do not constitute a part of the radiant heating system and, therefore, should not be figured in an estimate for such a heating system, nevertheless, the work and materials are necessary and must be paid for.

Good installation practice requires at least 6″ of gravel be placed over undisturbed earth. If the earth has been disturbed or if any fill has been made, the soil should be rolled so as to prevent any settlement after the installation is complete.

The amount of gravel necessary for the fill is found by multiplying the length of the floor panel area by the width, then multiplying by 6″ or 0.5′. Since gravel is usually estimated in cubic yards, it is necessary to divide the number of cubic feet by 27. The cost of gravel is found by multiplying the number of cubic yards by the prevailing cost per cubic yard.

Cinders are not used as fill material. Sulphur-bearing materials such as cinders will react with moisture to form sulphuric or sulphurous acids. Such acids, of course, are highly corrosive and would cause deterioration of the coil piping.

Moisture Barrier. A moisture barrier is placed over the gravel fill, particularly if there is any reason to believe that there will be inadequate drainage for the building site. The moisture barrier should be lapped at the edges up to 12″ and should be carried around the edge of the slab so that there is a barrier between the slab edge and the foundation, Fig. 15.

The amount of material required for the moisture barrier is found by determining the number of square feet to be covered. To this is added an additional 36½% for the material which will be "lost" due to the 12″ overlap which is recommended for each joint, and for spoilage or waste. Once the total amount of moisture barrier is known, the number of square feet is multiplied by the current price per square foot.

In order that each overlapped joint will be as resistant to water as the moisture barrier itself, each overlapped section must be covered with asphalt topping. In addition, the surfaces of the foundation over which the moisture barrier is laid (see Fig. 15) also must be covered with asphalt. The area of the overlapping sections is the same as that "lost" due to overlapping. Therefore, 36½% of the total area covered by the moisture barrier must be asphalted. Added to this is the area of the vapor barrier where it overlaps on the foundation. Assuming the foundation in Fig. 15 to be 8″ thick, the area covered by the moisture barrier is 10/12 or 5/6 feet multiplied by the length of the foundation perimeter. For ease in calculation, convert the 5/6 to a decimal (0.83′).

When the total area to be asphalted is determined, allow 2 gallons for every 100 square feet.

Insulation. In the past, it was con-

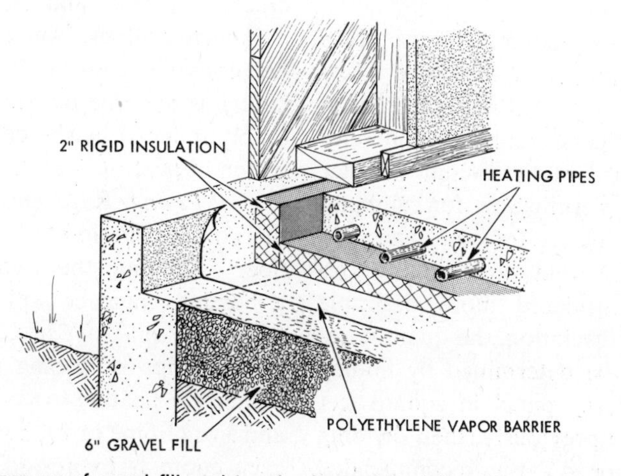

Fig. 15. The proper use of gravel fill, moisture barrier, and insulation for a concrete floor slab used as a radiant panel will minimize heat losses.

sidered necessary to insulate the slab perimeter only. Tests have indicated, however, that the flow of heat into the ground directly under the slab is great enough to merit the cost of the insulating material necessary to prevent this flow. The types of insulation which seem best adapted to this purpose include cellular glass, fiberglas, and insulating board with supplementary integral preservative treatment for protection against dry rot and decay. Whichever type of insulation is selected, it should be thoroughly coated on all sides and edges with asphalt or pitch before being applied.

The area of the floor panel has been previously determined in connection with the gravel fill and the moisture barrier. Therefore, the cost of the insulation is determined simply by multiplying the area in square feet by the cost of the insulation per square foot. The necessary asphalt is determined on the coverage basis previously established of 2 gallons per 100 square feet.

Insulating Concrete. The use of an insulating concrete in place of 2″ of insulation has proved advantageous in many instances. Insulating concretes are composed of lightweight cellular aggregates (Vermiculite, for example) instead of the customary sand and gravel. In the event the insulating concrete is considered more desirable than the 2″ insulation, the quantity required can be determined by finding the area of the panel in square feet as was done previously, then dividing by 3 (⅓′ or 4″, the recommended thickness of the insulating concrete). This figure is then divided by 27 to convert it into cubic yards.

The installation of a radiant heating system in the ceiling requires no elaborate preparation as was necessary for the floor panel. Once the ceiling joists are in place, the coils for the radiant panel can be formed and fixed in position. While some installations have been made in which an expanded metal lath was installed first, the ceiling coils then being placed in position, the better method is to place the metal lath *below* the coils. When the coils are placed below the metal lath, the plaster must cover the tubing with sufficient thickness to prevent cracking. Even with the tubing of small diameter, the plaster would have to be exceptionally thick. When the coils are placed above the expanded metal lath, the plaster can be carried over the whole of the ceiling area at an even thickness. This means that there is less weight for the expanded metal lath to support, less risk of sagging and cracked plaster, and less chance of darkened ceilings. Another factor to be considered is that the tensile strengths of some plasters are considerably reduced by the constant application of heat.

Coil Piping. Regardless of whether the coil installation is to go into the floor or ceiling, the architect and/or a heating engineer will have determined the quantity, size, and material for the pipe or tubing necessary to handle the heating load. The quantity and size of pipe or tubing required for each room is expressed in terms of inside diameter and linear feet. In addition, a detailed plan will have been prepared indicating the exact placement of the coils, the proper spacing, etc.

The total amount of pipe or tubing needed for both ceiling and floor installations is roughly estimated from the findings of the architect or engineer when the system was designed. (Each room will have been allowed a certain number of feet of piping, based on the known heat loss of the room.) A more accurate method would be to scale the drawing. In this way, an exact figure is possible.

When the total amount of piping is known (including piping around the boiler or heating unit and the supply and return mains, all of which will be indicated on the drawings), the number of linear feet is multiplied by the prevailing cost per foot and the material cost thus determined.

There is one other item of material cost which must be considered in connection with the coil piping for floor and/or ceiling installations. This is the quantity of balancing valves, shut-off valves, tees, elbows, and air vents which must be incorporated in the system at the time the coils are fabricated and installed. These fittings are taken from the drawings. The number of each type of fitting is multiplied by the unit cost per fitting in order to determine their cost.

To some extent, the location of the coil is a determining factor in the method by which the coil is formed. It has been found economical, for example, to fabricate coils of copper tubing, intended for ceiling installation, on forms built on top of benches. The entire completed coil is then raised into position at one time. Bench-formed coils are made either at the job or in the shop, whichever is more economical and practical. It is also possible to form ceiling coils of soft copper tubing in place on the ceiling joists without the use of forms or a bender. If steel or wrought-iron tubing is specified for a ceiling panel, the coil is formed on the floor and raised into position, or the coil could be prefabricated at the shop. Coils of either copper or steel or wrought iron, intended for use in a floor installation, will usually be formed in place, although some contractors find it preferable to prefabricate the coil in the shop.

If copper tubing is used instead of wrought iron or steel, the time which must be allowed for bending will be approximately the same since a bender is necessary for bending hard copper tubing, and desirable for soft copper tubing. While soft copper tubing can be bent by hand, a mechanical bender will produce even, uniform bends.

If steel or wrought iron is to be used, the approximate number of welds which must be made can be determined rapidly and easily by dividing the total number of linear feet of piping by 21. Steel and wrought-iron pipe come in uniform lengths of 21′-0″, Therefore, every 21 linear feet of coil piping will require a weld.

In addition to these welds, there are those which must be made for the installation of shut-off valves, air vents, balancing valves, etc. These items must be counted, using the plan prepared by the heating engineer and/or architect which gives the location and quantity of every valve and fitting to be used in the system. For each such fitting, two welds will be required.

It is possible to obtain copper tubing in lengths of 40, 60, and 100 feet. Therefore, if copper tubing is to be used, fewer joints are necessary because of the greater length in which the tubing is supplied. However, the same number of soldered joints must be made at balancing valves, air vents, shut-off valves, etc., as welds were made for the installation using steel or wrought-iron pipe.

Welding. Another item of material cost which must be considered is the amount of welding rod, oxygen, and acetylene gas necessary to make the welds that will be required. If copper tubing is to be used, the cost of the solder, flux, and gas used for making the soldered joints must be calculated.

Table II gives the materials *and* time required for making pipe welds. Time and materials for welding can be presented in tabular form (likewise for cutting) because their nature permits them to be reduced to an approximate average. A fairly definite quantity of gases is consumed by a blowpipe in operation with respect to the thickness and length of the steel welded or cut. Likewise, the amount of welding rod depends on the length and thickness of the joint; and actual time for welding and cutting is also approximately proportional to the amount of work to be done.

Estimates for the remaining time-consuming operations—laying out and

TABLE II
WELDING RATES AND MATERIAL CONSUMPTION BASED ON PIPE WALL THICKNESS
SINGLE FLAME WELDING*

Pipe Wall Thickness in Inches	Backhand Technique 70 Degree Vee, Rotation Welding			Forehand Technique † 90 Degree Vee, Rotation Welding		
	Linear Rate of Welding In. per Minute	Rod per In. of Weld in Lbs.	Oxygen per In. of Weld in Cu. Feet	Linear Rate of Welding In. per Minute	Rod per In. of Weld in Lbs.	Oxygen per In. of Weld in Cu. Feet
3/16	3.0	0.020	0.27	1.3	0.025	0.33
7/32	2.9	0.020	0.30	1.3	0.027	0.37
1/4	2.8	0.022	0.35	1.2	0.029	0.44
9/32	2.6	0.028	0.44	1.1	0.039	0.58
5/16	2.4	0.030	0.48	1.0	0.041	0.75
3/8	1.8	0.041	0.71	0.85	0.059	1.05
7/16	1.2	0.051	0.87	0.65	0.078	1.40
1/2	0.95	0.070	1.25	0.53	0.100	1.85

* The welding rates and material consumption in this table are based on continuous welding on or near the top of the pipe; work being done by a properly qualified pipe-welding operator; and actual welding only without allowance for tack-welding, starting, or finishing. Allowance for starting and finishing is approximately one-half to one minute per weld for pipe up to 10" in diameter.

These figures may vary 10 to 15% either way, depending on the size of the vee, the flame, the rod, and the diameter of the pipe.

† These figures will also apply for forehand welding with the neutral flame providing welds to be made are of minimum size.

Courtesy Linde Air Products Company

marking of cuts, joint alignment, tack-welding, etc.—as well as the unproductive time of the welder and his assistants can be determined only by experience, since these vary so much from one job to another.

While Table II is used to determine precisely what each welded joint should cost in terms of time and material, it is seldom necessary to estimate costs so closely. Rather, an estimate is made on the basis of previous experience. Moreover, the difficulty in arriving at an accurate estimate, due to uncontrollable and unforseeable events, makes such close figuring largely wasted effort.

If it is necessary or desirable to estimate welding costs as closely as possible, the first step is to determine the total number of welds. The length of each weld (pipe circumference) also must be found. The total amount of welding necessary (in terms of linear inches) is then multiplied by the prevailing cost, thus arriving at the materials cost for the welding.

The amount of material — solder, flux, and gas—necessary for a radiant panel job in which copper tubing is used is figured in the same manner as when steel or wrought-iron pipe and welding are used.

Boiler. The estimate for the boiler must be made on the basis of the schedule of heat losses and the specifications. The schedule of heat losses will have been prepared by the architect, a heating engineer, or some other competent person who knows every detail of the construction and insulation of the building. An allowance also must be made for heating domestic hot water, piping tax, and heating-up or starting-up.

The specifications will call for either a gas- or an oil-fired boiler, coal seldom being used for radiant heating installations. If a gas boiler is specified, stainless-steel flue liners, tees, elbows, and condensation drains are sometimes considered necessary for the chimney. Vitreous sewer tile with bell-mouth joints also make an acceptable lining. See Fig. 16. These special flue linings and other fittings are necessary because of the corrosive nature of the condensate from the combustion gases.

If the stainless-steel liner is used, its cost (which will be greater than any other type of flue lining) must be determined from manufacturers' or suppliers' literature and added to the estimate.

Whatever the type of flue lining used, it is customary to include a stainless-steel tee where the smoke pipe or breeching enters the chimney, and a drain fitting of stainless steel to which is attached a piece of plastic tubing. See Fig. 16. The condensation from the chimney is led off through this fitting and tubing to the nearest floor drain. These special fittings and the boiler are catalogued and their costs listed in manufacturers' literature.

A certain amount of piping is necessary in bringing the gas to the boiler in installations where gas is specified. The amount of piping is determined by the location of the meter and the boiler. This distance is scaled from the drawings. The various elbows,

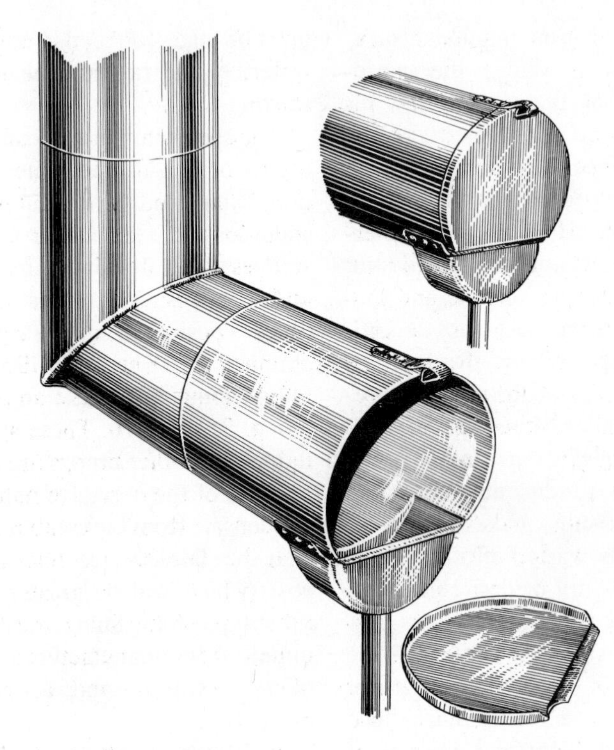

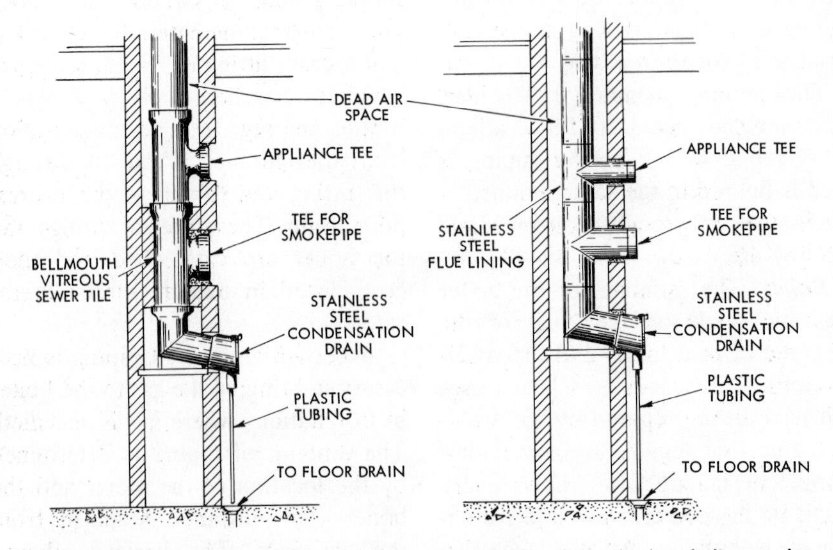

Fig. 16. Gas-fired boilers require stainless-steel condensation drains (top) and vitreous clay tile or stainless-steel flue liners (bottom).

valves, and other fittings also can be taken from the drawings, although the quantity needed, in terms of cost, can be approximated by allowing 100% of the cost of the necessary pipe.

If an oil-fired boiler is specified, the estimate would include a tank or tanks for the storage of oil, sized and placed according to the architect's specifications. The estimate is based on costs supplied by manufacturers or suppliers. The labor, as well as the material cost, is slightly higher for an oil-fired job than for a gas-fired installation since the storage tanks must be placed in addition to running the lines for the oil supply, setting and connecting the boiler, etc.

If the boiler is purchased as a packaged unit, all boiler controls are included in the price and will be an integral part of the unit. If the boiler is purchased in any other form, an electrical contractor, in most cases, is asked to bid on the job of installing all electrical materials. This bid should be added to the cost of the job and will include hooking up all the controls for the boiler, the controls for the panels, and the circulator.

The efficiency and the effectiveness of the radiant heating system is dependent to a large extent on automatic controls. There are a number of manufacturers producing automatic controls for radiant panel systems which are entirely satisfactory. The estimate of the cost of any particular control system is based on the cost for the material and the labor required to make the installation. The manufacturer of the equipment or one of his local distributors should quote prices on request.

When a complete cost list of the materials and labor needed for a job has been compiled, most contractors add 15% to the total cost estimate for overhead or operating expenses. To this new total is added about 10% for profit. Other additions could be as follows: 7% for employee insurance, 3% for employee welfare, 6% for testing, 2½% for cleaning up the job, and 10% for the handling of materials. The final or *grand total* is presented in the estimate.

Estimating labor

Labor estimating is done in any one of a number of acceptable forms. However, in many large cities the sheet metal contractors have formed associations which carry out studies of labor costs. The purpose of such studies is to find easy but accurate methods of estimating labor.

The following labor times are all average figures and will vary greatly among different regions, differing climatic conditions, etc.

For *radiant heating,* one laborer can place approximately 100 square feet of *moisture barrier* an hour. The total number of square feet of barrier, divided by 100, will give the number of man-hours required for its placing.

The labor required for mopping asphalt is determined by dividing the total area to be mopped by the rate at which a laborer works—approximately 60 square feet per hour. The total number of hours required for placing and mopping the barrier is then multiplied by the prevailing wage scale.

The cost of labor for *welding* is found by multiplying the length of each weld (pipe circumference) by the time necessary to complete the weld. To this must be added the time taken by the welder for starting and finishing. The total time required for making all welds is converted into hours, then multiplied by the prevailing hourly wage rate for welders.

The amount of labor necessary to make the connections in a *copper tubing radiant panel* installation is considerably less than for a wrought-iron or steel installation. An inexperienced journeyman and helper can jig, raise, and wire into place approximately 700 to 900 feet of copper tubing per 7 hour day. This includes all couplings bench soldered as part of the work.

For the floor panel installation, some time must be allowed for blocking up the coils to the level specified in the drawings. This operation will vary widely with each job, and experience alone will provide the best measure for estimating amount of time that will be necessary for this part of the job. However, approximately four hours' work by a journeyman fitter should be sufficient for the average six-room residence.

When the complete floor- or ceiling-panel system is in place and all soldered or welded connections have been made, a hydrostatic test must be made on the system. Where welding has been used, each joint should be rapped smartly with a hammer while the system is under pressure. A flow test also should be made in order to make certain there are no obstructions in any of the coils or other piping. Usually 6% of the total labor cost is figured for testing.

Supervisory time will be required during the pouring of the concrete for the floor slab. Since the heating contractor is responsible for the satisfactory operation of the completed system, he will want to make certain that the coils remain in place and that the piping is not dented or otherwise injured during the placement of the concrete. This cost is usually included in the 15% charged for overhead or operating expenses for the entire job.

Installation time for *boilers* varies greatly. Some sample times indicated that, on the average, two journeymen fitters will take one day (about 16 man-hours) to set and connect a boiler. Two journeymen fitters will take one-half day (about 8 man-hours) to install the boiler breeching. To run the gas piping and connect the burner will require two journeymen fitters about one day (16 man-hours).

For an *oil installation,* two journeymen fitters will carry in, set up, and pipe two 275 gallon storage tanks in 1½ days (about 24 man-hours).

The labor involved in the electrical work is usually included in the electrical contractor's bid.

Sample heating estimate

To demonstrate the itemizing methods

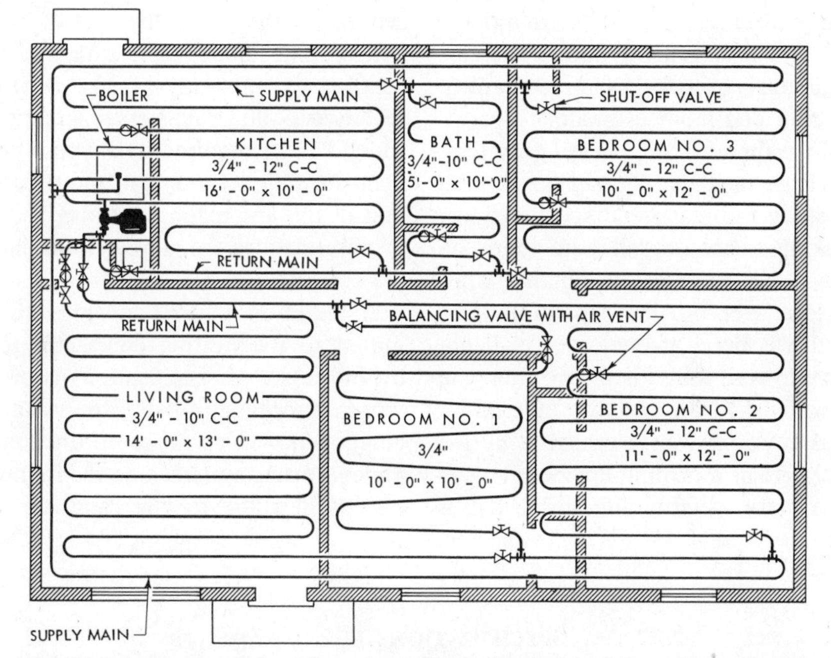

Fig. 17. A heating layout for a five-room residence. The layout of the radiant coils and fittings is accurate enough for estimating purposes.

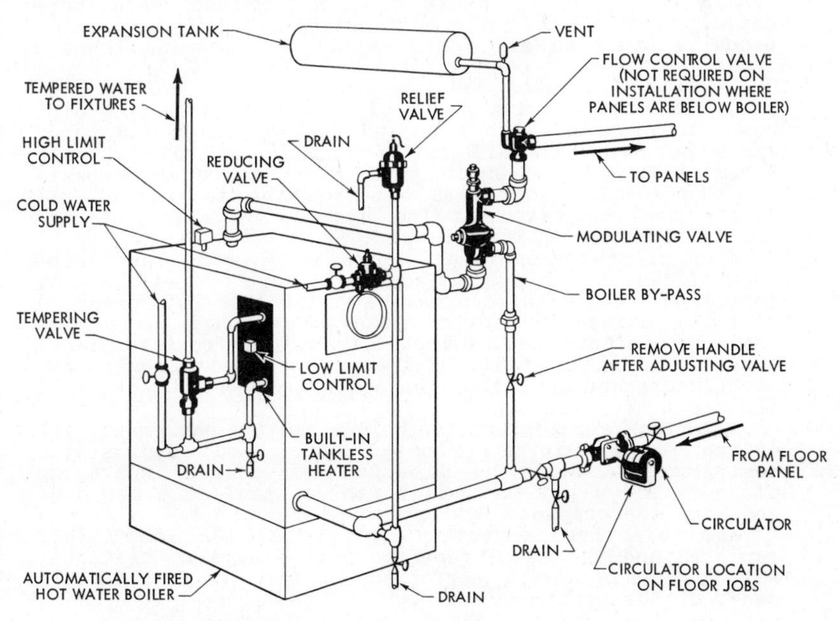

Fig. 18. Typical boiler details for the heating layout shown in Fig. 17.

and procedures described in this chapter, a typical radiant floor panel installation is used for a sample quantity take-off and labor estimate.

In estimating any heating system, the first thing that should be done is to study the plans and specifications, making a note of all equipment on which it is necessary to obtain manufacturer's or distributor's quotations. If this is done at once, you will not have to wait long for prices after you have finished the quantity take-off.

A set of drawings prepared by the architect or a heating engineer is necessary for determining just what is needed for the job in the way of materials. The floor plan and boiler detail shown in Figs. 17 and 18, respectively, illustrate every feature which will be required in a typical radiant heating system with the exception of the automatic controls. These two illustrations are more helpful than the conventional working drawings.

In addition to the drawings submitted to the heating contractor for his bid, a set of specifications is provided, describing in detail the various components which make up the completed heating system. Use the following specifications for this estimate.

TYPICAL SPECIFICATIONS FOR A RADIANT FLOOR PANEL HEATING SYSTEM

AIR VENTS: Automatic air vents shall be provided where indicated and shall be of the float-operated type. Small, manually operated air vents shall also be placed where indicated.

INSULATION: All exposed heating piping in the boiler room shall be covered with three-ply asbestos air cell pipe covering provided with a pasted canvas jacket and three bands per section. All fittings shall be covered with asbestos cement and enclosed in a pasted canvas jacket.

TESTS: The heating work shall include the testing of all piping and apparatus in the system for leaks, faulty joints, improper operation, etc.

When all heating piping is installed, with the exception of the boiler and its equipment, the system shall be tested to a hydrostatic pressure of 125 lbs. This pressure shall be maintained for at least four hours. This test shall be performed before the floors are poured into place.

Prior to installation of the boiler and its equipment, circulator, and controls, the contractor shall thoroughly flush out the system.

After installation of the boiler and its equipment, the entire system is to be filled with water, and a hydrostatic test pressure built up to equal the maximum operating pressure for which the boiler was designed as a means of checking the complete system for leaks.

Any defective joints or piping shall be immediately replaced and the tests repeated until found to be tight. All tests are to be performed in the presence of the owner or the architect.

Before final acceptance of the work, the contractor shall put the entire system into operation and demonstrate that all parts are in proper working order.

GUARANTEE: The contractor shall guarantee all material and workmanship for a period of one year.

MATERIAL

1 – Gas boiler complete . . . . $ ______
20' 0" – Stainless steel flue liner. . . . .
1 – Stainless steel drain fitting, tees, and plastic drip tubing . . . .
1 – Boiler breeching. . . . .
15' 0" – 1" gas piping with necessary fittings . .
1 – ¾" circulating pump . . . .
1 – ½" combination reducing and relief valve.
1 – ¾" modulating valve (including cost of automatic control system complete). . .
1 – Ten-gallon expansion tank . . . .
20 – ¾" gate valves. . . . .
6 – ¾" balancing valves . . . .
10 – ¾" tees . . . .
11 – Elbows. . . . .
900' 0" – ¾" black iron pipe. . . . . ______

Total $

LABOR

Set and connect boiler. . . . .$ ______
Boiler breeching. . . . .
Gas piping. . . . .
Circulator. . . . .
Supply and return piping. . . . .
Forming and installing coils. . . . . ______

Total $

RECAPITULATION

Material. . . . .$ ______
Labor . . . .

$ ______

Overhead (15% of total) . . . .

$ ______

Profit (10% of new total) . . . .

$ ______

Labor insurance (7%) and welfare (3%) (figured on labor cost) . . . .
Labor for testing (6% of total labor cost)
Labor for cleaning up (2½% of total labor cost) . . . .
Material Handling (10% of total material cost) . . . . ______

Grand total $

PLACEMENT OF COILS: A gravel fill of at least 6" thickness shall be placed over undisturbed earth or earth which has been rolled or tamped so as to preclude any possibility of settlement. In no instance shall cinders, slag, or other sulphur-bearing fill material be used.

Over the gravel fill shall be placed a moisture barrier of some suitable substance, lapped at the edges at least 6", and carried around to the edge of the slab. The overlapped sections and the surfaces of the foundation wall over which the moisture barrier is laid shall be mopped with pitch or asphaltum. A suitable type (Foamglas, Fiberglas, etc.) of insulation shall be placed over the moisture barrier and shall be thoroughly coated on all sides and edges with asphaltum or pitch before being applied.

The coils shall be blocked up in position approximately 1" above the surface of the insulation. The concrete for the floor shall then be poured directly over the heating pipes in accordance with the drawings. In no instance shall the concrete above the top of the piping be less than 3".

Fill and drainage shall be adequate to maintain piping free from surface drainage and subsurface water.

Coils shall be placed in accordance with the plans and shall be laid level throughout.

The contractor shall be responsible for maintaining all heating pipes in their proper positions during the pouring of the concrete floors. This shall include all necessary instruction, supervision, and inspection.

BOILER: The boiler shall be a gas- or oil-fired, cast-iron, hot water boiler capable of delivering 50,000 Btu to the system.

The boiler shall be equipped with an insulating jacket, combination pressure gauge and thermometer, flow and return tappings, etc., to make a complete unit.

CIRCULATING PUMP: The water circulating pump shall be capable of pumping five gallons per minute against a 2.5' head. The pump is to be equipped with flanged connections, vibration dampers, and motor as supplied by the manufacturer.

EXPANSION TANK: The expansion tank shall be an all-steel, airtight tank of ten-gallon capacity. The tank shall have all the necessary tappings as indicated on the drawings.

BOILER ACCESSORIES: The boiler accessories shall include the following which shall be properly connected to the boiler and piping system:

The water filling line shall be provided with a pressure reducing valve.

The pressure relief valve shall be of the diaphragm type.

The boiler shall be provided with a drain valve in the conventional manner.

CONTROLS: The control system shall be the outdoor-indoor type with continuous circulation. (If one particular type is favored, the manufacturer of the control system should be consulted for specification, installation, and operation details.)

VALVES: All gate, globe, and square-head valves should be designed for 125 lb. working pressure.

All balancing valves shall be ______ Company #324 iron body brass mounted cock for 1¼" size and ______ V-53 for 1" size, or approved equals, and one shall be installed with each coil as shown on the drawings.

SCOPE OF THE WORK: The work to be done by the contractor consists of the complete installation of a radiant heating system within the building, with the furnishing of all labor and material to provide a complete and satisfactory system. The radiant heating system will use a floor panel with forced hot water as the heating medium. The system shall be installed in accordance with the plans and details shown on sheet ______ ________________ of the drawings, and all work shall be done in a neat, thorough, and workmanlike manner.

CODES: All work in connection with the installation of this heating system must be performed in accordance with the applicable requirements of all pertinent building codes and shall meet the requirements of the fire underwriters.

CUTTING, PATCHING, AND EXCAVATING: All cutting and patching necessary for the installation of the heating system shall be done by this contractor. No cutting of the work of other contractors shall be done without the consent of the architect, and no structural member shall be cut or weakened in any way unless proper arrangement is made to reinforce the portion so weakened. In such matters it will be necessary to receive the permission of the architect before proceeding.

PIPE AND FITTINGS: All pipe for the heating system mains, coils, risers, etc., shall be standard weight, black, wrought-iron pipe in accordance with ASTM Designation A72.

All screwed joint fittings shall be black, cast, or malleable iron, beaded or banded fittings of 125 lbs. rating.

HANGERS AND SUPPORTS: All pipe hangers for the heating system shall be of the split-ring pattern with hanger rods. Hangers and rods shall be of a size and weight to safely support the load.

JOINTS: All joints in the heating system piping shall be welded except those which are immediately accessible in the vicinity of the boiler. The welding shall be done in accordance with the standards of the American Society for Pipe Welds. All ends of pipe shall be properly reamed before welding.

All boiler connections shall be made with screwed joints only.

Joints on all valves and cocks shall be either flanged or screwed.

COIL AND GRID FABRICATION: All coils shall be fabricated in accordance with the shape shown on the plans and shall contain not less than the pipe quantity and pipe sizes shown.

Elbows and return bends may be either shop fabricated or bent on the job to the radius shown on the drawings, but shall be free of crushing or deformation.

All joints and connections in the coils are to be welded, and the welding shall be smooth and sound with particular attention given to maintaining a flush, even interior surface.

Any prefabricated coils shall be adequately braced (temporarily) during transportation to the job to prevent springing, misalignment, etc.

All field welds shall be done in a manner to protect other workmen on the job from glare and injury, and every precaution shall be taken to avoid the possibility of fire.

Finally, there is a schedule of heat losses for each room of the house. See Table III. This table is prepared by the architect or some other qualified, competent designer.

Materials Estimate. Since the boiler is the largest single expense in this estimate, it will be the first item estimated.

Boiler. The specifications and the schedule of heat losses are the basis for the estimate of the boiler.

The total heat loss is found to be 32,400 *Btuh* (Btu's per hour) by add-

TABLE III
SCHEDULE OF HEAT LOSS FOR A RESIDENCE

Room	Room Size in Feet	Room Area in Square Feet	Heat Loss in Btu's
Living Room	14x13	182	9,600
Bedroom No. 1	10x10	100	3,500
Bedroom No. 2	11x12	132	5,800
Bedroom No. 3	10x12	120	5,700
Bath	5x10	50	2,100
Kitchen and Utility	10x16	160	5,700

ing the individual heat losses in Table III. To this is added 10% to compensate for heat loss to the ground under the slab, making a new total of 35,640 Btuh.

In average house heating systems, it is common practice to consider the *piping tax* (the estimated heat emission in Btuh of the piping connecting the radiators and other equipment to the boiler) as equal to approximately 25% of the net load *(radiation load).* However, since the supply and return mains in the example are, in a sense, a part of the panel system, a 10% figure is more reasonable. The piping tax, therefore, is approximately 3,564 Btuh.

The last item involved in selecting the boiler on the basis of its capacity is the allowance made for heating-up or picking-up. The usual allowance is about 20% of the *design load* (radiation load and the piping tax). This amounts to 7,841 Btuh.

The total capacity of the boiler, therefore, must be at least 47,045 Btuh in order to handle the heating load imposed upon it.

The specifications call for a gas- or oil-fired boiler. Since gas is mentioned first, the gas-fired boiler will be used for this estimate.

Flue Lining. The stack temperatures of a gas-fired boiler are lower than those where other fuels are used. Consequently, there is a greater amount of condensation present. The gases given off as a product of combustion combine with water vapor to form corrosive acids. It is therefore necessary to install a flue lining which will resist these acids. While a stainless-steel flue lining is usually used in new construction, vitreous, bellmouth sewer tile also gives a satisfactory installation. On old jobs, however, there is no substitute for the stainless-steel liner.

Assume that stainless steel is specified for the flue liner, and that the chimney height is 20′-0″.

Drain Fitting and Tee. A means must be provided for the disposal of the condensation which will accumulate at the bottom of the chimney. A stainless-steel elbow is obtainable which is a combination cleanout and drain. This elbow is attached at the bottom of the stainless-steel tee. The smoke pipe or breeching enters the tee from the side. To the top of the tee is attached the flue lining. The condensation is led off from the stainless-steel drain fitting through a piece of plastic tubing to the nearest drain. A second tee is necessary for the purpose of connecting the domestic hot water heater to the flue.

Boiler Breeching. Assume the boiler to be conveniently located with respect to the chimney so that no problem is encountered in installing the smoke pipe or breeching.

Gas Piping. It is usually customary for the local gas company to bring the gas piping from the street into the meter free of charge. While the meter location in Fig. 17 is not shown, you can assume that not more than 15′-0″ of 1″ pipe is needed. The cost of the fittings is determined usually by picking them off the drawings, but in this case by adding 100% to the total pipe cost.

Circulator. There are a variety of circulating pumps available, all of which will handle the needs of this heating installation.

Relief and Reducing Valves. Quite a few manufacturers produce relief valves and reducing valves. In some cases, these valves are combined as one unit.

Modulating Valve. The modulating valve is a part of the entire automatic control system. As such, it is sold as a complete unit and includes the outdoor anticipating bulb and the room thermostat.

Expansion Tank. The size of the expansion tank can be determined by the total Btu capacity of the installation. The heat loss schedule indicates that approximately 50,000 Btu's are required. If an expansion tank has a capacity of one gallon for every 5,000 Btu's, a tank having a ten gallon capacity is needed for this job.

Valves. A radiant panel heating system requires a greater number of valves than do other types of heating systems. Figs. 17 and 18 show that there are 20 gate valves of ¾″ size, 6 balancing valves (¾″), 10 tees (¾″), and 11 elbows required for the entire system.

A simpler and quicker, though not as accurate, method of determining the cost of such fittings is to assume they will amount to 100% of the cost of piping.

Coil Piping. The number of linear feet of piping required for each room and the pattern or way in which it should be placed will have been determined by the architect or some other qualified designer. You can determine pipe requirements from these figures. However, if the quantity of coil piping has not been indicated, it is easily determined from the schedule of heat losses, Table III, and from Table IV. For example, the living room has a heat loss of 9,600 Btuh. From Table IV you can see that it will take 200 linear feet of piping to overcome the loss. This table is based on a 120° F average water temperature and a 50 Btuh emission rate per square foot.

The necessary piping is determined for the rest of the rooms in the residence in the same manner. The lengths of the supply and return mains are scaled off the drawings. The total piping requirements are found to total ap-

TABLE IV
BTU REQUIREMENTS AND EQUIVALENT COIL SIZES

Btu* of Coil	Linear Feet of ¾" Pipe or Tubing Required	Btu* of Coil	Linear Feet of ¾" Pipe or Tubing Required
500	10′	5,500	110′
1,000	20′	6,000	120′
1,500	30′	6,500	130′
2,000	40′	7,000	140′
2,500	50′	7,500	150′
3,000	60′	8,000	160′
3,500	70′	8,500	170′
4,000	80′	9,000	180′
4,500	90′	9,500	190′
5,000	100′	*10,000	200′

* Maximum recommended output per coil is 10,000 Btu because of friction or resistance limitation. If more than 10,000 Btu is required, use two (2) or more coils. For example, if you have a room with a heat loss of 18,000 Btu, use two (2) 9,000 Btu coils, 180′ in each.

proximately 900 linear feet — 680 linear feet for coil piping, 200 linear feet for supply and return mains.

Material for the gravel fill, moisture barrier, asphalt topping, and slab insulation are not figured as a cost in this particular estimate since they are not actually a part of the heating plant. The amount of acetylene, oxygen, and welding rod used is not sufficient to merit an accurate estimate.

Labor Estimate. It was pointed out that the experienced estimator rarely itemizes each factor of a job as a means of determining its cost in advance. The estimate will more likely be made on the basis of previous installations which were similar. One of the main reasons for this condition is the difficulty in making accurate estimates of the amount of labor required for a particular job. While some of the following material on determining labor costs may seem to the experienced estimator as needlessly detailed, the information is intended only as a temporary substitute for experience.

Boiler. Experienced estimators have found that to set and connect a boiler for a typical, medium-sized residence such as this will take two journeymen fitters one day—about 16 man-hours.

Flue Lining. No additional labor is necessary as a result of using stainless-steel flue lining. A flue lining of some sort should be installed in every chimney. All that is needed, therefore, is a substitution of materials.

Drain Fitting and Tee. No additional labor is figured here. The stainless-steel parts are merely furnished to the mason who proceeds to lay up the chimney around them.

Boiler Breeching. Depending on the location and position of the chimney and boiler, the boiler breeching (smokepipe) may be simple or complicated. An average time for this installation is one-half day for two journeymen fitters. This is about 8 man-hours.

Gas Piping. The labor involved in

running the gas piping is determined to some extent by the distance from the meter to the boiler. As a general rule, it is safe to assume that the gas piping can be run and the boiler connected by two journeymen fitters working one day—16 man-hours.

Circulator. To connect the circulator requires the work of two fitters working one-half day—8 man-hours.

Relief and Reducing Valves. The installation of these valves is considered as a portion of the work done when connecting the circulator and setting the boiler. Likewise, the tempering valve, modulating valve, and expansion tank are installed as a part of the boiler labor and circulator labor.

Supply and Return Piping. It is customary to figure that two journeymen fitters require one day to run the supply and return piping—16 man-hours.

Coil Piping. Estimators allow two journeymen fitters four days in which to form, place, and weld the piping for the floor coils in a typical, average-sized residence such as this one. This is about 64 man-hours.

Finishing. Once the material and labor costs have been determined for a job, it is possible to complete the estimate by adding the customary percentages for overhead, profit, labor insurance and welfare, etc. These items are listed in their proper place on the estimating sheets.

A detailed estimate of even a single-unit residence may at first seem complicated. As you become more experienced you will be able to do estimates for apartment buildings and commercial buildings.

APPENDIX

CHECK LIST

The estimator must be systematic in his work. Check lists such as this one are especially valuable to the inexperienced estimator.

This list includes only those items which are discussed in the preceding chapters. You can use this list as the basis for your own check list. The best check list is often the specifications themselves.

Overhead Costs

Insurance
- Workmen's compensation
- Public liability
- Fire and extended coverage

Social security
Performance bond
Completion bond
Permits and inspection fees
Engineering services
Rubbish removal
Window cleaning
Winter protection
Watchmen
Inspection and testing
Field office
Project sign
Site fence
Temporary light and power
Temporary plumbing
Supervisory personnel
Timekeepers
Equipment

Surveying and Excavation

Site inspection
Surveying
Soil examination
Hauling distances
Piers
Ramps
Top soil
- Removal
- Storage
- Replacement

Sheet Piling
Underpinning
Cradling
Dewatering and well points
Slope of banks
Soil swell
Backfill

Compaction
Rough grading
General and special excavation
 Main excavation
 Footings
 Columns
 Areaways
 Sidewalks
 Driveways
 Utility trenches
 Boiler stack
 Incinerator foundation
Grubbing
Site layout
Batterboards
Equipment

Footings

Concrete
Other materials
Foundations
Partitions
Columns
Chimneys
Stairs
Porches

Foundations

Material
Thickness and depths
Anchor bolts

Concrete Work

Foundations—mix
Footings—mix
Partitions—mix
Areaway walls
Beams
Sidewalks
Driveways
Curbs
Floors
Chimneys
Stairways
Reinforcing steel
Retaining walls
Walls
Mortar
Waterproofing
Finishing

Formwork

Footings
Columns
Areaway walls
Floors
Sidewalks
Curbs
Driveways
Chimney footings
Outside basement stairs
Cisterns
Materials

Walls

Frame
Stucco
Brick
Brick veneer
Stone
Steel
Concrete
Tile and stucco

Framing and Millwork

Lumber
Studs
Wall sheathing

Plates
Siding and felt
Wall shingles
Sills
Braces
Interior walls
Wall insulation
Furring
Basement partitions
Soundproofing
Floor framing
  Joists
  Beams
  Bridging
  Headers
Rough floor
Finish floor
Creosoting
Doors and frames
Windows and frames
  Lintels
  Trim
Stairs
  Stringers
  Treads
  Basement
  Attic
Roofs
  Common rafters
  Hip rafters
  Valley rafters
  Jack rafters
  Purlins
  Collar beams
  Ridge pieces
  Board spacing
  Insulation
Dormers
Cornices and sills
  Fascia
  Plancier
  Frieze
Grounds
Chair rails
Base molding
Base board
Fireproofing
Closets
Cabinets
Shelves
Exterior trim
Interior trim
Plywood
Screens
Storm windows
Shutters

Masonry

Common brick
Modular brick
Firebrick
Face brick
SCR brick
Glass brick
Cinder block
Concrete block
Terra cotta
Hollow tile
Mortar
Anchors
Metal angles
Scaffolding
Tuckpointing
Fireplace
Firestops
Chimneys

Wall Coverings

Plaster
Stucco
Suction
Curing

Fireproofing
Lathing
  Metal
  Board
  Insulation
  Wood
Grounds
Corner beads
Metal trim
Metal arches
Furring
Scaffolding
Paneling
Nails
Taping
Spackling
Plasterboard
Insulation board

Electrical Wiring

Open conductors
Nonmetallic sheathed cable
Armored cable
Rigid conduit
Electrical metallic tubing
Service entrance wiring
  Switch box
  Fuses
  Circuit breakers
Ventilating fan
Entrance caps
Lock nuts
Bushings
Clamps
Receptacles
Outlet box
Fixtures

Special Interior Coverings

Plank floors
Block floors
Linoleum
Linoleum felt
Terrazzo
Asphalt tile
Vinyl, cork, and rubber tile
Ceramic tile
Mosaic tile
Wallpaper
Fabric wall covering
Presdwood tile
Small sheet tile
Wood paneling
Acoustical tile
Adhesive and paste
Caps
Bases
Special moldings
Furring
Troweling

Glass

Plate glass
Thermal glass
Special glass
Glazing
  metal sash
  putty
No. of lights
Hardware
Mailboxes
Nails
Hinges
Door locks
Push plates
Door knockers
Hooks
Towel racks
Knobs and fasteners
Shelf brackets
Window locks

Paint

Kind of paint
Putty
Cleaning
Work protection
Sanding
Scraping
Scaffolding
Calcimine
Glusize
Varnish
Stain
Shellac
Enamel
Metal paint

Window Coverings

Shades
Rollers
Venetian blinds
Curtain rods

Plumbing

Septic tanks
Sewers
Cisterns
Sump pumps
Fixtures
  Faucets
  Valves
Pipes
  Sewer
  Drain
  Catch basin
  Gas
Leaders and downspouts
Catch basins
Concrete
Permits
Excavation

Sheet Metal

Material
Barriers
Shields
Flashing
Drops and chutes
Stops
Ridging
Gutters
Leaders
Roofing
Dormer decks
Dormer pans
Drip caps
Accessories
  Brackets
  Hangers
  Clamps
  Elbows
  Felt
  Screws
  Nails
  Solder

Heating and Air Conditioning

Mechanical furnaces
Electrical heaters
Ducts
Fittings
Grilles
Dampers
Controls
Radiant heating
  Panels
  Coils
Snow melting
Insulation
Electrical work

Plumbing work
Welding
Boiler
Flue lining
Tees
Breeching
Gas piping
Circulator
Valves
Expansion tank

AVERAGE COSTS

The experienced estimator frequently estimates buildings by the square foot or cubic foot methods. These methods are not as precise as the detailed methods explained in the text, but they do save considerable time.

The basis for making such unit cost estimates is accurate records which are kept by the contractor and estimator. These are records of previous jobs involving the same kind of construction.

The following material is a compilation of average costs for constructing the average-size buildings and structures on an air base. The costs have been calculated on a square foot basis. A table is used to adjust the costs of proposed buildings which are larger or smaller than the typical size building.*

Office of Assistant Secretary of Defense for Installations and Land

Contract cost: complete facility to line 5 ft beyond foundation walls; includes all equipment permanently built in, attached or with fixed utility connections except technical, scientific, production and processing equipment.

Total cost: contract cost plus 8% for contingencies, 7% for engineering, supervision, inspection, administration, 6% for design.

Type of Structure	Size Gross sf	Contract Cost	Total Cost Less Dsgn	Total Cost	Type of Structure	Size Gross sf	Contract Cost	Total Cost Less Dsgn	Total Cost
Academic.............	25,000	$16.45	$19.00	$20.10	Messhall.............	15,000	$24.50	$28.30	$30.00
Administration.........	15,000	17.30	20.00	21.20	Operations, aircraft, with tower..........	13,000	28.00	32.35	34.15
Barracks, dormitories. .a	30,000	12.33	14.25	15.10					

*By permission of *Engineering News-Record,* copyrighted 1962 by McGraw-Hill Publishing Company, Inc.

Bachelor officers quarters	limit: $7,000 per officer			
Bowling alleys, with automatic pin spotters	8 alleys	29.00	33.50	35.35
Clubs, enlisted men	16,000	18.85	22.00	23.30
non-com officers	12,000	20.55	24.00	25.45
officers	10,000	23.55	27.50	29.15
Commissary	15,000	15.65	18.10	19.10
Communications	6,000	23.95	27.70	29.35
Exchange store	6,000	16.00	18.50	19.50
Fire & crash station	8,000	17.75	20.50	21.75
Guard house	4,000	23.20	26.80	28.40
Gymnasium	20,000	16.05	18.55	19.65
Hangar, gen or mainten.	50,000	15.80	18.25	19.35
Hospitals	500 beds	26.05	29.00	30.10
	200–300	26.95	30.00	31.10
	75–100	28.75	32.00	33.20
	50	26.65	33.00	34.25
Dental clinics	24 dors	27.40	30.50	31.65
	18–22	28.30	31.50	32.70
	8–16	28.75	32.00	33.20
	6	29.20	32.50	33.70
Dispensary	8,000	26.95	30.00	31.10
without tower	10,000	23.00	26.55	28.05
Readiness crew bldg.	23,000	20.35	23.55	24.95
Chapel	8,000	24.70	28.55	30.25
Religious education	4,000	19.40	22.40	23.75
Reserve training	20,000	16.30	18.85	20.00
Shops: auto mainten.	5,000	17.30	20.00	21.20
arm & elec.	12,000	16.30	18.85	20.00
instlns mainten.	15,000	14.80	17.10	18.10
para & dinghy	8,000	21.40	24.75	26.20
rocket assem & checkout	5,000	22.45	25.95	27.50
Storage: auto w/ht.	20,000	11.75	13.55	14.35
auto w/o ht.	20,000	10.95	12.65	13.40
cold . . . b	6,000	27.70	32.00	33.90
gen purpose . . . c	50,000	6.90	8.00	8.50
Swimming pools, outdoor:				
82′ 2″ x 63′ 0″	5,176	16.75	19.35	20.45
164′ 4″ x 54′ 0″	8,874	15.35	17.75	18.75
164′ 4″ x 75′ 0″	12,325	13.85	16.00	16.90
Bathhouse	3,000	22.00	25.40	27.85
Theater, stage & seats incl.	10,000	23.00	26.55	28.05

•a Limit: $850 per occupant •b Limit: $32.00 per sf •c Limits: $8.00 per sf

How to Adjust Unit Cost for Size of Building

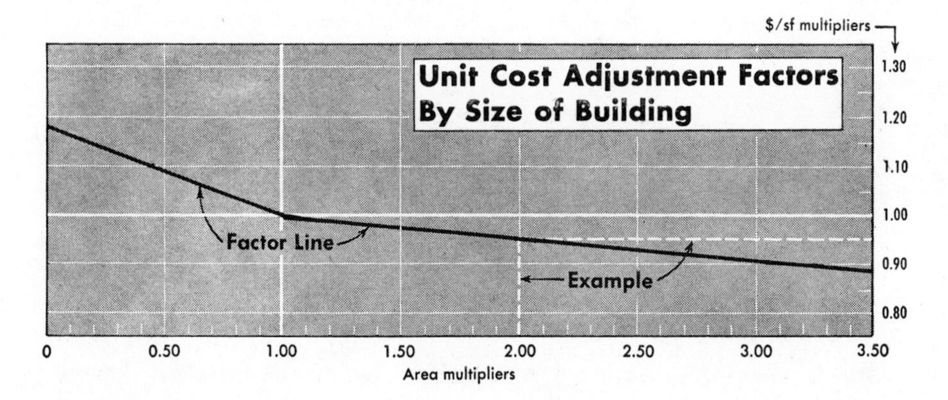

To find the adjusted unit cost of a proposed building larger or smaller than the typical size given for that type building in above table:

(1) Divide gross area of proposed building by typical size to find area multiplier.
(2) Locate multiplier on horizontal scale. Trace vertically to Factor Line then across to vertical scale to find $/sf multiplier.
(3) Multiply typical unit cost given on table by $/sf multiplier to determine the adjusted unit cost for the proposed building.

Example: typical dormitory, area = 30,000 sf
cost = $15.10/sf
proposed dormitory, area = 60,000 sf

60,000 ÷ 30,000 = 2.00 (area multiplier)
$/sf multiplier = 0.95
$15.10 × .95 = $14.35/sf (adjusted unit cost for proposed building)

FUNCTIONS OF NUMBERS

Functions of Numbers

1 to 59

No.	Square	Cube	Square Root	Cubic Root
1	1	1	1.0000	1.0000
2	4	8	1.4142	1.2599
3	9	27	1.7321	1.4422
4	16	64	2.0000	1.5874
5	25	125	2.2361	1.7100
6	36	216	2.4495	1.8171
7	49	343	2.6458	1.9129
8	64	512	2.8284	2.0000
9	81	729	3.0000	2.0801
10	100	1000	3.1623	2.1544
11	121	1331	3.3166	2.2240
12	144	1728	3.4641	2.2894
13	169	2197	3.6056	2.3513
14	196	2744	3.7417	2.4101
15	225	3375	3.8730	2.4662
16	256	4096	4.0000	2.5198
17	289	4913	4.1231	2.5713
18	324	5832	4.2426	2.6207
19	361	6859	4.3589	2.6684
20	400	8000	4.4721	2.7144
21	441	9261	4.5826	2.7589
22	484	10648	4.6904	2.8020
23	529	12167	4.7958	2.8439
24	576	13824	4.8990	2.8845
25	625	15625	5.0000	2.9240
26	676	17576	5.0990	2.9625
27	729	19683	5.1962	3.0000
28	784	21952	5.2915	3.0366
29	841	24389	5.3852	3.0723
30	900	27000	5.4772	3.1072
31	961	29791	5.5678	3.1414
32	1024	32768	5.6569	3.1748
33	1089	35937	5.7446	3.2075
34	1156	39304	5.8310	3.2396
35	1225	42875	5.9161	3.2711
36	1296	46656	6.0000	3.3019
37	1369	50653	6.0828	3.3322
38	1444	54872	6.1644	3.3620
39	1521	59319	6.2450	3.3912
40	1600	64000	6.3246	3.4200
41	1681	68921	6.4031	3.4482
42	1764	74088	6.4807	3.4760
43	1849	79507	6.5574	3.5034
44	1936	85184	6.6332	3.5303
45	2025	91125	6.7082	3.5569
46	2116	97336	6.7823	3.5830
47	2209	103823	6.8557	3.6088
48	2304	110592	6.9282	3.6342
49	2401	117649	7.0000	3.6593
50	2500	125000	7.0711	3.6840
51	2601	132651	7.1414	3.7084
52	2704	140608	7.2111	3.7325
53	2809	148877	7.2801	3.7563
54	2916	157464	7.3485	3.7798
55	3025	166375	7.4162	3.8030
56	3136	175616	7.4833	3.8259
57	3249	185193	7.5498	3.8485
58	3364	195112	7.6158	3.8709
59	3481	205379	7.6811	3.8930

Table from *How to Estimate for the Bldg. Trades*

60 to 119

No.	Square	Cube	Square Root	Cubic Root
60	3600	216000	7.7460	3.9149
61	3721	226981	7.8102	3.9365
62	3844	238328	7.8740	3.9579
63	3969	250047	7.9373	3.9791
64	4096	262144	8.0000	4.0000
65	4225	274625	8.0623	4.0207
66	4356	287496	8.1240	4.0412
67	4489	300763	8.1854	4.0615
68	4624	314432	8.2462	4.0817
69	4761	328509	8.3066	4.1016
70	4900	343000	8.3666	4.1213
71	5041	357911	8.4261	4.1408
72	5184	373248	8.4853	4.1602
73	5329	389017	8.5440	4.1793
74	5476	405224	8.6023	4.1983
75	5625	421875	8.6603	4.2172
76	5776	438976	8.7178	4.2358
77	5929	456533	8.7750	4.2543
78	6084	474552	8.8318	4.2727
79	6241	493039	8.8882	4.2908
80	6400	512000	8.9443	4.3089
81	6561	531441	9.0000	4.3267
82	6724	551368	9.0554	4.3445
83	6889	571787	9.1104	4.3621
84	7056	592704	9.1652	4.3795
85	7225	614125	9.2195	4.3968
86	7396	636056	9.2736	4.4140
87	7569	658503	9.3274	4.4310
88	7744	681472	9.3808	4.4480
89	7921	704969	9.4340	4.4647
90	8100	729000	9.4868	4.4814
91	8281	753571	9.5394	4.4979
92	8464	778688	9.5917	4.5144
93	8649	804357	9.6437	4.5307
94	8836	830584	9.6954	4.5468
95	9025	857375	9.7468	4.5629
96	9216	884736	9.7980	4.5789
97	9409	912673	9.8489	4.5947
98	9604	941192	9.8995	4.6104
99	9801	970299	9.9499	4.6261
100	10000	1000000	10.0000	4.6416
101	10201	1030301	10.0499	4.6570
102	10404	1061208	10.0995	4.6723
103	10609	1092727	10.1489	4.6875
104	10816	1124864	10.1980	4.7027
105	11025	1157625	10.2470	4.7177
106	11236	1191016	10.2956	4.7326
107	11449	1225043	10.3441	4.7475
108	11664	1259712	10.3923	4.7622
109	11881	1295029	10.4403	4.7769
110	12100	1331000	10.4881	4.7914
111	12321	1367631	10.5357	4.8059
112	12544	1404928	10.5830	4.8203
113	12769	1442897	10.6301	4.8346
114	12996	1481544	10.6771	4.8488
115	13225	1520875	10.7238	4.8629
116	13456	1560896	10.7703	4.8770
117	13689	1601613	10.8167	4.8910
118	13924	1643032	10.8628	4.9049
119	14161	1685159	10.9087	4.9187

Functions of Numbers (continued)

120 to 179

No.	Square	Cube	Square Root	Cubic Root
120	14400	1728000	10.9545	4.9324
121	14641	1771561	11.0000	4.9461
122	14884	1815848	11.0454	4.9597
123	15129	1860867	11.0905	4.9732
124	15376	1906624	11.1355	4.9866
125	15625	1953125	11.1803	5.0000
126	15876	2000376	11.2250	5.0133
127	16129	2048383	11.2694	5.0265
128	16384	2097152	11.3137	5.0397
129	16641	2146689	11.3578	5.0528
130	16900	2197000	11.4018	5.0658
131	17161	2248091	11.4455	5.0788
132	17424	2299968	11.4891	5.0916
133	17689	2352637	11.5326	5.1045
134	17956	2406104	11.5758	5.1172
135	18225	2460375	11.6190	5.1299
136	18496	2515456	11.6619	5.1426
137	18769	2571353	11.7047	5.1551
138	19044	2628072	11.7473	5.1676
139	19321	2685619	11.7898	5.1801
140	19600	2744000	11.8322	5.1925
141	19881	2803221	11.8743	5.2048
142	20164	2863288	11.9164	5.2171
143	20449	2924207	11.9583	5.2293
144	20736	2985984	12.0000	5.2415
145	21025	3048625	12.0416	5.2536
146	21316	3112136	12.0830	5.2656
147	21609	3176523	12.1244	5.2776
148	21904	3241792	12.1655	5.2896
149	22201	3307949	12.2066	5.3015
150	22500	3375000	12.2474	5.3133
151	22801	3442951	12.2882	5.3251
152	23104	3511808	12.3288	5.3368
153	23409	3581577	12.3693	5.3485
154	23716	3652264	12.4097	5.3601
155	24025	3723875	12.4499	5.3717
156	24336	3796416	12.4900	5.3832
157	24649	3869893	12.5300	5.3947
158	24964	3944312	12.5698	5.4061
159	25281	4019679	12.6095	5.4175
160	25600	4096000	12.6491	5.4288
161	25921	4173281	12.6886	5.4401
162	26244	4251528	12.7279	5.4514
163	26569	4330747	12.7671	5.4626
164	26896	4410944	12.8062	5.4737
165	27225	4492125	12.8452	5.4848
166	27556	4574296	12.8841	5.4959
167	27889	4657463	12.9228	5.5069
168	28224	4741632	12.9615	5.5178
169	28561	4826809	13.0000	5.5288
170	28900	4913000	13.0384	5.5397
171	29241	5000211	13.0767	5.5505
172	29584	5088448	13.1149	5.5613
173	29929	5177717	13.1529	5.5721
174	30276	5268024	13.1909	5.5828
175	30625	5359375	13.2288	5.5934
176	30976	5451776	13.2665	5.6041
177	31329	5545233	13.3041	5.6147
178	31684	5639752	13.3417	5.6252
179	32041	5735339	13.3791	5.6357

180 to 239

No.	Square	Cube	Square Root	Cubic Root
180	32400	5832000	13.4164	5.6462
181	32761	5929741	13.4536	5.6567
182	33124	6028568	13.4907	5.6671
183	33489	6128487	13.5277	5.6774
184	33856	6229504	13.5647	5.6877
185	34225	6331625	13.6015	5.6980
186	34596	6434856	13.6382	5.7083
187	34969	6539203	13.6748	5.7185
188	35344	6644672	13.7113	5.7287
189	35721	6751269	13.7477	5.7388
190	36100	6859000	13.7840	5.7489
191	36481	6967871	13.8203	5.7590
192	36864	7077888	13.8564	5.7690
193	37249	7189057	13.8924	5.7790
194	37636	7301384	13.9284	5.7890
195	38025	7414875	13.9642	5.7989
196	38416	7529536	14.0000	5.8088
197	38809	7645373	14.0357	5.8186
198	39204	7762392	14.0712	5.8285
199	39601	7880599	14.1067	5.8383
200	40000	8000000	14.1421	5.8480
201	40401	8120601	14.1774	5.8578
202	40804	8242408	14.2127	5.8675
203	41209	8365427	14.2478	5.8771
204	41616	8489664	14.2829	5.8868
205	42025	8615125	14.3178	5.8964
206	42436	8741816	14.3527	5.9059
207	42849	8869743	14.3875	5.9155
208	43264	8998912	14.4222	5.9250
209	43681	9129329	14.4568	5.9345
210	44100	9261000	14.4914	5.9439
211	44521	9393931	14.5258	5.9533
212	44944	9528128	14.5602	5.9627
213	45369	9663597	14.5945	5.9721
214	45796	9800344	14.6287	5.9814
215	46225	9938375	14.6629	5.9907
216	46656	10077696	14.6969	6.0000
217	47089	10218313	14.7309	6.0092
218	47524	10360232	14.7648	6.0185
219	47961	10503459	14.7986	6.0277
220	48400	10648000	14.8324	6.0368
221	48841	10793861	14.8661	6.0459
222	49284	10941048	14.8997	6.0550
223	49729	11089567	14.9332	6.0641
224	50176	11239424	14.9666	6.0732
225	50625	11390625	15.0000	6.0822
226	51076	11543176	15.0333	6.0912
227	51529	11697083	15.0665	6.1002
228	51984	11852352	15.0997	6.1091
229	52441	12008989	15.1327	6.1180
230	52900	12167000	15.1658	6.1269
231	53361	12326391	15.1987	6.1358
232	53824	12487168	15.2315	6.1446
233	54289	12649337	15.2643	6.1534
234	54756	12812904	15.2971	6.1622
235	55225	12977875	15.3297	6.1710
236	55696	13144256	15.3623	6.1797
237	56169	13312053	15.3948	6.1885
238	56644	13481272	15.4272	6.1972
239	57121	13651919	15.4596	6.2058

Functions of Numbers (continued)

240 to 299

No.	Square	Cube	Square Root	Cubic Root
240	57600	13824000	15.4919	6.2145
241	58081	13997521	15.5242	6.2231
242	58564	14172488	15.5563	6.2317
243	59049	14348907	15.5885	6.2402
244	59536	14526784	15.6205	6.2488
245	60025	14706125	15.6525	6.2573
246	60516	14886936	15.6844	6.2658
247	61009	15069223	15.7162	6.2743
248	61504	15252992	15.7480	6.2828
249	62001	15438249	15.7797	6.2912
250	62500	15625000	15.8114	6.2996
251	63001	15813251	15.8430	6.3080
252	36504	16003008	15.8745	6.3164
253	64009	16194277	15.9060	6.3247
254	64516	16387064	15.9374	6.3330
255	65025	16581375	15.9687	6.3413
256	65536	16777216	16.0000	6.3496
257	66049	16974593	16.0312	6.3579
258	66564	17173512	16.0624	6.3661
259	67081	17373979	16.0935	6.3743
260	67600	17576000	16.1245	6.3825
261	68121	17779581	16.1555	6.3907
262	68644	17984728	16.1864	6.3988
263	69169	18191447	16.2173	6.4070
264	69696	18399744	16.2481	6.4151
265	70225	18609625	16.2788	6.4232
266	70756	18821096	16.3095	6.4312
267	71289	19034163	16.3401	6.4393
268	71824	19248832	16.3707	6.4473
269	72361	19465109	16.4012	6.4553
270	72900	19683000	16.4317	6.4633
271	73441	19902511	16.4621	6.4713
272	73984	20123648	16.4924	6.4792
273	74529	20346417	16.5227	6.4872
274	75076	20570824	16.5529	6.4951
275	75625	20796875	16.5831	6.5030
276	76176	21024576	16.6132	6.5108
277	76729	21253933	16.6433	6.5187
278	77284	21484952	16.6733	6.5265
279	77841	21717639	16.7033	6.5343
280	78400	21952000	16.7332	6.5421
281	78961	22188041	16.7631	6.5499
282	79524	22425768	16.7929	6.5577
283	80089	22665187	16.8226	6.5654
284	80656	22906304	16.8523	6.5731
285	81225	23149125	16.8819	6.5808
286	81796	23393656	16.9115	6.5885
287	82369	23639903	16.9411	6.5962
288	82944	23887872	16.9706	6.6039
289	83521	24137569	17.0000	6.6115
290	84100	24389000	17.0294	6.6191
291	84681	24642171	17.0587	6.6267
292	85264	24897088	17.0880	6.6343
293	85849	25153757	17.1172	6.6419
294	86436	25412184	17.1464	6.6494
295	87025	25672375	17.1756	6.6569
296	87616	25934336	17.2047	6.6644
297	88209	26198073	17.2337	6.6719
298	88804	26463592	17.2627	6.6794
299	89401	26730899	17.2916	6.6869

300 to 359

No.	Square	Cube	Square Root	Cubic Root
300	90000	27000000	17.3205	6.6943
301	90601	27270901	17.3494	6.7018
302	91204	27543608	17.3781	6.7092
303	91809	27818127	17.4069	6.7166
304	92416	28094464	17.4356	6.7240
305	93025	28372625	17.4642	6.7313
306	93636	28652616	17.4929	6.7387
307	94249	28934443	17.5214	6.7460
308	94864	29218112	17.5499	6.7533
309	95481	29503629	17.5784	6.7606
310	96100	29791000	17.6068	6.7679
311	96721	30080231	17.6352	6.7752
312	97344	30371328	17.6635	6.7824
313	97969	30664297	17.6918	6.7897
314	98596	30959144	17.7200	6.7969
315	99225	31255875	17.7482	6.8041
316	99856	31554496	17.7764	6.8113
317	100489	31855013	17.8045	6.8185
318	101124	32157432	17.8326	6.8256
319	101761	32461759	17.8606	6.8328
320	102400	32768000	17.8885	6.8399
321	103041	33076161	17.9165	6.8470
322	103684	33386248	17.9444	6.8541
323	104329	33698267	17.9722	6.8612
324	104976	34012224	18.0000	6.8683
325	105625	34328125	18.0278	6.8753
326	106276	34645976	18.0555	6.8824
327	106929	34965783	18.0831	6.8894
328	107584	35287552	18.1108	6.8964
329	108241	35611289	18.1384	6.9034
330	108900	35937000	18.1659	6.9104
331	109561	36264691	18.1934	6.9174
332	110224	36594368	18.2209	6.9244
333	110889	36926037	18.2483	6.9313
334	111556	37259704	18.2757	6.9382
335	112225	37595375	18.3030	6.9451
336	112896	37933056	18.3303	6.9521
337	113569	38272753	18.3576	6.9589
338	114244	38614472	18.3848	6.9658
339	114921	38958219	18.4120	6.9727
340	115600	39304000	18.4391	6.9795
341	116281	39651821	18.4662	6.9864
342	116964	40001688	18.4932	6.9932
343	117649	40353607	18.5203	7.0000
344	118336	40707584	18.5472	7.0068
345	119025	41063625	18.5742	7.0136
346	119716	41421736	18.6011	7.0203
347	120409	41781923	18.6279	7.0271
348	121104	42144192	18.6548	7.0338
349	121801	42508549	18.6815	7.0406
350	122500	42875000	18.7083	7.0473
351	123201	43243551	18.7350	7.0540
352	123904	43614208	18.7617	7.0607
353	124609	43986977	18.7883	7.0674
354	125316	44361864	18.8149	7.0740
355	126025	44738875	18.8414	7.0807
356	126736	45118016	18.8680	7.0873
357	127449	45499293	18.8944	7.0940
358	128164	45882712	18.9209	7.1006
359	128881	46268279	18.9473	7.1072

Functions of Numbers (continued)

360 to 419

No.	Square	Cube	Square Root	Cubic Root
360	129600	46656000	18.9737	7.1138
361	130321	47045881	19.0000	7.1204
362	131044	47437928	19.0263	7.1269
363	131769	47832147	19.0526	7.1335
364	132496	48228544	19.0788	7.1400
365	133225	48627125	19.1050	7.1466
366	133956	49027896	19.1311	7.1531
367	134689	49430863	19.1572	7.1596
368	135424	49836032	19.1833	7.1661
369	136161	50243409	19.2094	7.1726
370	136900	50653000	19.2354	7.1791
371	137641	51064811	19.2614	7.1855
372	138384	51478848	19.2873	7.1920
373	139129	51895117	19.3132	7.1984
374	139876	52313624	19.3391	7.2048
375	140625	52734375	19.3649	7.2112
376	141376	53157376	19.3907	7.2177
377	142129	53582633	19.4165	7.2240
378	142884	54010152	19.4422	7.2304
379	143641	54439939	19.4679	7.2368
380	144400	54872000	19.4936	7.2432
381	145161	55306341	19.5192	7.2495
382	145924	55742968	19.5448	7.2558
383	146689	56181887	19.5704	7.2622
384	147456	56623104	19.5959	7.2685
385	148225	57066625	19.6214	7.2748
386	148996	57512456	19.6469	7.2811
387	149769	57960603	19.6723	7.2874
388	150544	58411072	19.6977	7.2936
389	151321	58863869	19.7231	7.2999
390	152100	59319000	19.7484	7.3061
391	152881	59776471	19.7737	7.3124
392	153664	60236288	19.7990	7.3186
393	154449	60698457	19.8242	7.3248
394	155236	61162984	19.8494	7.3310
395	156025	61629875	19.8746	7.3372
396	156816	62099136	19.8997	7.3434
397	157609	62570773	19.9249	7.3496
398	158404	63044792	19.9499	7.3558
399	159201	63521199	19.9750	7.3619
400	160000	64000000	20.0000	7.3681
401	160801	64481201	20.0250	7.3742
402	161604	64964808	20.0499	7.3803
403	162409	65450827	20.0749	7.3864
404	163216	65939264	20.0998	7.3925
405	164025	66430125	20.1246	7.3986
406	164836	66923416	20.1494	7.4047
407	165649	67419143	20.1742	7.4108
408	166464	67917312	20.1990	7.4169
409	167281	68417929	20.2237	7.4229
410	168100	68921000	20.2485	7.4290
411	168921	69426531	20.2731	7.4350
412	169744	69934528	20.2978	7.4410
413	170569	70444997	20.3224	7.4470
414	171396	70957944	20.3470	7.4530
415	172225	71473375	20.3715	7.4590
416	173056	71991296	20.3961	7.4650
417	173889	72511713	20.4206	7.4710
418	174724	73034632	20.4450	7.4770
419	175561	73560059	20.4695	7.4829

420 to 479

No.	Square	Cube	Square Root	Cubic Root
420	176400	74088000	20.4939	7.4889
421	177241	74618461	20.5183	7.4948
422	178084	75151448	20.5426	7.5007
423	178929	75686967	20.5670	7.5067
424	179776	76225024	20.5913	7.5126
425	180625	76765625	20.6155	7.5185
426	181476	77308776	20.6398	7.5244
427	182329	77854483	20.6640	7.5302
428	183184	78402752	20.6882	7.5361
429	184041	78953589	20.7123	7.5420
430	184900	79507000	20.7364	7.5478
431	185761	80062991	20.7605	7.5537
432	186624	80621568	20.7846	7.5595
433	187489	81182737	20.8087	7.5654
434	188356	81746504	20.8327	7.5712
435	189225	82312875	20.8567	7.5770
436	190096	82881856	20.8806	7.5828
437	190969	83453453	20.9045	7.5886
438	191844	84027672	20.9284	7.5944
439	192721	84604519	20.9523	7.6001
440	193600	85184000	20.9762	7.6059
441	194481	85766121	21.0000	7.6117
442	195364	86350888	21.0238	7.6174
443	196249	86938307	21.0476	7.6232
444	197136	87528384	21.0713	7.6289
445	198025	88121125	21.0950	7.6346
446	198916	88716536	21.1187	7.6403
447	199809	89314623	21.1424	7.6460
448	200704	89915392	21.1660	7.6517
449	201601	90518849	21.1896	7.6574
450	202500	91125000	21.2132	7.6631
451	203401	91733851	21.2368	7.6688
452	204304	92345408	21.2603	7.6744
453	205209	92959677	21.2838	7.6801
454	206116	93576664	21.3073	7.6857
455	207025	94196375	21.3307	7.6914
456	207936	94818816	21.3542	7.6970
457	208849	95443993	21.3776	7.7026
458	209764	96071912	21.4009	7.7082
459	210681	96702579	21.4243	7.7138
460	211600	97336000	21.4476	7.7194
461	212521	97972181	21.4709	7.7250
462	213444	98611128	21.4942	7.7306
463	214369	99252847	21.5174	7.7362
464	215296	99897344	21.5407	7.7418
465	216225	100544625	21.5639	7.7473
466	217156	101194696	21.5870	7.7529
467	218089	101847563	21.6102	7.7584
468	219024	102503232	21.6333	7.7639
469	219961	103161709	21.6564	7.7695
470	220900	103823000	21.6795	7.7750
471	221841	104487111	21.7025	7.7805
472	222784	105154048	21.7256	7.7860
473	223729	105823817	21.7486	7.7915
474	224676	106496424	21.7715	7.7970
475	225625	107171875	21.7945	7.8025
476	226576	107850176	21.8174	7.8079
477	227529	108531333	21.8403	7.8134
478	228484	109215352	21.8632	7.8188
479	229441	109902239	21.8861	7.8243

Functions of Numbers (continued)

480 to 539

No.	Square	Cube	Square Root	Cubic Root
480	230400	110592000	21.9089	7.8297
481	231361	111284641	21.9317	7.8352
482	232324	111980168	21.9545	7.8406
483	233289	112678587	21.9773	7.8460
484	234256	113379904	22.0000	7.8514
485	235225	114084125	22.0227	7.8568
486	236196	114791256	22.0454	7.8622
487	237169	115501303	22.0681	7.8676
488	238144	116214272	22.0907	7.8730
489	239121	116930169	22.1133	7.8784
490	240100	117649000	22.1359	7.8837
491	241081	118370771	22.1585	7.8891
492	242064	119095488	22.1811	7.8944
493	243049	119823157	22.2036	7.8998
494	244036	120553784	22.2261	7.9051
495	245025	121287375	22.2486	7.9105
496	246016	122023936	22.2711	7.9158
497	247009	122763473	22.2935	7.9211
498	248004	123505992	22.3159	7.9264
499	249001	124251499	22.3383	7.9317
500	250000	125000000	22.3607	7.9370
501	251001	125751501	22.3830	7.9423
502	252004	126506008	22.4054	7.9476
503	253009	127263527	22.4277	7.9528
504	254016	128024064	22.4499	7.9581
505	255025	128787625	22.4722	7.9634
506	256036	129554216	22.4944	7.9686
507	257049	130323843	22.5167	7.9739
508	258064	131096512	22.5389	7.9791
509	259081	131872229	22.5610	7.9843
510	260100	132651000	22.5832	7.9896
511	261121	133432831	22.6053	7.9948
512	262144	134217728	22.6274	8.0000
513	263169	135005697	22.6495	8.0052
514	264196	135796744	22.6716	8.0104
515	265225	136590875	22.6936	8.0156
516	266256	137388096	22.7156	8.0208
517	267289	138188413	22.7376	8.0260
518	268324	138991832	22.7596	8.0311
519	269361	139798359	22.7816	8.0363
520	270400	140608000	22.8035	8.0415
521	271441	141420761	22.8254	8.0466
522	272484	142236648	22.8473	8.0517
523	273529	143055667	22.8692	8.0569
524	274576	143877824	22.8910	8.0620
525	275625	144703125	22.9129	8.0671
526	276676	145531576	22.9347	8.0723
527	277729	146363183	22.9565	8.0774
528	278784	147197952	22.9783	8.0825
529	279841	148035889	23.0000	8.0876
530	280900	148877000	23.0217	8.0927
531	281961	149721291	23.0434	8.0978
532	283024	150568768	23.0651	8.1028
533	284089	151419437	23.0868	8.1079
534	285156	152273304	23.1084	8.1130
535	286225	153130375	23.1301	8.1180
536	287296	153990656	23.1517	8.1231
537	288369	154854153	23.1733	8.1281
538	289444	155720872	23.1948	8.1332
539	290521	156590819	23.2164	8.1382

540 to 599

No.	Square	Cube	Square Root	Cubic Root
540	291600	157464000	23.2379	8.1433
541	292681	158340421	23.2594	8.1483
542	293764	159220088	23.2809	8.1533
543	294849	160103007	23.3024	8.1583
544	295936	160989184	23.3238	8.1633
545	297025	161878625	23.3452	8.1683
546	298116	162771336	23.3666	8.1733
547	299209	163667323	23.3880	8.1783
548	300304	164566592	23.4094	8.1833
549	301401	165469149	23.4307	8.1882
550	302500	166375000	23.4521	8.1932
551	303601	167284151	23.4734	8.1982
552	304704	168196608	23.4947	8.2031
553	305809	169112377	23.5160	8.2081
554	306916	170031464	23.5372	8.2130
555	308025	170953875	23.5584	8.2180
556	309136	171879616	23.5797	8.2229
557	310249	172808693	23.6008	8.2278
558	311364	173741112	23.6220	8.2327
559	312481	174676879	23.6432	8.2377
560	313600	175616000	23.6643	8.2426
561	314721	176558481	23.6854	8.2475
562	315844	177504328	23.7065	8.2524
563	316969	178453547	23.7276	8.2573
564	318096	179406144	23.7487	8.2621
565	319225	180362125	23.7697	8.2670
566	320356	181321496	23.7908	8.2719
567	321489	182284263	23.8118	8.2768
568	322624	183250432	23.8328	8.2816
569	323761	184220009	23.8537	8.2865
570	324900	185193000	23.8747	8.2913
571	326041	186169411	23.8956	8.2962
572	327184	187149248	23.9165	8.3010
573	328329	188132517	23.9374	8.3059
574	329476	189119224	23.9583	8.3107
575	330625	190109375	23.9792	8.3155
576	331776	191102976	24.0000	8.3203
577	332929	192100033	24.0208	8.3251
578	334084	193100552	24.0416	8.3300
579	335241	194104539	24.0624	8.3348
580	336400	195112000	24.0832	8.3396
581	337561	196122941	24.1039	8.3443
582	338724	197137368	24.1247	8.3491
583	339889	198155287	24.1454	8.3539
584	341056	199176704	24.1661	8.3587
585	342225	200201625	24.1868	8.3634
586	343396	201230056	24.2074	8.3682
587	344569	202262003	24.2281	8.3730
588	345744	203297472	24.2487	8.3777
589	346921	204336469	24.2693	8.3825
590	348100	205379000	24.2899	8.3872
591	349281	206425071	24.3105	8.3919
592	350464	207474688	24.3311	8.3967
593	351649	208527857	24.3516	8.4014
594	352836	209584584	24.3721	8.4061
595	354025	210644875	24.3926	8.4108
596	355216	211708736	24.4131	8.4155
597	356409	212776173	24.4336	8.4202
598	357604	213847192	24.4540	8.4249
599	358801	214921799	24.4745	8.4296

INDEX

Numerals in **bold face** indicate illustrations

Q

R